T0139114

Multi-sensor System Applications in the Everglades Ecosystem

Remote Sensing Applications
Series Editor
Qihao Weng

Remote Sensing of Land Use and Land Cover
Principles and Applications
Chandra P. Giri

Remote Sensing of Natural Resources
edited by Guangxing Wang and Qihao Weng

Global Urban Monitoring and Assessment through Earth Observation
edited by Qihao Weng

Remote Sensing of Impervious Surfaces in Tropical and Subtropical Areas
Hui Lin, Yuanzhi Zhang, and Qihao Weng

Remote Sensing Applications for the Urban Environment
George Z. Xian

Remote Sensing for Sustainability
Qihao Weng

Integrating Scale in Remote Sensing and GIS
Dale A. Quattrochi, Elizabeth A. Wentz, Nina Siu-Ngan Lam, and Charles W. Emerson

Hyperspectral Remote Sensing
Fundamentals and Practices
Ruiliang Pu

LiDAR Remote Sensing and Applications
Pinliang Dong, Qi Chen

Urban Remote Sensing, Second Edition
edited by Qihao Weng, Dale Quattrochi, and Paolo E. Gamba

Multi-sensor System Applications in the Everglades Ecosystem
Caiyun Zhang

For more information about this series, please visit: www.crcpress.com/Remote-Sensing-Applications-Series/book-series/CRCREMSENAPP

Multi-sensor System Applications in the Everglades Ecosystem

Caiyun Zhang

CRC Press
Taylor & Francis Group
Boca Raton London New York

CRC Press is an imprint of the
Taylor & Francis Group, an **informa** business

CRC Press
Taylor & Francis Group
6000 Broken Sound Parkway NW, Suite 300
Boca Raton, FL 33487-2742

First issued in paperback 2022

© 2020 by Taylor & Francis Group, LLC

CRC Press is an imprint of Taylor & Francis Group, an Informa business

No claim to original US Government works

ISBN 13: 978-1-03-247487-8 (pbk)
ISBN 13: 978-1-4987-1177-7 (hbk)

DOI: 10.1201/9780429075872

This book contains information obtained from authentic and highly regarded sources. Reasonable efforts have been made to publish reliable data and information, but the author and publisher cannot assume responsibility for the validity of all materials or the consequences of their use. The authors and publishers have attempted to trace the copyright holders of all material reproduced in this publication and apologize to copyright holders if permission to publish in this form has not been obtained. If any copyright material has not been acknowledged, please write and let us know so we may rectify in any future reprint.

Except as permitted under US. Copyright Law, no part of this book may be reprinted, reproduced, transmitted, or utilized in any form by any electronic, mechanical, or other means, now known or hereafter invented, including photocopying, microfilming, and recording, or in any information storage or retrieval system, without written permission from the publishers.

For permission to photocopy or use material electronically from this work, please access www.copyright.com (http://www.copyright.com/) or contact the Copyright Clearance Center, Inc. (CCC), 222 Rosewood Drive, Danvers, MA 01923, 978-750-8400. CCC is a not-for-profit organization that provides licenses and registration for a variety of users. For organizations that have been granted a photocopy license by the CCC, a separate system of payment has been arranged.

Trademark Notice: Product or corporate names may be trademarks or registered trademarks, and are used only for identification and explanation without intent to infringe.

Publisher's Note
The publisher has gone to great lengths to ensure the quality of this reprint but points out that some imperfections in the original copies may be apparent.

Visit the Taylor & Francis Web site at
http://www.taylorandfrancis.com

and the CRC Press Web site at
http://www.crcpress.com

Contents

Part I Florida Everglades and Remote Sensing

Part II   Multispectral Remote Sensing Applications

9. Applying Landsat Products to Assess the Damage and Resilience of Mangroves from Hurricanes

Part III   Hyperspectral Remote Sensing Applications

10. Applying Point Spectroscopy Data to Assess the Effects of Salinity and Sea Level Rise on Canopy Water Content of *Juncus roemerianus*

Part IV Lidar Remote Sensing Applications

Series Foreword

Comprising the southern half of a large drainage basin within the Neotropic ecozone, the Florida Everglades is a natural treasure of tropical wetlands. Within this region, we can find many one-of-a-kind national parks, wetlands, seashores, natural preserves, and wildlife refuges, interspersed among a bounty of other ecosystems. Conservation and restoration of such a diverse ecoregion is challenging yet critical in the context of continued population growth, economic and ecological conflict, and climate change. Remote sensing provides one of the most effective technologies for inventorying and monitoring protected areas as well as assessing their conditions for subsequent resource planning and management. However, traditional multispectral remote sensors are often not sufficient for all these tasks. Confronting many ecological and environmental issues facing the Everglades requires the ability to reconcile, integrate, and analyze data from other remote sensors. Though the task is no short one, I am pleased that Professor Caiyun Zhang has the vision, energy, and knowledge to develop a suite of methodologies using contemporary remote sensing technology that will be essential to the conservation and management of this threatened frontier on our planet Earth.

Multi-sensor System Applications in the Everglades Ecosystem is the first comprehensive study of the Everglades ecoregion and its ecosystems. It examines and illustrates how airborne and spaceborne multispectral and hyperspectral sensors, light detection and ranging (lidar), Unmanned Aircraft Systems (UAS), and handheld spectroradiometers can be applied to study many key issues of this ecoregion, ranging from vegetation mapping, biomass, and water quality assessment to hurricane monitoring and coastal and coral reef preservation. As the intellectual bridge between wetland and coastal ecosystems and remote sensing science and technology, I cannot overstate this book's importance—Professor Zhang is the first to span this knowledge gap crucial to protecting the Everglades. By reading this book, readers will surely be amazed by the pure depth of knowledge that Zhang possesses; however, I believe the book's true value lies in Zhang's skill and dexterity in applications of such knowledge. From her years of integrating her research on the Everglades, Zhang has written this book such that it is not only a research treatise at the front edge of remote sensing; many chapters are also excellent teaching materials, as already demonstrated at Florida Atlantic University.

This volume adds to the already-rich contents of the *Taylor & Francis Series in Remote Sensing Applications* that I started to edit in 2007. The book series contributes to advancements in theories, methods, techniques, and applications of remote sensing, while serving professionals, researchers, professors, and students as references and textbooks around the globe. I hope that the publication of this book will promote a wider and deeper appreciation and application of multi-sensor remote sensing technology—and I am confident it will contribute to the Comprehensive Everglades Restoration Plan that is currently underway. Finally, my sincere congratulations go to Professor Zhang for achieving a new milestone in the literature of remote sensing.

Qihao Weng, Ph.D., IEEE Fellow
Hawthorn Woods, Indiana
August 22, 2019

Preface

The Florida Everglades, the largest subtropical wetland in the USA, is regionally, nationally, and internationally significant to both humans and their surrounding natural environments but has been largely modified by human activities in the past century leading to a range of environmental issues. To restore, preserve, and protect the Everglades ecosystem while providing for other water-related needs of the region, the largest hydrological restoration project in the USA, known as the Comprehensive Everglades Restoration Plan (CERP), is underway. In CERP, information related to vegetation patterns and changes, water quality, biomass, coral reefs, beaches, Digital Elevation Models (DEM) and tracking effects from disasters (e.g., sea level rise and hurricanes), as well as others, are in demand for different regions to guide restoration and conservation. Acquisition of this type of information is often from field methods or other time-consuming and labor-intensive procedures. In this book, we explore the applicability of multiple remote sensors to indirectly acquire these types of information using data collected from airborne and spaceborne multispectral/hyperspectral sensors, light detection and ranging (lidar), Unmanned Aircraft System (UAS), and a handheld spectroradiometer. These remote sensing datasets were combined with data from the field, laboratory, or other sources to understand the patterns, changes, and disaster effects at a larger scale. We apply modern digital data processing techniques to identify accurate models and generate informative maps such as Object-Based Image Analysis (OBIA), machine learning data mining, texture analysis, and multi-sensor data fusion approaches. We applied these remote sensing techniques in multiple ecosystems in the Florida Everglades such as sawgrass, mangroves, Florida Bay, Florida Reef Tract, and upper Everglades (Lake Okeechobee and Kissimmee River watersheds). We provide a comprehensive application of remote sensing techniques in the Florida Everglades and its coastal ecosystems. It is expected that the techniques used in the book can benefit the restoration and conservation of the Florida Everglades in particular and global wetland ecosystems in general.

The book is organized into four parts based on the sensor systems used. Part I (Chapters 1 to 3) introduces the Florida Everglades, remote sensing, and three vegetation classification systems used in the Everglades. Part II (Chapters 4-9) shows multispectral systems (aerial photography, UAS products, and Landsat data) for vegetation mapping, biomass modeling, water quality modeling, and mangrove damage and resilience analyses from hurricanes. Part III (Chapters 10-14) applies hyperspectral systems for plant stress analysis, coastal sand characterization, and vegetation and coral reef mapping. Part IV (Chapters 15-17) features airborne lidar data for coastal vulnerability analysis, DEM improvement, and mangrove structure analysis. Part V (Chapters 18 and 19) explores a fusion of multiple sensors for vegetation and coral reef benthic habitat mapping.

We acknowledge Dr. Qihao Weng at Indiana State University for the encouragement to write this book.

Caiyun Zhang
Department of Geosciences
Florida Atlantic University
Boca Raton, Florida, USA

Author

Dr. Caiyun Zhang is a professor in the Department of Geosciences at Florida Atlantic University (FAU). Dr. Zhang received her PhD in Geospatial Information Sciences from University of Texas, Dallas. Her research at FAU focuses on vegetation characterization in the Florida Everglades using multiple sensors, biomass modeling and mapping, water quality monitoring and mapping, and the analysis of coastal vulnerability to sea level rise and hurricanes. She has developed innovative methodology frameworks to monitor and map the Greater Everglades by combining multiple sensors and GIS techniques, which can assist with the restoration and conservation of the Florida Everglades ecosystem. She applies modern machine learning and advanced remote sensing image processing techniques in the coastal environments to understand the effects of human activities and natural disasters on the modification of coastal landscapes. She teaches five remote sensing courses at FAU including *Remote Sensing of Environment, Digital Image Analysis, Hyperspectral Remote Sensing, Lidar Remote Sensing and Applications,* and *Photogrammetry and Aerial Photo Interpretation.*

Contributors

Dr. Zhang's former and current PhD students contributed to this book, including Sara Denka Durgan (Chapter 5), David Brodylo (Chapter 9), Molly E. Smith (Chapter 11), Dr. Donna Selch (Chapters 10 and 11), and Dr. Hannah Cooper (Chapters 15 and 16). In addition, Cara J. Abbott made a minor contribution to Chapter 10.

Cara J. Abbott
The Institute for Regional Conservation
Delray Beach, Florida

David Brodylo
Department of Geosciences
Florida Atlantic University
Boca Raton, Florida

Hannah Cooper
Department of Geography, Planning,
    and Environment
East Carolina University
Greenville, North Carolina

Sara Denka Durgan
Department of Geosciences
Florida Atlantic University
Boca Raton, Florida

Donna Selch
School of Marine and Atmospheric
Sciences
Stony Brook University
Stony Brook, New York

Molly E. Smith
Department of Geosciences
Florida Atlantic University
Boca Raton, Florida

Abbreviations

ANN	Artificial Neural Network
ARD	Analysis Ready Data
ARIES	Australian Resource Information and Environment Satellite
ASPRS	American Society for Photogrammetry and Remote Sensing
AVIRIS	Airborne Visible/Infrared Imaging Spectrometer
BRDF	Bidirectional Reflectance Distribution Function
BV	Brightness Value
CASI	Compact Airborne Imaging Spectrometer
CERP	Comprehensive Everglades Restoration Plan
CHM	Canopy Height Model
CHRIS	Compact High Resolution Imaging Spectrometer
DAAC	Distributed Active Archive Center
DEM	Digital Elevation Model
DOQQ	Digital Orthophoto Quarter-Quadrangles
DSM	Digital Surface Model
EAA	Everglades Agricultural Areas
ENP	Everglades National Park
ENVI	ENvironment for Visualizing Images
EO-1	Earth Observing-1
ESA	European Space Agency
ETM+	Enhanced Thematic Mapper Plus
FDEM	Florida Division of Emergency Management
FDOT	Florida Department of Transportation
FLAASH	Fast Line-of-sight Atmospheric Analysis of Spectral Hypercubes
FLUCCS	Florida Land Use, Cover, and forms Classification System
FPAR	Fraction of Photosynthetically Active Radiation
FPL	Florida Power & Light
FWC	Florida Fish and Wildlife Conservation Commission
FWRI	Fish and Wildlife Research Institute
GCP	Ground Control Point
G-LiHT	Goddard's Lidar, Hyperspectral & Thermal Imager
GPP	Gross Primary Productivity
GPS	Global Positioning System
HAB	Harmful Algal Bloom
HICO	Hyperspectral Imager for the Coastal Ocean
HISUI	Hyperspectral Imager SUIte
HLS	Harmonized Landsat and Sentinel-2
HRIS	High Resolution Imaging Spectrometer
HRS	Hyperspectral Remote Sensing
IDW	Inverse Distance Weighted
INS	Inertial Navigation System
ISODAT	Iterative Self-Organizing Data Analysis Technique
JPL	Jet Propulsion Laboratory

k-NN	k-Nearest Neighbor
KRREP	Kissimmee River Restoration Evaluation Program
LABINS	Land Boundary Information System
LAI	Leaf Area Index
LCLU	Land Cover Land Use
LEISA	Linear Etalon Imaging Spectral Array
Lidar	Light detection and ranging
LWIR	Long-Wave Infrared
MAE	Mean Absolute Error
MAP	Monitoring and Assessment Plan
MCC	Mapir Camera Control
MLR	Multiple Linear Regression
MMU	Minimum Mapping Unit
MNF	Minimum Noise Fraction
MODIS	Moderate Resolution Imaging Spectroradiometer
MSI	Multispectral Instrument
MSS	Multispectral Scanner System
MVS	Multiview Stereopsis
NAIP	National Agriculture Imagery Program
NASA	National Aeronautics and Space Administration
NAWCA	North American Wetland Conservation Act
NBCD	National Biomass and Carbon Dataset
NDVI	Normalized Difference Vegetation Index
NHAP	National High Altitude Photography
NIR	Near INfrared
NOAA	National Oceanic and Atmospheric Administration
NPP	Net Primary Productivity
OA	Overall Accuracy
OLI	Operational Land Imager
ORNL	Oak Ridge National Laboratory
PA	Producer's Accuracy
PCA	Principle Component Analysis
PIF	Pseudo-Invariant Feature
PRISMA	Hyperspectral Precursor of the Application Mission
RECOVER	REstoration COordination and VErification
RF	Random Forest
RGB	Red-Green-Blue
RMSE	Root Mean Square Error
SeaWiFS	Sea-viewing Wide Field-of-view Sensors
SfM	Structure from Motion
SFWMD	South Florida Water Management District
SLFMR	Scanning Low-Frequency Microwave Radiometer
SOFIA	South Florida Information Access
SRTM	Shuttle Radar Topography Mission
SVM	Support Vector Machine
SWIR	Short-Wavelength Infrared
TIRS	Thermal Infrared Sensor
TM	Thematic Mapper

UA	User's Accuracy
UAS	Unmanned Aerial Systems
UAV	Unmanned Aerial Vehicles
USACE	US Army Corps of Engineers
USDA	US Department of Agriculture
USGS	US Geological Survey
UTM	Universal Transverse Mercator
VNIR	Visible and Near-Infrared
WCA	Water Conservation Area
WEKA	Waikato Environment for Knowledge Analysis

Part I

Florida Everglades and Remote Sensing

1

Florida Everglades and Restoration

Caiyun Zhang

1.1 Florida Everglades

The Florida Everglades is the largest subtropical wetland in the USA. It is recognized both nationally and internationally as one of the world's most unique natural and cultural resources, as it supports many threatened and endangered species. Historically, the system was characterized by a continuum of clean, flowing fresh water from just south of present-day Orlando, down the Kissimmee River, through Lake Okeechobee (a 1900 km^2 lake with an average depth of 2.7 m), across "River of Grass" with 60 miles (97 km) wide and 100 miles (160 km) long, and ultimately into Florida Bay and the Gulf of Mexico (USGS/Florida Everglades, 2019). The Kissimmee River, Lake Okeechobee, and Everglades National Park (ENP) together form the valuable Greater Everglades ecosystem, encompassing about 10,000 km^2 (4000 square miles) and taking up a third of the Florida peninsula in the 1800s before drainage attempts (Figure 1.1). The Everglades is home to around 11,000 species of seed-bearing plants and 400 species of land or water vertebrates.

The name "River of Grass" for the Everglades is from Marjory Stoneman Douglas, who noticed the thread of the Everglades, helped establish the ENP, and saved the Everglades from the destruction of drainage (Richardson, 2010). The name is used to describe the vast sawgrass marshes (icon of the Everglades), which are part of the interdependent Everglades ecosystem including cypress swamps, hardwood hammocks, pine forests, and coastal environment. The major landscape types in the Greater Everglades prior to human intervention are displayed in Figure 1.2a. Water is the heart of the Everglades, and its landscape is mainly controlled by the water flow. The slow-moving water flows downstream through the low-gradient wetland landscape; it formed a vast expanse of sawgrass prairie south of Lake Okeechobee, slough and ridge in the middle of the flow, pine flatwoods over most of the eastern boundary of the flow, and relatively higher wetlands in the western boundary, now known as Big Cypress National Preserve. Much of the flow discharged into the Gulf of Mexico through Shark River Slough and mangrove estuaries. South of the flow, the pine flatwoods were absent, and the Atlantic Coastal Ridge became discontinuous, forming a series of islands separated by coastal rivers. These rivers resulted in a portion of the flow being discharged eastward into Biscayne Bay and the Atlantic Ocean. The remainder of the flow discharged southward through Taylor Slough into Florida Bay. The current system of the Greater Everglades is displayed in Figure 1.2 (b); they are highly managed and impacted by a network of pumps, gates, culverts, canals, and levees. The landscape is dominated by state-owned Water Conservation Areas (WCAs), Everglades Agricultural Areas (EAA), Big Cypress National Preservation, and ENP. The surrounding areas are Florida Bay, developed urban areas, and Biscayne Bay. Two Indian Reservations (the Seminole Indian Reservation and the Miccosukee Indian Reservation) are also part of the current Everglades.

3

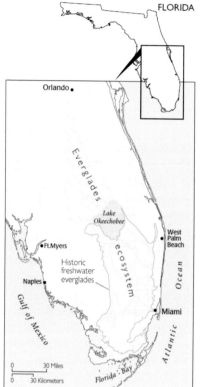
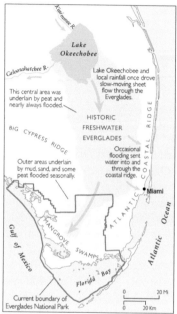

FIGURE 1.1
The original Greater Everglades ecosystem and its natural flow pattern around 1900 (from Galloway et al., 1999, USGS).

1.2 Geology and Landforms of the Everglades

All of Florida resides within the Atlantic Coastal Plain province, and most of Florida is underlain by a thick sequence of carbonate sediment (limestone and dolostone) known as the Florida Platform, which was formed during Pleistocene or Ice Age (from 2.588 million to 11,700 years before the present time) (Davis, 1943). South Florida was sea floor for most of the last 150 million years. There are five geological formations in south Florida: the Tamiami Formation, Caloosahatchee Formation, Anastasia Formation, Miami Limestone, and the Fort Thompson Formation. The Tamiami Formation is a compression of highly permeable light-colored fossiliferous sands and pockets of quartz. It is named for the Tamiami Trail that follows the upper bedrock of the Big Cypress Swamp and underlies the southern portion of the Everglades. Between the Tamiami Formation and Lake Okeechobee is the Caloosahatchee Formation, named for the river over it. Both the Tamiami and Caloosahatchee Formations developed during the Pliocene Epoch (5.33–2.58 million years before the present). The Fort Thompson Formation surrounds the southern part of Lake Okeechobee. Dense, hard limestone,

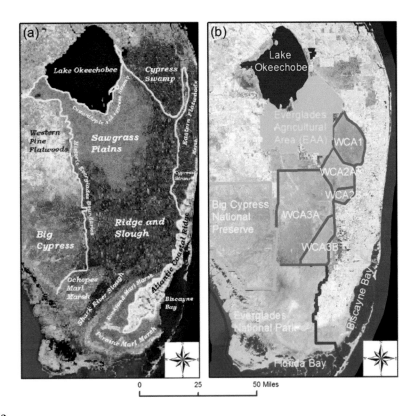

FIGURE 1.2
The predicted major landscapes of the Everglades in the 1800s before the drainages. (a) A synthetic satellite image reconstructed by the South Florida Water Management District (SFWMD) Hydrological Systems Modeling Division using the Natural System Model; it simulates the hydrologic response of the pre-drainage Everglades system to historical meteorological data. This view of the landscape was likely to have existed about 1850. (b) The current system of the Everglades on a mosaic of 2004 Landsat 7 ETM.

shells, and sand were formed. Rock beds are generally impermeable in this formation. The Anastasia Formation is underneath Palm Beach County and is much more permeable and filled with pocks and solution holes. The Fort Thompson and Anastasia Formations and Miami Limestone were formed during the Sangamon interglacial period (about 11,500 years before the present).

The biggest geological influence on the development of the Everglades is the Miami Limestone, which forms the floor of the lower Everglades. Miami Limestone is made up tiny egg-shaped concentric shells and calcium carbonate, called ooids. It has an extraordinary ability to store water and affects the hydrology, plants, and wildlife above it. The deposited limestone layer is known as the Biscayne Aquifer formed in the last glacial minimum around 110,000 years ago. Around the same time, the Atlantic Coastal Ridge (Figure 1.3(a)), a several mile-wide strip of limestone 20 feet higher than the current sea level, was deposited. This built the eastern edge of the Everglades. Meanwhile, along the western border of the Big Cypress Swamp, the Immokolee Ridge, also known as southwestern Pine Flatwoods, was formed as a slight rise with compressed sands. The rises in elevation along the eastern and western sides created a basin forcing water out of Lake Okeechobee to creep southward. When sea levels subsided, the

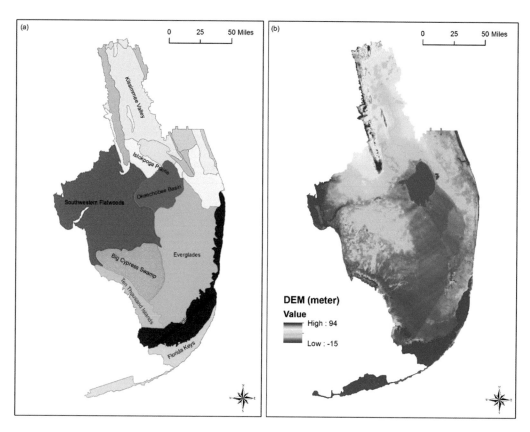

FIGURE 1.3
(a) physiographic provinces and (b) digital elevation model (DEM) of the Greater Everglades. The physiographic data is from St. Johns River Water Management District, and the DEM data is from USGS's Global Multi-resolution Terrain Elevation Data 2010, 7.5-arc-second products.

Atlantic Coastal Ridge kept much of its water flowing south into Florida Bay though some flowed east into Biscayne Bay. Changes in these freshwater flows have changed both bays. The formation of the Everglades is a joint result of geology, climate, and geography. About 5000 years ago, south Florida's climate took on its current subtropical and monsoonal character of dry winters followed by hot and moist summers with large amounts of rain (on average 50–60 inches per year), as seas surround it on three sides. The geographic location, along with the heavy precipitation in a wet season and the geological basin, the vast Everglades was formed from the water runoff of Lake Okeechobee. The ecosystems evolved with time, and the living habitats adapted to one another and to the nonliving environment. The stable warm weather, abundant fresh water, and sunlight in the past thousand years, consequently, led to a biologically productive Everglades ecosystem that abounded with fish and other aquatic animals. This in turn supported large number of reptiles, birds, and mammals.

The bedrock geology of the Everglades has been modified over time to respond to the processes of weathering, erosion, compaction of organic sediments, unique hydrologic conditions, and episodes of sea-level rise and fall. Locally, the Atlantic Coastal Plain province in the southern Everglades is divided into seven subprovinces: Okeechobee Basin, Everglades,

Southwestern Flatwoods, Big Cypress Swamp, Southern Atlantic Coastal Strip, Ten Thousand Islands, and the Florida Keys (National Park Service/Geology of Everglades, 2018) (Figure 1.3a). This division is closely related to the water flow in the Everglades. The Okeechobee Basin subprovince is the headwaters of the Everglades. Water slowly trickled south down to the sea from central Florida. During the summer wet season (April to October), fresh water overflowed the south shore of Lake Okeechobee and formed the *sheetflow* downstream 100 miles, dropping only 12 to 14 feet in elevation, to reach Florida Bay, as demonstrated on the digital elevation model (DEM) in Figure 1.3(b). Dams, canals, ditches, and levees were built in the basin, which largely dammed the natural *sheetflow* from Lake Okeechobee to Florida Bay, leading to a severe degradation of the original Everglades ecosystem.

In the long water journey, an elongate, south-dipping, spoon-shaped, and low-lying area was developed into the Everglades subprovince between the Southern Atlantic Coastal strip to the east and the Big Cypress Swamp to the west. The basin has very low relief, as demonstrated by the DEM in Figure 1.3(b). The low-lying basin of this subprovince and its slight slope from north to south provided a perfect geological condition to develop the Everglades. The elevation change is only 3.6 to 4.3 m (12–14 ft) from the maximum near Lake Okeechobee to sea level. Prior to the digging of canals and ditches and building of dams, flow in this drainage system was slow and steady from north to south. Two major types of soils appear in this subprovince, marl and peat. Marl is a product of periphyton, a complex assemblage of algae, cyanobacteria, microbes, and detritus. During the dry season (November to April), organic material in the periphyton oxidizes, leaving calcium carbonate behind as light-colored soil. Marl is common in the short-hydroperiod (short-term flooding) wet prairies of ENP, where bedrock lies close to the surface. Peat soil is a product of long-hydroperiod (long-term flooding) wetlands and typically occurs in areas of deeper bedrock. Peat is composed of the organic remains of dead plants and has a very low bulk density (often exceeding 85% pore space by volume) (Craft and Richardson, 1998; Chambers et al., 2015). The Everglades is a patterned peat-land formed from sawgrass and other aquatic plant materials that has accumulated over millennia (Hohner and Dreschel, 2015). Peat started to accumulate in the northern Everglades when it was formed 5000 years ago, and further south around 2000 to 3000 years ago (Gleason and Stone, 1994). The presence of peat deposits is evidence of the widespread flooding of the Everglades since it was formed. Studies have shown that peat accretion rate is heterogeneous across the Everglades with a higher rate in the northern region than in the southern part, leading to greater depth in the northern than the southern part in the Everglades (Richardson and Huvane, 2008; Richardson, 2010). In the late 1800s, field surveys identified 3 or more meters of highly organic peat soils in the Everglades, which was reduced to only 24% of its original volume, 17% of its mass, and 19% of its carbon after the drainage. A comparison of peat depth between pre-drainage (around 1885) and post-drainage (2005) is displayed in Figure 1.4. Peat soils accumulated to 4 meters in the north but are often less than 20 cm deep in portions of ENP. The deepest peats in the southern Everglades are found in depressions and major water flows, such as Shark River slough. Studies of peat accumulation and degradation in the Everglades have been of interest to scientists for decades. Accumulation of peat requires anaerobic (lack of oxygen) conditions. Without oxygen, microorganisms cannot decompose plant material as fast as it accumulates. Abundant precipitation in south Florida during the summer rainy season causes flooding of vast low-lying areas, which prevents oxygen in the air from touching soils and allows the organic material to transform into peat. If left undisturbed over long periods of time, increasingly thick layers of peat accumulate until the surface can dry sufficiently to allow either decay or fire. Marl and peat soils are opposites

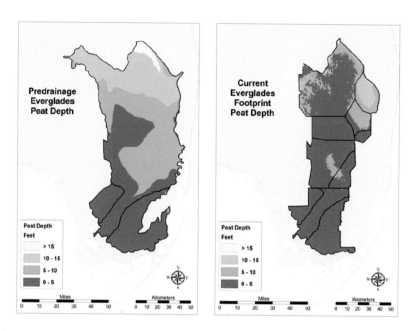

FIGURE 1.4
Peat depth comparison between pre-drainage (around 1885) and post-drainage (2005) based on the Natural Systems Regional Simulation Model (Hohner and Dreschel, 2015, from SFWMD).

that cannot coexist. Marl requires aerobic conditions; in contrast, peat requires anaerobic conditions. Peat does not accumulate in the short-hydroperiod marshes in which marl accumulates, and the acidic conditions in which a peat soil thrives would dissolve marl.

Current peat degradation in the Everglades is mainly caused by climate change and sea level rise, which leads to saltwater intrusion and "domino effects" to the entire coastal ecosystem. Salt water and fresh water beneath the land surface struggle for dominance in the coastal Everglades. Salt water moves landward due to sea level rise and drives the brackish zone farther inland. The saltier water causes peat soil to collapse and sawgrass to die, leaving a pockmark pattern of brackish pools of water in the wetland (Figure 1.5). Peat collapse was first described for Cape Sable, ENP, where canals were dug in 1920s. This allowed salt water to infiltrate freshwater marshes, leading to a dramatic modification in habitat over a timeframe of 10 to 20 years (SERES Project, 2016). The latest science shows that the process of peat collapse begins when freshwater marshes and peat soils are exposed to salt water. A cascade of events then takes place leading to the creation of a "hole" in the landscape. Depending on the duration of exposure, freshwater plants (e.g., sawgrass) are stressed or die back, increasing the exposure and temperature of the soil. This reduces or stops the production of roots that contribute to the formation of peat soil. Salt water also provides an enhanced supply of sulfate that increases bacterial activity. The combination of plant dieback and increased bacterial activity tips the peat soil balance toward a rapid breakdown and ultimate collapse. The long-term implication of peat collapse is that these habitats do not transition to mangrove forest. They remain as open water and contribute to a net loss of land. Scientists have noticed the impacts of climate change and sea level rise on peat collapse and

FIGURE 1.5
Photos showing evidence of peat collapse in a sawgrass marsh surrounded by expanding mangrove forest in lower Shark River Slough, ENP (From SFWMD).

relevant "domino effects" on the coastal Everglades and are making efforts to enhance the resilience of Everglades by a range of restoration projects.

Karst is also observed in the Everglades subprovince. It describes terrain produced by the chemical erosion of carbonate rocks such as limestone and dolomite. Acidic water dissolves the carbonate rock along cracks and fractures in the bedrock. Most precipitation is of relatively neutral pH but becomes increasingly acidic as it infiltrates live plant tissue, decaying plant debris and soils before seeping into the ground. Over thousands of years, dissolution within pore spaces and along fractures creates increasingly larger voids (National Park Service/Geology of Everglades, 2018).

The Southwestern Flatwoods subprovince lies to the northwest of the Everglades subprovince and west of the Okeechobee Basin subprovince. Miocene and Pliocene sedimentary rocks and sediments underlie this subprovince. Recent deposits are thin or nonexistent. Landforms in this subprovince include flatwoods, cypress swamps, rocklands, and marl plains. The Big Cypress Swamp subprovince defines the western boundary of the Everglades. Big Cypress National Preserve is within this subprovince, characterized by a great biodiversity including sawgrass prairies, estuaries, mangrove forests, pinelands, cypress swamps (strands and domes), hardwood tree islands and hammocks, slow-flowing freshwater sloughs, and coastal marshes. Currently, as much as 90% of the preserve is inundated from May to October. The rock under this subprovince is among the oldest in South Florida and is composed of silt, sand, and carbonate minerals. This area is slightly higher in elevation than the Everglades basin because it is underlain primarily by the coral-rich limestones of the Pliocene (3 to 5 million years ago) Tamiami Formation, which is exposed in large areas of Big Cypress National Preserve. Elevation in Big Cypress Swamp ranges from 3.6 to 12 meters (12 to 39 feet) above mean sea level. Drainage in the province is primarily to the south and southwest (Thornberry-Ehrlich, 2008). The Southern Atlantic Coastal strip subprovince consists of Pleistocene Miami limestone. The subprovince ranges in elevation from 1.5 to 6 m (5 to 20 feet) in the southernmost parts. The width of the linear ridge ranges from 16 km (10 miles) in southern Miami–Dade County to 5 to

8 km (3–5 miles) farther north. The southern parts of the ridge are breached in places by sloughs (transverse marshes) oriented perpendicularly to the trend of the ridge. The subprovince wraps around the southern end of Florida containing vast tracts of coastal marshes and mangrove swamps. Freshwater runoff and tidal fluxes cause the salinity to vary dramatically. The mangrove, capable of enduring such salinity changes, thrives in this area. Ten Thousand Islands subprovince is characterized by a vast labyrinth of mangrove forest, oyster bars, tidal sloughs, and lagoons. It extends for about 60 miles along the southwestern edge of the Big Cypress Swamp and Everglades subprovinces on Florida's Gulf Coast. The complex of small islands that make up the Ten Thousand Islands protects inland areas from the destructive and powerful winds, large breaking waves, storm surges, and flooding caused by tropical storms and hurricanes that rampage from time to time in the Gulf of Mexico. Florida Keys subprovince consists of long, narrow islands that stretch in an arcuate pattern from the southeastern tip of Florida to the Dry Tortugas. Coral rock underlies the northern keys. The islands that make up the northern, upper keys are the exposed remnants of coral reefs that fossilized and were exposed as the sea level declined. North of Elliott Key in Biscayne National Park lie several small transitional keys that are composed of sand built up around areas of exposed ancient coral reefs. Further north in the Miami metropolitan area, Key Biscayne is composed of sand, as are the barrier islands that protect much of the entire east coast of Florida.

1.3 Everglades Ecosystem

The current South Florida is a regional system of humanity and nature. The original Everglades ecosystem before 1900 was characterized by the wildness of bays, marshes, swamps, pine forests, and prairies. Each of these natural features was defined by its physical and chemical conditions of the land, by climate, and by representative plants, animals, and microbes. This jointly constituted an integrated ecosystem of the original Everglades (McPherson et al., 1976). Each ecosystem within the Everglades was sustained by sunlight, wind, rain, nutrients, and along the coast by tides and currents. Each, however, was not isolated, but rather dependent on nearby ecosystems for some of its nutrients, water, or wildlife. Each of the interdependent ecosystems, then, was a subsystem of the regional ecosystem. After 1900, due to human activities, two components were added into the natural ecosystem of the Everglades, an agricultural and an urban component. Some of the natural ecosystems have remained to the present. This book presents a brief review of the major natural subsystems in the Greater Everglades, the freshwater ecosystem and the coastal ecosystem. A detailed description of the ecology of South Florida can be found in McPherson et al. (1976).

1.3.1 Freshwater Ecosystem

Natural fresh water in the Greater Everglades lies within four physiographic regions (Figure 1.3(a)): (1) the Okeechobee basin; (2) the Everglades; (3) the Southwestern Flatwoods; and (4) the Big Cypress Swamp. The Everglades and the Big Cypress Swamp are predominantly freshwater marsh and swamp systems. The Flatwoods are composed of swamp, marsh, and terrestrial systems. All of these natural systems are affected to some degree by human activities and interact with man-made ecosystems.

The freshwater ecosystem in the Greater Everglades is characterized by lakes, sawgrass marsh, wet prairies, sloughs, pine forests, cypress forests, and hardwood hammock.

1) Lakes

The largest lake in the Everglades is Lake Okeechobee, which serves as the "water heart" of the Everglades (Figures 1.1 and 1.2). It covers 730 square miles (1,900 km^2) with a shallow depth. Lake Okeechobee is fed from the north mainly by the Kissimmee River and has been flanked by pumping stations along its south shore, which during wet periods, pump surplus water from the north part of the Everglades agricultural area into the lake. Generally, during dry periods, water is released from the lake to sustain the agricultural areas just south of the lake (Figure 1.2(b)) and to supply water to the urban lower east coast. When levels in the lake exceed scheduled elevations during some rainy seasons, releases are made through the St. Lucie Canal and the Caloosahatchee River. Channels, gates, and levees were constructed in the lake area for flood control due to the historical catastrophes caused by hurricanes. Reduced water quality and relevant ecological damages such as algal bloom are the main concerns in this region.

2) Sawgrass marsh or sawgrass prairie

Sawgrass is the icon of the Everglades (Figure 1.6). It usually occurs on land slightly higher than that of the sloughs and wet prairies. Sawgrass communities are characterized by the large sedge, sawgrass, which comprises approximately 65 to 70% of the total vegetation cover of the remaining Everglades spanning from Lake Okeechobee to Florida Bay. This is the titular "River of Grass" popularized by Marjory Stoneman Douglas, a conservationist known for her

FIGURE 1.6
Sawgrass in the Everglades (from USGS).

staunch defense of the Everglades against efforts to drain it. Some authors refer to the sawgrass and water combination as the "true Everglades" or just "Glades" (McPherson and Halley, 1996). Sawgrass communities thrive in slowly moving water but may die in unusually deep floods if oxygen is unable to reach their roots. They appear as the dominant plant, either in almost pure stands or mixed with a wide variety of other sedges, grasses, herbs, and attached emergent or floating leafed aquatic plants. The composition of sawgrass marsh is affected by prevailing water conditions. Abundance and density in the community are greatly influenced by water depth, period of inundation, and rate of rise and fall of water levels. A longer hydroperiod (a period during which a wetland is covered by water) along with increased water depths produces taller, thicker stands of sawgrass, while a short hydroperiod and shallow waters result in limited growth. Dense areas of sawgrass have low species diversity, with alligators often nesting in these habitats. During most of the year, wet soils protect sawgrass roots from damaging fires. While the above-ground plant tissues are burned, the wet roots can survive, allowing the plant to make a complete recovery. Fires play an important role in sawgrass habitats by limiting the invasion of woody vegetation that would eventually change the sawgrass marsh into the next successional habitat (Florida Museum, 2018).

The sawgrass-flag-maidencane type is the most common sawgrass community type and covers vast areas throughout the entire marsh. Sawgrass of this type may be either sparse or dense, sometimes occurring in communities of almost pure stands. It grows in association with flag, maidencane, pickerel, and cattails. Small trees and shrubs also commonly present including buttonbush and willow. The sawgrass-myrtle-dahoon holly, uncommon in the Everglades prior to the initiation of drainage, has become increasingly important because of the general lowering of water levels and perhaps fire (Loveless, 1959). It often appears in a drier site and covers extensive areas in some sections in the Everglades. Sawgrass-maidencane occurs irregularly over much of the region and is like the sawgrass-flag-maidencane in its species composition. During dry years, several incipient plants can appear in sawgrass communities such as dog fennel, plume-grass pigweed, and goldenrod.

Based on a land cover land use dataset interpreted from USGS Color Infrared Digital Orthophotographic Quarter Quadrangles (DOQQs) acquired from November 2004 through March 2005, sawgrass marshes cover about 8676 km^2 in the Greater Everglades, and their distribution is displayed in Figure 1.8. Remaining sawgrass marshes are mainly distributed in the Water Conservation Areas and ENP. Sawgrass once covered the northern portion of the Everglades (south of Lake Okeechobee), growing to heights of over 9 feet (2.7 m) tall. The rich, dark peat soil was converted into agricultural areas to grow sugarcane (Figure 1.2).

3) Wet prairie

Wet prairie is also common in the freshwater ecosystem. It is dominated by emergent plants including grasses and other low growing plants, supporting a greater diversity of species than other types of marshes. Wet prairies are found on continuously wet, but not inundated soils, on somewhat flat or gentle slopes between lower lying depression marshes, shrub bogs, or dome swamps and slightly higher wet or mesic flatwoods or dry prairies. There are two types of wet prairies in the Everglades: marl prairie and water-marsh community (Figure 1.7). Marl prairies are located where marl covers limestone that may protrude as pinnacles or erode into solution holes, which are depressions in the Earth's surface caused by dissolving substrate composed primarily of calcium carbonate. Located along both the eastern and western portions of the Everglades, marl prairies have

FIGURE 1.7
Wet prairies in the Everglades: marl prairie (top) and water-marsh prairie (bottom) (from USGS and SFWMD).

a short hydroperiod of only 3 to 7 months a year. The water-marsh community of wet prairies is found over rich peat soils with a longer hydroperiod than marl prairies. This community is often located between sawgrass marshes and sloughs with less plant diversity. It provides habitat for fish, waterfowl, and other wildlife that require a permanent source of water. During the wet season, animals of the wet prairies include an assemblage of aquatic and semiaquatic species like those of the sloughs. During the dry season, aquatic animals such as the freshwater prawns and fish are forced into the deeper sloughs and ponds. Water levels rarely recede more than a foot below the land surface except in abnormally dry years (Loveless, 1959; McPherson et al., 1976). Alligators have created an ecological niche in wet prairies; they dig at low spots with their claws and snouts and create ponds free of vegetation that remain submerged throughout the dry season. Alligator holes are integral to the survival of aquatic invertebrates, turtles, fish, small mammals, and birds during extended drought periods. The 2004 land cover land use dataset shows that wet prairies cover approximate 1012 km^2 in the current Greater Everglades. They occur throughout the Everglades, in parts of the Big Cypress Swamp and Flatwoods, as shown in Figure 1.8.

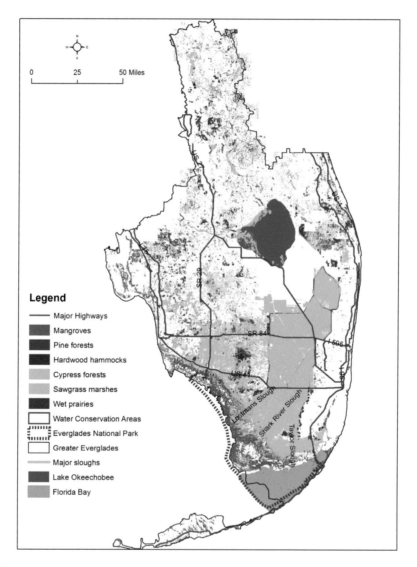

FIGURE 1.8
Distribution of sawgrass marshes, wet prairies, pine forests, cypress forests, and hardwood hammock in the Greater Everglades. Map is based on the land cover land use data of 2004 interpreted from USGS Color Infrared Digital Orthophotographic Quarter Quadrangles (DOQQs) acquired from November 2004 through March 2005 (SFWMD). Sawgrass marshes, wet prairies, pine forests, cypress forests, and hardwood hammock cover about 8676 km^2, 1012 km^2, 1560 km^2, 2305 km^2, and 228 km^2, respectively, in the current Greater Everglades. Mangrove data is from USGS (Giri et al., 2011).

4) Sloughs

Sloughs are channels of free-flowing water between the sawgrass marshes and play important roles to deliver fresh water to bays and estuaries, a function like rivers and canals in the Everglades (Figure 1.9). However, flow in sloughs is slower than in a river. Sloughs are deeper than sawgrass marshes and stay flooded for at least 11 months out of the year if not multiple years in a row. The peat beds that support sawgrass are slightly elevated and may begin abruptly creating ridges of grass. The elevated sawgrass lands and deep sloughs between the borders are called the "ridge-and-slough" landscape in the Everglades. The "ridge-and-slough" once was the major landscape in the Everglades (Figure 1.2(a)), but only three major sloughs remain, including Lostmans Slough bordering the Big Cypress Preservation, Shark River Slough draining to Gulf of Mexico, and Taylor Slough located on the east side of the southern Everglades draining water to Florida Bay (Figure 1.8). In the pristine state, soil surface of ridges is 2 to 3 feet higher than that of the sloughs. Despite the elevation difference between ridges and sloughs, ridges are also covered with water for most of the year. Ridges can dry out entirely during the dry season. Tree islands are a third element in the "ridge-and-slough" landscape. They are higher in elevation than ridges and tend to expose soil except during periods of unusually high water (Figures 1.10 and 1.11). A diverse assemblage of plants, animals, and microorganisms is found in the sloughs. Plants are usually submerged or floating such as bladderwort, waterlily, and spatterdock. Aquatic animals such as turtles, young alligators, snakes, and fish live in sloughs, and they usually feed on aquatic invertebrates. Periphyton, an assemblage of algae, and other microorganisms, also thrive in sloughs. Periphyton often forms dense calcareous mats several inches thick that are attached to the bottom or to vascular plants. Because of its concentration of microorganisms, which require oxygen and other nutrients, periphyton has a pronounced effect on water quality.

FIGURE 1.9
Freshwater sloughs in the Everglades (from National Park Service).

FIGURE 1.10
The "ridge-and-slough" with tree islands landscape in the Everglades (from SFWMD).

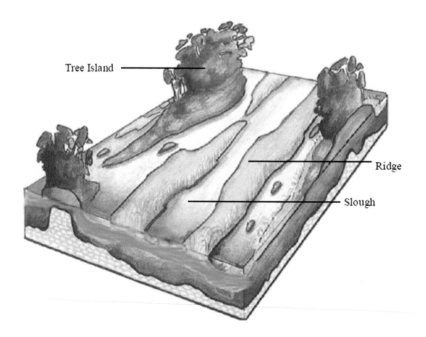

FIGURE 1.11
Artist's reconstruction of the landscape of "ridge-and-slough" and tree islands in the Everglades (from National Academies Press, 2003).

FIGURE 1.12
Pine forests and fires in the Everglades (from USGS).

5) Pine forests

Pine forests (Figure 1.12) are common in the Everglades, and they are mainly distributed in the Big Cypress Swamp and Southwestern Flatwoods, north of the Big Cypress Swamp, and in the Southern Atlantic Coastal Strip (Figure 1.8) (McPherson et al., 1976). They are also known as pine rocklands or pinelands and cover the highest part of the Everglades with little to no hydroperiod, although some floors may have flooded solution holes or puddles for a few months at a time (McPherson and Halley, 1996). In much of the Big Cypress Swamp, the pine forests grow on small islands of limestone several inches to several feet higher than surrounding cypress forest land. In the northern Big Cypress Swamp, however, pine forests occur extensively on a relatively high sandy soil. Interspersed with the pines are prairies and depressions or low areas of marsh and swamp. In the western Big Cypress Swamp, pines and cypress grow together over large areas of poorly drained soil. On the Southern Atlantic Coastal Strip, generally 3 to 7.5 m (10 to 25 feet) above sea level, pines grow on extremely rough and solution-pitted oolitic limestone. The characteristic plants of pine forests are the slash pine and a variety of hardwood trees and shrubs, palms, grasses, and other plants. Slash pine can reach a height of 22 feet (6.7 m); it is used in paper and pulp industries. Fire is important to maintain pine forest ecosystems. It eliminates competing vegetation on the forest floor and opens pinecones to germinate seeds. A period without significant fire can turn pineland into hardwood hammock as larger trees overtake the slash pines. Today prescribed fires occur in ENP in pine forests every 3 to 7 years. Wildlife in pine forests is diverse including 15 species of birds and more than 20 species of reptiles and amphibians. Based on the land cover land use data of 2004, pine forests cover about 1560 km^2 in the Greater Everglades.

6) Cypress forests

Cypress forests (Figure 1.13) are mainly distributed in the Big Cypress Swamp, which is slightly elevated at 6.7 m (22 feet) at its highest point and slopes gradually to the coastline for approximately 56 km (35 miles) (Figure 1.8). Cypress forests are well adapted to the water-logged soils of the Everglades. Their roots produce "knees" that protrude above the soil, ranging from a few inches to 6 feet (several centimeters to 1.8 meters) tall, which aid in respiration and provide required oxygen to the root tissues and structural support in water-logged soils. Forests in the swamp include open areas of small cypress trees and a scattered sparse growth of herbaceous plants, growing on a thin layer of marl soil or sand over limestone. Cypress domes and strands of larger trees grow over much of the forest. Domes are circular or egg-shaped features that are dome shaped in profile on the horizon. Strands are elongate areas of large trees that follow depressions. Shrubs and small swamp trees such as wax myrtle, coco plum, and pond apple are common understory species within the domes and strands. They grow in areas that are covered in water for longer periods than the surrounding marshlands; they require abundant moisture for 1 to 3 months after seedfall for germination. Water facilitates germination by allowing the hard seed coats to swell and soften. Seeds covered by water as long as 30 months may germinate if the water recedes. After germination, the seedlings require dry conditions for a time, and to survive they must grow high enough to stay above the seasonal floods of the next rainy season. Once established, larger trees can grow in the absence of seasonal inundation. Lowered water levels, however, make cypress susceptible to fire (McPherson et al., 1976). Two species of cypress occur in the Everglades, the bald cypress and the pond cypress. Cypress trees can live for hundreds of years, and some can grow up to 130 feet (40 m) and 500 years old. Current cypress forests are preserved and protected in the Big Cypress National Preserve, Corkscrew Swamp Sanctuary, Fakahatchee Strand State Preserve, and two Indian reservations (McPherson and Halley, 1996). Based on the land cover land use data of 2004, cypress forests cover about 2305 km^2 in the Greater Everglades.

7) Hardwood hammocks

Hardwood hammocks (Figure 1.14) are areas of dense vegetation including hardwood trees and shrubs, palms, ferns, and epiphytes. They grow on tree islands in freshwater sloughs, sawgrass marshes, or pine forests. They are found throughout the Everglades (Figure 1.8) but only cover about 228 km^2 based on the 2004 land cover land use data. Hammocks in the northern Everglades consist of more temperate plant species, including the live oak and the hackberry; closer to Florida Bay, the trees are tropical and smaller shrubs are more prevalent (McPherson and Halley, 1996). These tropical species were transported to the Everglades by migrating birds, winds, and water

FIGURE 1.13
Cypress forests in the Big Cypress Swamp (from National Park Service).

currents over a long period of time. Near the bases of hammocks, saw palmettos flourish, making the hammocks very difficult to penetrate. Water in sloughs flow around the islands, creating moats. Hardwood hammocks are fire resistant habitats due to their high humidity and open understory. However, they are susceptible to fire damage under certain conditions. During drought conditions, hammock habitats can be destroyed by natural fires. It may take decades for a hammock to recover after suffering major fire damage (Florida Museum, 2018). Moats are therefore essential for protection. Hardwood hammocks are home to many threatened and endangered plant species. Today more than 120 species of tropical plants including trees and shrubs are found in the Everglades. Out of these 120 species, 36 are listed as endangered or threatened.

FIGURE 1.14
Hardwood hammock in ENP (from National Park Service).

1.3.2 Coastal Ecosystem

The coastal ecosystems in the Everglades include sandy beaches, mangroves and salt marshes, shallow bays, and Florida reef tract, which blend from one to another (McPherson et al., 1976). Water is the medium connecting these coastal ecosystems. Salt water from offshore currents moves over the reef tract (Figure 1.15), through the bays and mangroves (Figures 1.16), and back offshore carrying physical and biological products from one coastal environment to another. Fresh water, moving overland from the mainland through the mangroves, into the bays, and out to sea, transports terrestrial products to marine habitats. Where moving water acquires sufficiently different characteristics, it supports different plant and animal communities. The Florida Current, moving around the tip of Florida, provides clean, warm, and saline water to the coastal areas, forming coral reef ecosystems and supporting a diverse tropical flora and fauna. It also transports and distributes juvenile fish and invertebrates to coastal waters. The interaction of fresh water and tidal saline water creates the extensive grass beds with highly diverse and productive habitats. This book provides more details about two coastal ecosystems: reef tract and seagrass beds, and mangrove forests. We applied remote sensing data to characterize them.

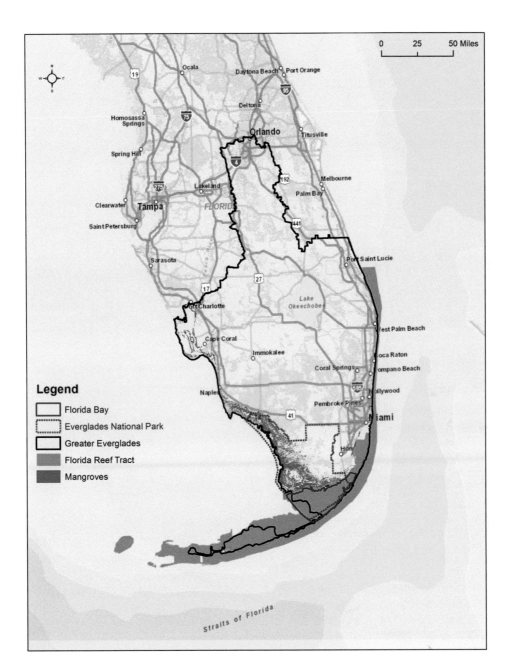

FIGURE 1.15
Florida reef tract and mangroves in the coastal ecosystems of the Everglades. Reef tract data is from Florida Fish and Wildlife Conservation Commission-Fish and Wildlife Research Institute (Unified Florida Patch Reef Map v1.0, 2014); mangrove data is from Giri et al. (2011).

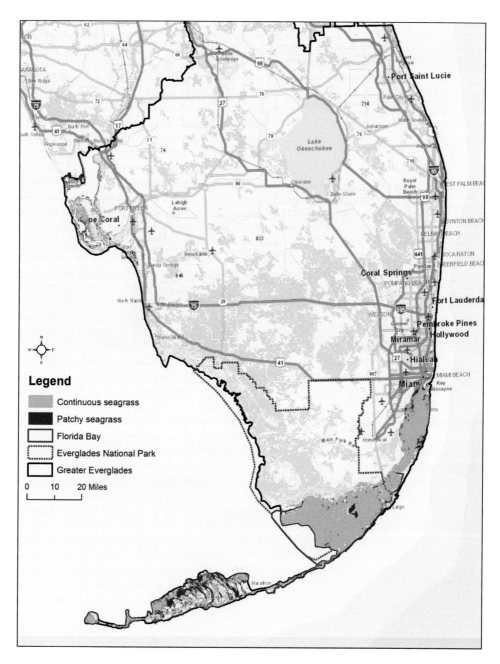

FIGURE 1.16
Seagrass distribution in the Greater Everglades. Seagrass data is from Florida Fish and Wildlife Conservation Commission, which are compiled from various source agencies and scales in date from 1987 to 2016.

FIGURE 1.17
Riverine mangrove forest at the mouth of the Shark River Slough in the Everglades. (from National Science Foundation. Credit: Victor H. Rivera-Monroy).

1) Mangrove forests

Mangroves thrive in the transitional zone where fresh water meets salt water in the coastal Everglades (Figure 1.17). The brackish mixture of water and mangroves, crisscrossed by many tidal creeks in the southern tip of the Everglades, harbors a very productive ecosystem. The Everglades have the most extensive contiguous system of mangroves in the world. The mangrove forests of the Ten Thousand Islands cover almost 200,000 acres (810 km^2) (McPherson and Halley, 1996). A total 2097 km^2 of mangroves are present in the Greater Everglades (Figures 1.8), based on a mangrove dataset from USGS (Giri et al., 2011). Three mangrove species are found: red mangroves, black mangroves, and white mangroves. They are all tolerant of salt, brackish, and fresh water, grow in oxygen-poor soils, and can survive drastic water level changes. Mangroves provide a range of benefits to humans and the natural environment, such as stabilizing coastlines, protecting communities from storms, and storing vast amounts of carbon (Zhang et al., 2012; Beck et al., 2018). They are vulnerable to sea level rise and tropical storms such as hurricanes (Doyle et al., 2003; Ellison, 2015).

2) Reef tract and seagrass beds

Florida is the only state in the continental United States to have extensive shallow coral reef formations near its coasts. These reefs extend from St. Lucie Inlet in Martin County to the Dry Tortugas in the Gulf of Mexico (Figure 1.15). The most prolific reef development occurs seaward of the Florida Keys, and the most extensive living coral reef is adjacent to the island chain of the Florida Keys (NOAA/Coral Reef Information System, 2018). These reefs, nearly 100,000 years old, form a tract almost 240 km (150 miles) long and about 6 km (4 miles) wide sloping to the edge of the Florida Straits. The Florida Reef Tract is the third largest barrier reef ecosystem in the world. Discontinuous and less biologically diverse coral reef communities continue northward to St. Lucie Inlet. Nearly 25% of all ocean life thrives on coral reefs, making these fragile habitats a necessity to ocean ecosystems. The reef tract is threatened by pollution, overfishing, and climate change. A joint reef-monitoring program conducted by the Environmental Protection Agency (EPA), Florida Marine Research Institute, and National Oceanic and Atmospheric Administration (NOAA) recorded a loss of 6% to 10% living corals at 40 sampling stations from 1996 to 2000. Currently, the Florida reef is protected by federal and state governments, including the establishment of John Pennekamp Coral Reef State Park, Biscayne National Park, and the Florida Keys National Marine Sanctuary (Figure 1.18). Reef restoration efforts have also been made through many projects such as the Coral Reef Evaluation and Monitoring Project, which has monitored the condition of coral reef and hard-bottom habitats annually throughout the Florida Keys since 1996.

The Reef Tract is co-distributed with seagrasses, which are submerged vascular plants and can form dense vegetative communities in shallow water estuaries. These "grass-like" plants grow in a highly variable salinity environment. In the coastal Everglades, they are mainly distributed in Florida Bay, Biscayne Bay, and Florida Keys, with an approximate coverage of 2854 km^2 (Figure 1.16). Three seagrass species commonly occur in the coastal Everglades: turtle grass, shoal grass, and manatee grass (Figure 1.19). Like coral reef and

FIGURE 1.18
Coral reefs in Florida Keys National Marine Sanctuary (from NOAA).

FIGURE 1.19
Three major seagrass species occur in the coastal Everglades. Top: turtle grass, the largest of the Florida seagrasses, has deeper root structures than other species and forms extensive beds in Florida Bay; middle: shoal grass, an early colonizer of vegetated areas, usually grows in very shallow water; bottom: manatee grass, usually found in mixed seagrass beds or small, dense monospecific patches (from Florida Department of Environmental Protection).

mangrove ecosystems, seagrasses are an important component of the coastal ecosystems in the Everglades. They provide food and habitat to numerous marine species, stabilize the ocean bottom, maintain water quality, and help support local economies (Florida Fish and Wildlife Conservation Commission/Seagrass, 2018). They are also threatened by humans, storm activity, and climate change. The wave energy introduced by storms can uproot seagrasses and cause extensive damage. Other climate conditions, such as floods and droughts, can affect seagrasses by changing the salinity of the water, which can affect their distribution. Grazing is another natural factor affecting seagrasses. Some marine animals, such as the endangered Florida manatee and green sea turtle, feed directly on seagrasses. Many smaller organisms forage for food within seagrass beds: crabs, fish, skates, and rays can disturb seagrasses during their search for food. In addition to these natural threats, humans can damage seagrass communities through activities such as dredging and boating. Run-off is another major problem because it can change water quality and reduce the amount of light reaching these plants. Docks and boats can shade seagrass beds, causing them to die from lack of light. To protect seagrasses, many projects have been conducted including seagrass habitat characterization, restoration and monitoring methods.

1.4 Everglades Restoration

In the last century, the Greater Everglades was largely modified due to human activities. Levees and canals were built for agriculture development, urban development, tourism, and real estate development in the state. The 1926 Miami Hurricane and 1928 Okeechobee Hurricane caused widespread devastation and flooding, which resulted in the construction of a dike around Lake Okeechobee. Flood from hurricanes further pushed the construction of 2,300 km (1,400 miles) of canals and levees, hundreds of pumping stations, and other water control devices under the Central and Southern Florida Flood Control Project (C&SF) in 1947. Urban development and agricultural use in 1960s further decreased the size of the original Everglades due to the altered freshwater flow through the system. In sum, 30% of the original Everglades has been converted to agricultural and urban development, and 35% has been converted as Water Conservation Areas (WCAs) 1, 2, and 3 for flood protection, water supply, and allied purposes of navigation and fish and wildlife protection (Figure 1.2). The remaining 25% of the Everglades in its original state is protected in ENP, which is listed as a World Heritage Site in 1979 by the United Nations Educational, Scientific and Cultural Organization (UNESCO) and included on the list of Wetlands of International Importance in 1987 by Ramsar Convention. The current Greater Everglades is a mosaic of inter-connected freshwater wetlands and estuaries located primarily to the east and south of the Everglades Agricultural Area. In total, approximately 50% of the Everglades' habitat has been lost. The "ridge-slough" and tree islands landscape extends through the Water Conservation Area (WCA) 1, 2, 3A, and 3B into Shark River Slough within ENP. The "ridge-slough" wetlands drain into tidal rivers that flow through mangrove estuaries into the Gulf of Mexico and Florida Bay. Higher elevation wetlands in either side of Shark River Slough are characterized by marl substrates and exposed limestone bedrock. The marl wetland areas located to the east of Shark River Slough form the drainage basin for Taylor Slough, which flows through an estuary of dwarf mangrove forests into northeastern Florida Bay. The Everglades marshes merge with the forested wetlands of Big Cypress National Preserve to the west of WCA 3 and

ENP (CERP, Program-level Adaptive Management Plan, 2015). The drainage also modified the vegetation community, as shown in Figure 1.20.

In the 1970s, as environmental protection became a national priority, protection activities in the Everglades occurred such as the declaration of Big Cypress National Preserve and Fakahatchee Strand State Preserve. Though protections were established, the degraded ecosystems resulting from changes to the natural hydrologic regime in the original Everglades were visible. Biology and hydrology are closely linked. The Everglades now supports 90% fewer nesting birds than it did in the 1930s. ENP is still threatened by human activities and the modified hydrologic regime. In the 1980s, projects were conducted in the park to change water quantity and quality to sustain the park's resources, but more restoration efforts were needed to accomplish restoration of the Greater Everglades ecosystem, which includes the Kissimmee River, Lake Okeechobee, and ENP. In 2000, the Comprehensive Everglades Restoration Plan (CERP) was authorized by Congress to restore, preserve, and protect the south Florida ecosystem while providing for other water-related needs of the region, including water supply and flood protection (Kloor, 2000; CERP, 2018). CERP is the largest hydrologic restoration project ever undertaken in the United States, and the restoration efforts will take decades and require the resolution of complex science, engineering, management, and policy issues by many federal, Native

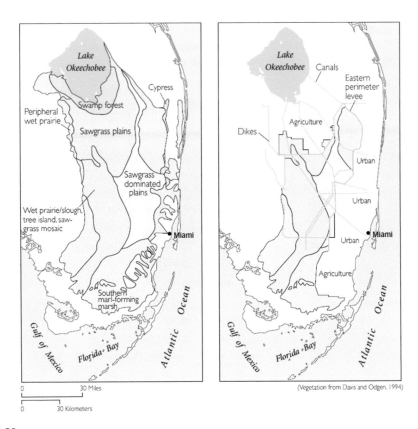

FIGURE 1.20
Water management has changed the vegetation patterns in the Greater Everglades. Left: the historical vegetation pattern around 1900; right: the current vegetation pattern around 1990 (Galloway et al., 1999, USGS).

American, state, and local organizations. Its goal is to use the best available science to restore the "right quantity, quality, timing, and distribution" of freshwater in the Everglades. It will increase freshwater storage capacity and distribute this fresh water for urban, agricultural, and environmental purposes while reducing damaging floodwater discharges in the coastal estuaries. CERP ensures that water for the natural systems remains its priority. The historical, current, and planned water flows under CERP are displayed in Figure 1.21. CERP is expected to reduce 20% water flow per year to the Atlantic Ocean with new environmental water flow to ENP. Flow to Lake Okeechobee is expected to be near 1994 levels, while outflows from the lake to the Caloosahatchee River are expected to double. Note that CERP was never intended to return the Everglades to its pre-drainage state but rather to a state of enhanced functionality by minimizing or removing impacts caused by the operation of the current C&SF infrastructure.

CERP contains 68 project components and associated operational regimes such as water quality programs, northern Everglades initiatives, and invasive exotic species strategic action. A Florida-based working group has been established to support the tasks in CERP, representing tribal, local, state. and federal entitles. With CERP's approval, the US Congress also requested adaptive management principles during Everglades restoration to improve the Plan based on new information, improved modeling, new technology, and changed circumstances (CERP/Program-level Adaptive Management Plan, 2015). The major science and research components in CERP include hydrological restoration to store, clean, and redirect water back to the Everglades, water quality improvement by reducing phosphorus to the Everglades (SERES Project, 2016), fresh water amount to estuaries, restoring peat, restoring "ridge-slough" and tree islands landscape, restoring food web, impact of climate change and sea level rise on restoration, and reducing invasive species.

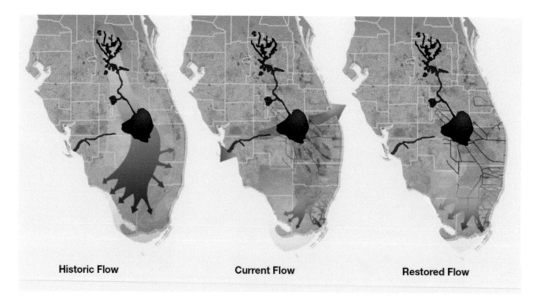

FIGURE 1.21
Left: the historical flow in the Greater Everglades; center: the current flow showing much of the flow going to the Caloosahatchee (west) and St. Lucie Rivers (east); right: the restored flow reflecting the intended pattern and magnitude of flows in the current Greater Everglades (from National Park Service).

References

Beck, M.W., Narayan, S., Trespalacios, D., et al., 2018. *The Global Value of Mangroves for Risk Reduction*, Summary Report, The Nature Conservancy, Berlin.

CERP/Program-level Adaptive Management Plan, 2015. https://evergladesrestoration.gov/. Accessed on 31 October 2018.

Chambers, L.G., Davis, S.E., and Troxler, T.G., 2015. Sea level rise in the Everglades: Plant-soil-microbial feedbacks in response to changing physical conditions. In: Entry, J.A., Gottlieb, A.D., Jayachandrahan, K., and Ogram, A., (eds.) *Microbiology of the Everglades Ecosystem*, CRC Press, Boca Raton, FL, 89–112.

Comprehensive Everglades Restoration Plan (CERP), 2018. https://evergladesrestoration.gov/. Accessed on 31 October 2018.

Craft, C.B., and Richardson, C.J., 1998. Recent and long-term organic soil accretion and nutrient accumulation in the Everglades. *Soil Science Society of America Journal*, 62, 834–843.

Davis, J.H., 1943. The natural features of southern Florida, especially the vegetation, and the Everglades Florida. *Geological Survey Bulletin*, 25, 1–311.

Doyle, T.W., Girod, G.F., and Books, M.A., 2003. Modeling mangrove forest migration along the southwest coast of Florida under climate change. In: Ning, Z.H., Turner, R.E., Doyle, T.W., and Abdollahi, K. (eds.) *Integrated Assessment of the Climate Change Impacts on the Gulf Coast Region*, GCRCC, Baton Rouge, LA, 211–221.

Ellison, J., 2015. Vulnerability assessment of mangroves to climate change and sea-level rise impacts. *Wetlands Ecology and Management*, 23, 115–137.

Florida Fish and Wildlife Conservation Commission/Seagrass, 2018. http://myfwc.com/research/habitat/seagrasses/. Accessed on October 2018.

Florida Museum, 2018. www.floridamuseum.ufl.edu/. Accessed on 31 October 2018.

Galloway, D., Jones, D.R., and Ingebritse, S.E., 1999. *Land Subsidence in the United States*. US Geology Survey, Circle 1182, Reston, VA.

Giri, C., Ochieng, E., Tieszen, L.L., et al., 2011. Status and distribution of mangrove forests of the world using earth observation satellite data. *Global Ecology and Biogeography*, 20, 154–159.

Gleason, P.J., and Stone, P., 1994. Age, origin, and landscape evolution of the Everglades peatland. In: Davis, S.M. and Ogden, J.C. (eds.) *Everglades: The Ecosystem and Its Restoration*, St. Lucie Press, Delray Beach, FL, 149–197.

Hohner, S.M., and Dreschel, T.W., 2015. Everglades peats: Using historical and recent data to estimate predrainage and current volumes, masses and carbon contents. *Mires and Peat*, 16, 1–15.

Kloor, K., 2000. Everglades restoration plan hits rough waters. *Science*, 288, 1166–1167.

Loveless, C.M., 1959. A study of the vegetation in the Florida Everglades. *Ecology*, 40, 1–9.

McPherson, B.F., and Halley, R.B., 1996. *The South Florida Environment: A Region under Stress*, US Geology Survey, Circular 1134, Reston, VA.

McPherson, B.F., Hendrix, C.Y., Klein, H., and Tyus, H.M., 1976. The environment of south Florida, a summary report. Geological Survey Professional Paper 1011.

National Academies Press, 2003. *Does Water Flow Influence Everglades Landscape Pattern?* National Academies Press, Washington, DC, USA.

National Park Service/Geology of Everglades, 2018. www.nps.gov/ever/learn/nature/evergeology.htm. Accessed on 31 October 2018.

NOAA/Coral Reef Information System, 2018. www.coris.noaa.gov/. Accessed on 31 October 2018.

Richardson, C.J., 2010. The Everglades: North America's subtropical wetland. *Wetlands Ecology and Management*, 18, 517–542.

Richardson, C.J., and Huvane, J.K., 2008. Ecological status of the Everglades: Environmental and human factors that control the peatland complex on the landscape. In: Richardson, C.J. (ed.) *The Everglades Experiments: Lessons for Ecosystem Restoration*, Springer, New York, 13–58.

SERES Project, 2016. Management-driven science synthesis: An evaluation of Everglades restoration trajectories. Technical Report by Davis, S., Naja, G., Lent, T., Wetzel, P., Davis, S., et al., Palmetto Bay, FL, 60.

Thornberry-Ehrlich, T., 2008. *Big Cypress National Preserve Geologic Resource Evaluation Report.* Natural Resource Report NPS/NRPC/GRD/NRR—2008/021. National Park Service, Denver, CO.

USGS/Florida Everglades, 2019. https://archive.usgs.gov/archive/sites/sofia.usgs.gov/publications/circular/1182/. Accessed on 25 June 2019.

Zhang, K., Liu, H., Li, Y., et al., 2012. The role of mangroves in attenuating storm surges. *Estuarine, Coastal and Shelf Science*, 102–103, 11–23.

2

Introduction to Remote Sensing

Caiyun Zhang

2.1 Overview of Remote Sensing

Remote sensing is the science, technology, and art of collecting information of an object (e.g., a building or a tree) or phenomenon (e.g., El Niño) without being in direct physical contact with it (Jensen, 2007). To collect the information, a recording device known as a sensor is required. To generate useful information, various techniques, such as preprocessing, interpretation, and analysis, are applied to the collected data. Remote sensing data can be acquired from different sensors, and these sensors can be mounted on different platforms such as aircraft, spacecraft, and ships. Remote sensors can be grouped into two types: passive instruments and active instruments. Passive sensors detect natural/solar energy that is reflected or emitted from the observed scene, while active sensors measure the reflection or emission from the observed scene using their own energy sources (Figure 2.1). Only a limited spectral range can be used in remote sensing because the radiation from the targets passes through the atmosphere, which absorbs most of the spectral regions. The commonly used spectra include visible, infrared, thermal, and microwave for Earth observations, as shown in Figure 2.2.

Both passive and active remote sensing can be used in the Everglades. Several passive imaging sensors have been developed for mapping and monitoring Earth, but only a few can be used for mapping the Everglades due to the varying spatial resolution of these sensors. The spatial resolution, also known as ground resolution, is the ground area that an image pixel covers. It describes how much detail in an image is visible to human eyes. For example, when we say 30 meters Landsat data, we are saying that the spatial resolution of Landsat is 30 meters and an object with a length and width of at least 30 meters in 2 dimensions can be observed. Targets less than 30 meters will be challenging to observe using Landsat. Because the Everglades has a high degree of heterogeneity, only sensors with relatively high spatial resolution can discriminate the targets over an area of interest for Everglades restoration and management. The primary passive remote sensor data in the Everglades include aerial photography with a spatial resolution of 5 meters or less, drone imagery and products with a spatial resolution of 1 meter or less, Landsat imagery with a spatial resolution of 30 meters, and hyperspectral imagery with a spatial resolution of 30 meters or less. Active sensors such as light detection and ranging (lidar) have also been used in the Everglades. In this book, these primary passive and active sensors and their relevant products are detailed and organized into three sections: multispectral remote sensing, hyperspectral remote sensing, and lidar. They are presented in the following sections.

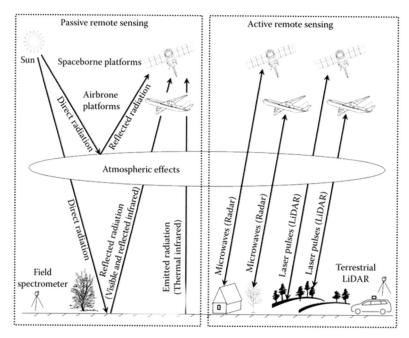

FIGURE 2.1
Illustration of passive (left) and active (right) remote sensing (from Dong and Chen, 2018).

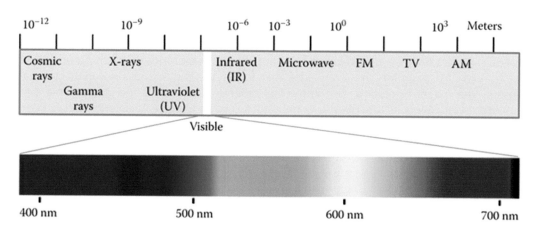

FIGURE 2.2
The commonly used spectral range in remote sensing: visible, infrared, and microwave (from Dong and Chen, 2018).

2.2 Multispectral Remote Sensing

Passive multispectral sensors collect images using visible, infrared, and/or thermal wavelengths with several bands. Sensors can be mounted on aircraft to get high-resolution aerial photography, on Unmanned Aerial vehicles (UAVs) or drones to get very high-resolution digital imagery or on spacecraft to collect satellite imagery. The unique feature of multispectral sensors is that they collect data using a few spectral channels, resulting in a coarse spectral resolution of data. Thus, multispectral instruments are also known as broadband sensors. The spectral resolution is the number of bands used by a sensor. It describes the ability of a sensor to define wavelength intervals. The finer the spectral resolution, the narrower the wavelength range for a specific channel or band. For example, the sensor of Landsat 8 Operational Land Imager (OLI) has 9 bands collecting data over visible, infrared, and thermal spectral channels. In this section, multispectral data (from aircraft, drone, and spacecraft platforms) are provided in more detail.

2.2.1 Aerial Photography and Products

Multispectral aerial photography has often been used for fine scale mapping and regional monitoring purposes. Modern remote sensing for Earth observation has evolved from early photography techniques, which collected imagery using ornithopters, balloons, kites, pigeons, and gliders (Jensen, 2007). Today, multispectral aerial photography using aircraft is still frequently collected by federal, state, and local governments for various applications.

In the US, the National Agriculture Imagery Program (NAIP) administered by the US Department of Agriculture (USDA) acquires aerial imagery at a 5-year cycle beginning 2003 and a 3-year cycle from 2009, with 2008 as a transition year (USGS/NAIP, 2018). NAIP collects aerial imagery at a spatial resolution of 1 meter and a spectral resolution of 3 bands (Red, Green, and Blue) or 4 bands (Red, Green, Blue, and Near Infrared). It produces imagery products either as Digital Ortho Quarter Quad tiles (DOQQs) or as Compressed County Mosaics. Each individual image tile within the mosaic covers a 3.75 × 3.75 minute quarter quadrangle plus a 300-meter buffer on all four sides. The DOQQs are GeoTIFFs, and the area corresponds to the USGS topographic quadrangles. The USGS distributes NAIP products in GeoTIFF and JPEG2000 formats at USGS's Earth Explorer (https://earthexplorer.usgs.gov/). NAIP quarter quads are projected to the Universal Transverse Mercator (UTM) coordinate system and referenced to NAD 83. Compressed data in the JPEG 2000 format acquired after 2013 are referenced to WGS 84 and utilize the WGS 1984 Web Mercator Auxiliary Sphere projection. In addition to the NAIP program, the National High Altitude Photography (NHAP) operational from 1980 to 1989 (USGS/NHAP, 2018) and the National Aerial Photography Program (NAPP) operational from 1987 to 2007 also collected black-white and color infrared aerial photography over the conterminous United States. Both NHAP and NAPP were coordinated by USGS and data are available at USGS' Earth Explorer web portal.

Regionally, the Florida Department of Revenue collects aerial photography and distributes ortho imagery of approximately one-third of the state each year. The imagery is collected using a multispectral digital camera with four bands (Red, Green, Blue, and Near Infrared), and imagery products are distributed with a spatial resolution of 0.5 feet. The specified flight season is from October 1 to March 15 every year. The state imagery is available from LAnd Boundary INformation System (LABINS) (www.labins.org/). Locally, due to extreme events such as hurricanes, aerial photography is often acquired after the episodic disturbance by federal agencies such as the National Aeronautics and Space Administration (NASA) and National Oceanic and Atmospheric Administration (NOAA). These images are available to the public for relevant applications and research. Figure 2.3 shows aerial photography collected from NAIP in 2015 and 2017 (after Hurricane Irma),

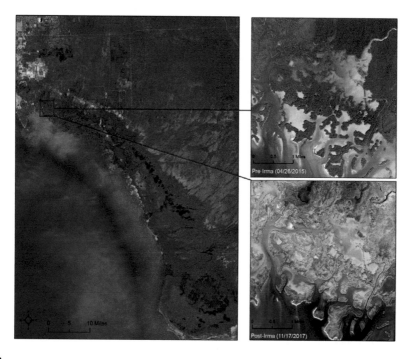

FIGURE 2.3
Aerial photography collected by the National Agriculture Imagery Program (NAIP) in 2015 and 2017 reveals the impact of Hurricane Irma (a Category 3 hurricane that made landfall on 10 September 2017) on the mangrove forests in Ten Thousand Islands of the coastal Everglades. The aerial imagery is displayed in a color infrared composite. The healthy mangroves showed brighter red on the pre-Irma imagery, while damaged mangroves illustrated darker red on the post-Irma imagery.

which revealed the impact of Hurricane Irma on the mangrove forests in the coastal Everglades.

Main advantages of aerial photography include its flexibility for data collection, high spatial resolution, and stereo capability. Compared with satellite remote sensing, aerial acquisition offers the flexibility of being fitted with a wide range of sensors such as multispectral, hyperspectral/thermal, lidar and other survey sensors. The newly developed sensor technologies can be adapted very quickly to an aircraft. It is also flexible for planning data acquisition according to local weather conditions and collecting data in cloud-free conditions. It is possible to fly under cloud cover with minor corrections applied during post-processing. This guarantees cloud-free data delivery. Due to low altitude acquisition, aerial data doesn't suffer from atmospheric effects that can impact the quality of the data in satellite imagery. Aerial photography has high spatial resolution, which can be used to produce high spatial resolution maps using image tone, size, shape, and other image-interpretation keys. Although high spatial resolution satellite data are now available, the flexibility by airborne sensors is more suitable for a wide range of applications. That is the main reason NAIP continues the acquisition of aerial photography nationwide. Aerial surveys can also capture imagery with full stereo mapping capability (60–80% forward overlap between images) and 30% side overlap between runs. This enables a wide range of other products to be generated with a high degree of accuracy, including Digital Elevation Models (DEMs), Digital Surface Models (DSMs), contours, orthophotos, and 3-Dimension (3-D) GIS feature data capture. Major limitations of aerial photography include its temporal resolution, limited spectral bands, and large data volume. The temporal resolution of remote sensing refers to the revisit period or how often the sensor will collect data for the same area. NAIP has collected aerial imagery at 3-year cycles since 2009, which may not meet the needs for a higher frequency monitoring purpose. Collecting aerial photography with an increased temporal resolution is costly. Also, aerial photography has limited spectral channels (3 or 4 bands), while modern satellite sensors can provide data with more bands. In some cases, satellite data collection is not restricted by time of a day, weather conditions, or other environmental anomalies. Lastly, the high spatial resolution of aerial photography consumes enormous amounts of storage space. Mapping a broad area using aerial photography involves processing a large volume of data, which is a challenge using traditional mapping software packages such as ArcGIS and Erdas Imagine.

2.2.2 Drone Remote Sensing and Products

UAVs, also known as Unmanned Aerial Systems (UASs) or simply drones, is an airborne remote sensing system or an aircraft operated remotely by a human operator or autonomously by an onboard computer. UAVs open a new opportunity for collecting remote sensing data for various applications with largely reduced cost compared with manned or satellite acquisitions (e.g., González-Jorge et al., 2017). Research and application of UAVs has been thriving, especially in the past decade, as evidenced by thousands of referenced UAVs (Colomina and Molina, 2014) and thousands of UAV peer-reviewed papers (Singh and Frazier, 2018). UAVs are now widely used for environmental applications due to miniaturized sensors, improved navigation and georeferencing algorithms, and increased sophistication of analytical methods (Milas et al., 2018).

Two types of drones (fixed-wing UAVs and rotary-wing UAVs) are mainly used in environmental remote sensing, which concerns the collection and interpretation of information about land, ocean, and the atmosphere from a remote vantage point (Figure 2.4). Fixed-wing UAVs are mini-airplanes, while rotary-wing UAVs are mini-helicopters. Payloads can be small imaging sensors covering visible, infrared, and thermal spectra or active sensors such as lidar and radar. Current UAVs are mainly used to collect ultra-high spatial resolution imagery down to centimeters and to generate DEMs and DSMs using Structure from Motion (SfM) photogrammetric techniques, which estimate 3-D structures from 2-D image sequences and generate 3-D point cloud data like lidar. Drone-based SfM is attractive in environmental remote sensing because it provides a low-cost alternative to lidar with reliable accuracies for certain applications. Figure 2.5 provides an example of drone products generated from a rotary-wing drone with a 4-band camera. UAVs fill a gap between aerial photography and satellites. They provide flexibility in timely data collection and can be tailored to required spatial and temporal resolutions. Klemas (2015) provides a review of UAVs in coastal and environmental remote sensing, but so far there are no published papers of drone applications in the Everglades. Personal communications with South Florida Water Management District (SFWMD), a key agency involved in CERP, have shown that the District has been exploring drone techniques for data collection in the Everglades. Our team has also been working on

FIGURE 2.4
Top: a fixed-wing drone; bottom: a rotary-wing drone ready for collecting data in the coastal Everglades.

identifying the best configuration of drone systems (fixed-wing and rotary-wing) for generating wetland products to be used in the Everglades, increasing integrity of drone products for coastal vulnerability analyses and species mapping using drone products.

2.2.3 Spaceborne Sensors and Products

Multispectral satellite sensors applicable in the Everglades can be grouped into three types in terms of their spatial resolution: high spatial resolution satellites such as QuickBird and Ikonos with a resolution of 5 meters or less, moderate spatial resolution sensors such as Landsat and Sentinel-2 with a resolution of between 5 and 30 meters, and coarse spatial resolution sensors such as Moderate Resolution Imaging Spectroradiometer (MODIS) with a resolution of 30 to 500 meters. They are detailed below.

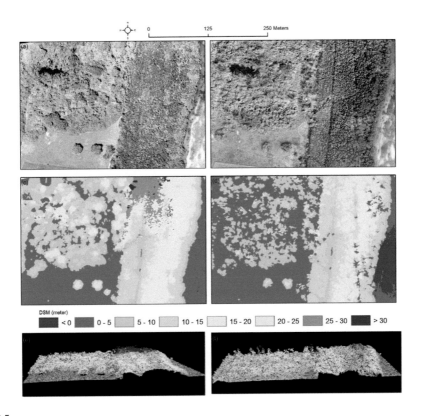

FIGURE 2.5
Drone products generated from a rotary-wing UAV over the wetlands of Biscayne National Park in the Greater Everglades. A drone-mosaic image collected (a) before Hurricane Irma; (b) after Hurricane Irma; DSM generated from the SfM technique and (c) pre-Irma photographs; (d) post-Irma photographs; (e) pre-Irma 3-D point cloud; and (f) post-Irma point cloud from drone photographs. Hurricane Irma made landfall in the coastal Everglades on September 10, 2017 as a Category 3 hurricane. Damages from Irma are revealed from both the drone-mosaic images and DSM products.

1) High spatial resolution satellite sensors and their products

In the US, currently, high spatial resolution satellite sensors are mainly operated and owned by the commercial company DigitalGlobe, which launched Ikonos in 1999, QuickBird in 2001, WorldView-1 in 2001, GeoEye-1 in 2008, WorldView-2 in 2009, WorldView-3 in 2014, and WorldView-4 in 2016 (DigitalGlobe, 2018). All these satellites have multispectral sensors and collect imagery with a spatial resolution less than 5 meters except for WorldView-1. The spatial, spectral, and temporal resolution of these multispectral sensors is listed in Table 2.1. Ikonos, QuickBird, GeoEye-1, and WorldView-4 have four spectral bands (Blue, Green, Red, and Near Infrared), while WorldView-2 has eight spectral channels. WorldView-3 is the industry's first multi-payload, super-spectral, high-resolution commercial satellite. Operating at an altitude of 617 km, WorldView-3 provides 31 cm panchromatic resolution, 1.24 m multispectral resolution, 3.7 m short-wave infrared resolution, and 30 m CAVIS resolution. It has an average revisit time of <1 day and can collect up to 680,000 km^2 per day. These high spatial

TABLE 2.1

Spatial, spectral, and temporal configuration of high-resolution satellite sensors

Sensors	Spatial Resolution or Ground Sample Distance (GSD)	Spectral Resolution	Temporal Resolution	Temporal Coverage
Ikonos	Panchromatic: 0.82 m Multispectral: 3.2 m	4 Multispectral: Blue: 445–526 nm Green: 506–595 nm Red: 632–698 nm NIR: 757–853 nm	Approximately 3 days	1999–2015
QuickBird	Panchromatic: 55 cm/61cm at nadir Multispectral: 2.16/2.44 m at nadir	4 Multispectral: Blue: 430–545 nm Green: 466–620 nm Red: 590–710 nm NIR: 715–918 nm	from 2 to 12 days depending on target location as the orbit decays	2001–2015
GeoEye–1	Panchromatic 0.41 m at nadir Multispectral 1.65 m at nadir	4 Multispectral: Blue: 450–510 nm Green: 510–580 nm Red: 655–690 nm NIR: 780–920 nm	2.6 days at 30° off–nadir	2008–
WorldView–2	Panchromatic: 0.46 m at nadir 0.52 m at 20° off–nadir Multispectral: 1.85 m at nadir 2.07 m GSD at 20° off–nadir	8 Multispectral: Coastal: 400–450 nm Blue: 450–510 nm Green: 510–580 nm Yellow: 585–625 nm Red: 630–690 nm Red Edge: 705–745 nm NIR1: 770–895 nm NIR2: 860–1040 nm	1.1 days at 1 m 3.7 days at 20° off–nadir	2009–
WorldView–3	Panchromatic nadir: 0.31m 20° off–nadir: 0.34 m Multispectral nadir: 1.24 m 20° off–nadir: 1.38 m SWIR nadir: 3.70 m 20° off–nadir: 4.10 m CAVIS nadir: 30.00 m 20° off–nadir: 30 m	8 Multispectral: Coastal: 397–454 nm Blue: 445–517 nm Green: 507–586 nm Yellow: 580–629 nm Red: 626–696 nm Red Edge: 698–749 nm NIR1: 765–899 nm NIR2: 857–1039 nm 8 SWIR Bands: SWIR–1: 1184–1235 nm SWIR–2: 1546–1598 nm SWIR–3:1636–1686 nm SWIR–4: 1702–1759 nm SWIR–5: 2137–2191 nm SWIR–6: 2174–2232 nm SWIR–7: 2228–2292 nm SWIR–8: 2285–2373 nm 12 CAVIS Bands: Desert Clouds: 405–420 nm Water–3: 930–965 nm Aerosol–1: 459–509 nm NDVI–SWIR: 1220–1252 nm Green: 525–585 nm	1 m GSD: < 1.0 day 4.5 days at 20° off nadir	2014–

(Continued)

TABLE 2.1 (Cont.)

Sensors	Spatial Resolution or Ground Sample Distance (GSD)	Spectral Resolution	Temporal Resolution	Temporal Coverage
		Aerosol–2: 635–685 nm Water–1: 845–885 nm Water–2: 897–927 nm Cirrus: 1365–1405 nm Snow: 1620–1680 nm Aerosol–1: 2105–2245 nm Aerosol–2: 2105–2245 nm		
WorldView–4	Panchromatic Nadir: 0.31 m 20° Off–Nadir: 0.34 m 56° Off–Nadir: 1.00 m Multispectral Nadir: 1.24 m 20° Off–Nadir: 1.38 m 56° Off–Nadir: 4.00 m	4 Multispectral: Red: 655–690 nm Green: 510–580 nm Blue: 450–510 nm NIR: 780–920 nm	1 m GSD: < 1.0 day Total constellation >4.5 accesses/day	2016–

resolution images are available by purchase. NOAA purchased Ikonos images for mapping coral reef ecosystems in the Florida Keys (NOAA/National Centers for Coastal Ocean Science, 2018), and the footprints of these images are displayed in Figure 2.6. These Ikonos images are free to the public. The high spatial resolution satellite sensors provide imagery like aerial photography and are useful for detailed mapping, but for south Florida, heavy cloud is a big issue for acquiring such satellite imagery. Their high temporal resolution can mitigate this issue to produce a cloud-free mosaic to improve the application. Note that the application of such type of satellite imagery is limited in the Everglades because they are not free to the public. Only limited images are provided for research experiments.

2) Moderate spatial resolution satellite sensors and products

2.1) Landsat series
Landsat provides the world's longest continuously acquired collection of moderate resolution land remote sensing data. Four decades of imagery offers a unique resource for various applications in wetlands, agriculture, geology, forestry, regional planning, education, mapping, and global change research. Landsat images are also invaluable for emergency response and disaster relief (USGS/Landsat, 2018). As a joint initiative between the USGS and NASA, the Landsat Project and the data it collects support government, commercial, industrial, civilian, military, and educational communities throughout the United States and worldwide. The spectral, spatial, temporal configuration of Landsat series is displayed in Table 2.2 starting from Landsat 1 launched in 1972 to current Landsat 8 launched in 2013. Land 9 has a launch readiness date of December 2020. Figure 2.7 shows the chronology of Landsat series. Landsat has a suitable spectral and spatial resolution for Earth observation and has been widely applied in wetland monitoring and mapping.

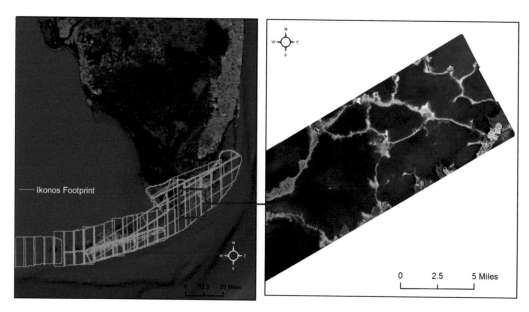

FIGURE 2.6
Ikonos imagery purchased by NOAA for mapping coral reef ecosystems in the Florida Keys. The images were collected between 2005 and 2006 (NOAA/National Centers for Coastal Ocean Science, 2018). Left: footprint of Ikonos images; right: a color infrared composite of Ikonos imagery.

Currently, USGS produces three major Landsat products: Landsat Collection Level-1 and Level-2 products and Landsat Analysis Ready Data (ARD) products, which are available to users at no cost. Level-1 products are processed to standard parameters and distributed as scaled and calibrated digital numbers that can be further processed to calibrated radiance or reflectance values using the provided metadata. In 2016, USGS reorganized the Landsat archive into a tiered collection structure to ensure the consistency of Level-1 products for time-series analyses and data "stacking." The collection consists of three categories: Tier 1, Tier 2, and Real-Time. Tier 1 meets formal geometric and radiometric quality criteria. Tier 2 does not meet the Tier 1 criteria. The Real-Time Tier contains data immediately after acquisition that use estimated parameters. Real-Time data are reprocessed and assessed for inclusion into Tier 1 or Tier 2 as soon as final parameters are available. Level-1 data products are generated from Landsat 8 Operational Land Imager (OLI) and Thermal Infrared Sensor (TIRS), Landsat 7 Enhanced Thematic Mapper Plus (ETM+), Landsat 4–5 Thematic Mapper ™, and Landsat 1–5 Multispectral Scanner System (MSS) instruments.

Landsat Collection Level-2 products are surface reflectance data to support land surface change studies. Surface reflectance is the fraction of incoming solar radiation that is reflected from Earth's surface to the Landsat sensor. Landsat 4–5 TM and Landsat 7 ETM+ Surface Reflectance data are generated using the Landsat Ecosystem Disturbance Adaptive Processing System, specialized software originally developed by NASA and the University of Maryland. The software applies MODIS atmospheric correction routines to Landsat Level-1 data products. Landsat 8 OLI Surface Reflectance data are generated from the Landsat Surface Reflectance Code, which makes use of the coastal

TABLE 2.2

Summary of Landsat series revised from USGS/Landsat (2018)

Instrument	Sensors	Spectral and Spatial Configuration	Temporal Coverage
Landsat 1	MSS	Landsat 1–3 MSS, 60 meters, 18 days revisit	1972–1978
Landsat 2	MSS	Band 4: Green, 0.5–0.6µm	1975–1982
Landsat 3	MSS	Band 5: Red, 0.6–0.7µm	1978–1983
Landsat 4	MSS and TM	Band 6: NIR, 0.7–0.8µm	1982–1993
		Band 7: NIR, 0.8–1.1µm	
		Landsat 4–5 MSS, 60 meters, 16 days revisit	
Landsat 5	MSS and TM	Band 1: Green, 0.5–0.6µm	1984–2013
		Band 2: Red, 0.6–0.7µm	
		Band 3: NIR, 0.7–0.8µm	
		Band 4: NIR, 0.8–1.1µm	
		Landsat 4–5 TM, 30 meters, 16 days revisit	
		Band 1: Blue, 0.45–0.52µm	
		Band 2: Green, 0.52–0.60µm	
		Band 3: Red, 0.63–0.69µm	
		Band 4: NIR, 0.76–0.90µm	
		Band 5: SWIR 1, 1.55–1.75µm	
		Band 6: Thermal, 10.40–12.50µm (120 m)	
		Band 7: SWIR 2, 2.08–2.35µm	
Landsat 6	MSS		1993/failed
Landsat 7	ETM+	ETM+, 30 meters, 16 days revisit	1999–present
		Band 1: Blue, 0.45–0.52µm	
		Band 2: Green, 0.52–0.60µm	
		Band 3: Red, 0.63–0.69µm	
		Band 4: NIR, 0.77–0.90µm	
		Band 5: SWIR 1, 1.55–1.75µm	
		Band 6: Thermal, 10.40–12.50 µm, 60 m	
		Band 7: SWIR 2, 2.09–2.35 µm	
		Band 8: Panchromatic, 0.52–0.90 µm, 15 m	
Landsat 8	OLI and TIRS	OLI, 30 meters, 16 days revisit	2013–present
		Band 1: Ultra Blue, 0.435–0.451µm	
		Band 2: Blue, 0.452–0.512 µm	
		Band 3: Green, 0.533–0.590µm	
		Band 4: Red, 0.636–0.673µm	
		Band 5: NIR, 0.851–0.879µm	
		Band 6: SWIR 1, 1.566–1.651µm	
		Band 7: SWRI 2: 2.107–2.294 µm	
		Band 8: Panchromatic, 0.503–0.676 µm, 15 m	
		Band 9: Cirrus, 1.363–1.384	
		Band 10: Thermal (TIRS), 10.60–11.19 µm, 100 m	
		Band 11: Thermal (TIRS), 11.50–12.51 µm, 100 m	
Landsat 9	A rebuild of its predecessor Landsat 8		2020/expected

MSS: Multispectral Scanner; TM: Thematic Mapper; ETM: Enhanced Thematic Mapper; OLI: Operational Land Imager; TIRS: Thermal InfraRed Sensor; NIR: Near Infrared; SWIR: Shortwave Infrared.

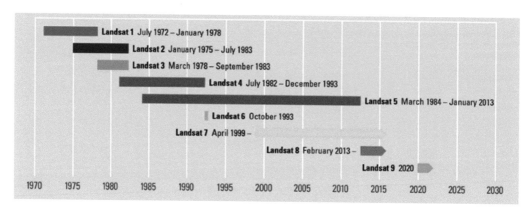

FIGURE 2.7
Chronology of Landsat series starting with Landsat 1 from July 1972 (USGS/Landsat, 2018).

aerosol band to perform aerosol inversion tests and uses auxiliary climate data from MODIS and a unique radiative transfer model.

Landsat ARD products are produced to the highest scientific standards and level for direct use in monitoring and assessing landscape change. A fundamental goal for Landsat ARD is to significantly reduce the magnitude of data processing for application scientists, especially when large amounts of Landsat scenes are requested for time-series analysis. The product consists of top of atmosphere reflectance, top of atmosphere brightness temperature, surface reflectance, surface temperature, and quality assessment data that are consistently processed using per pixel solar zenith angle corrections gridded to a common cartographic projection and accompanied by appropriate metadata to enable further processing while retaining data provenance. Landsat ARD products are processed to a common tiling scheme of uniform dimensions bounded by static corner points in a defined grid system.

In addition to these three products, metadata and relevant quality assessment results are also provided to users. Datasets of mosaics of Landsat products are also provided, such as Global Land Survey, TM and ETM+ mosaics. Figures 2.8 and 2.9 show some examples of these products in the Everglades. Since 2018, USGS has also been producing Level-3 science products derived from Level-2, including surface temperature to represent the temperature of the earth's surface, dynamic surface water extent to describe the existence and condition of surface water, fractional snow covered area to indicate the percentage of pixels covered by snow and burned area to represent per pixel burn classification and burn probability. All these data products are available at USGS' Earth Explorer (https://earthexplorer.usgs.gov/).

2.2) Sentinel-2
Landsat-type multispectral satellite sensors such as Sentinel-2 can be applied in the Everglades. Sentinel-2 was developed by the European Space Agency (ESA) (ESA/Sentinel Online, 2018). It consists of two identical satellites, Sentinel-2A and Sentinel-2B, with a multispectral instrument (MSI) onboard to collect visible, near infrared, and shortwave infrared imagery using 13 bands. It has a temporal resolution of 5 days under the same viewing angles, and spatial resolution of 10 meters, 20 meters, and 60 meters. Sentinel-2 also has a free and open data policy. Sentinel-2A was launched in June of 2015, and Sentinel-2B was launched in March of 2017. The spatial and spectral configuration of Sentinel-2 is displayed in Table 2.3.

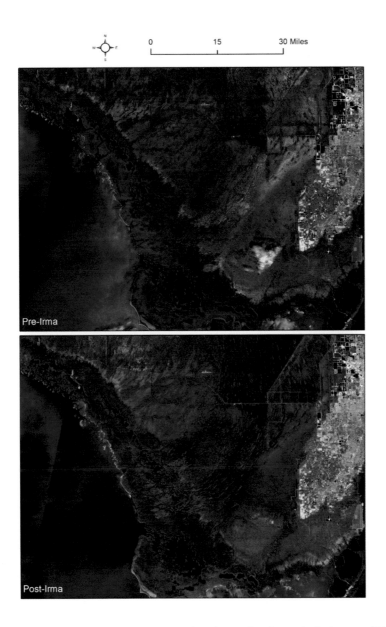

FIGURE 2.8
A color infrared composite of Landsat 8 OLI imagery (Level-2 products) reveals the impact of Hurricane Irma on the coastal Everglades. Top: pre-Irma mosaic produced from Landsat 8 OLI imagery collected in October and December 2016; bottom: post-Irma mosaic produced from Landsat 8 Imagery collected in November 2017. Irma made landfall on September 10, 2017.

Currently, three products are generated from Sentinel-2: Level-1B, Level-1C, and Level-2A. Level-1B provides top of atmosphere radiances in sensor geometry. It is composed of granules with one granule/tile having a data volume of approximately 27 MB. Level-1B products require expert knowledge of orthorectification techniques. Level-1C provides top of atmosphere reflectance in fixed cartographic geometry framed into tiles (also named as granules). Level-1C products contain applied radiometric and geometric

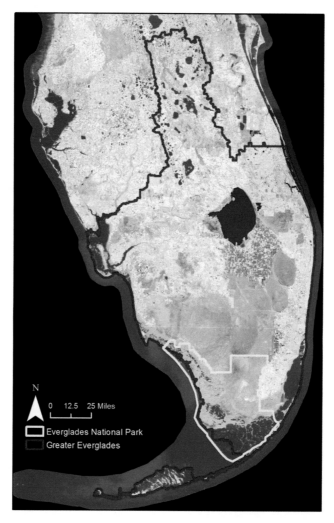

FIGURE 2.9
A natural color mosaic of Landsat 7 ETM+ imagery collected in 2004 shows the Greater Everglades.

corrections (including orthorectification and spatial registration). An example of Sentinel-2 Level-1C products in the Everglades is displayed in Figure 2.10. Level-2A offers bottom of atmosphere reflectance in cartographic geometry. This product is currently processed on the user side. All these products are distributed by ESA Open Access Hub.

2.3) Harmonized Landsat and Sentinel-2 (HLS) products
The HLS project is a NASA initiative aiming to produce a virtual constellation of surface reflectance data acquired by Landsat 8 OLI and Sentinel-2 MSI (Claverie et al., 2018; NASA/Harmonized Landsat Sentinel-2, 2018). Merging multiple Earth observations can improve the temporal resolution, resulting in a broader application of the products. Combining data from two or more sensors into a single data set is called "virtual constellation." HLS products are based on a set of algorithms to obtain seamless products from both on-orbit sensors: atmospheric correction, cloud and cloud-shadow masking, spatial co-registration

TABLE 2.3

Spatial and spectral configuration of Sentinel-2 (ESA/Sentinel Online, 2018)

Band Number	Sentinel-2A		Sentinel-2B		Spatial resolution (m)
	Central wavelength (nm)	Bandwidth (nm)	Central wavelength (nm)	Bandwidth (nm)	
1	443.9	27	442.3	45	60
2	496.6	98	492.1	98	10
3	560.0	45	559	46	10
4	664.5	38	665	39	10
5	703.9	19	703.8	20	20
6	740.2	18	739.1	18	20
7	782.5	28	779.7	28	20
8	835.1	145	833	133	10
8a	864.8	33	864	32	20
9	945.0	26	943.2	27	60
10	1373.5	75	1376.9	76	60
11	1613.7	143	1610.4	141	20
12	2202.4	242	2185.7	238	20

and common gridding, bidirectional reflectance distribution function normalization, and spectral band pass adjustment. Three products have been derived from the HLS processing chain: (1) S10: full resolution MSI SR at 10 m, 20 m, and 60 m spatial resolutions; (2) S30: a 30 m MSI Nadir Bidirectional Reflectance Distribution Function (BRDF)-Adjusted Reflectance; and (3) L30: a 30 m OLI Nadir BRDF Adjusted Reflectance. All three products are processed for every Level-1 input product from Landsat 8 OLI and Sentinel-2 MSI Level-2C products. In addition, a fourth product of the surface reflectance dataset named M30 with a spatial resolution of 30 meters and temporal resolution of 5-day over the Sentinel-2 tiling system is also underway. This project prototypes the harmonization for all of North America and other globally distributed test sites, and data products are available at https://hls.gsfc.nasa.gov/. A summary of HLS products is displayed in Table 2.4.

3) Coarse spatial resolution satellite sensors and products

The MODIS is a spaceborne imaging instrument mounted on the *Terra* (launched in 1999) and *Aqua* (launched in 2002) satellites that view the entire Earth's surface every 1 to 2 days. It acquires data in 36 spectral bands ranging in wavelength from 0.4 μm to 14.4 μm. Bands 1 and 2 are imaged at a nominal resolution of 250 m at nadir, while bands 3 to 7 acquire data at 500 m, and the remaining 29 bands at 1 km. The specifications of MODIS are listed in Table 2.5. MODIS plays a vital role in the development of validated, global, interactive Earth system models. It can predict global change accurately enough to assist policymakers in making sound decisions concerning the protection of our environment (NASA/MODIS, 2018).

Products from MODIS include 1) level-1 data; 2) land products; 3) atmosphere products; 4) ocean products; and 5) cryosphere products (https://modis.gsfc.nasa.gov/data/dataprod/). MODIS land products are suitable for research in the Everglades: surface reflectance, land surface temperature and emissivity, land cover products, vegetation index products, Fraction of Photosynthetically Active Radiation (FPAR)/Leaf Area Index (LAI), evapotranspiration, Gross Primary Productivity (GPP)/Net Primary Productivity

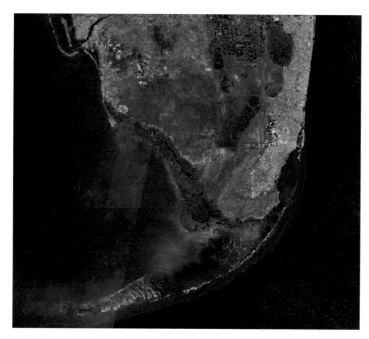

FIGURE 2.10

A screenshot of color infrared mosaic (Bands 8, 4, and 3) of Sentinel-2 Level-1C products with cloud pixels removed in the Everglades. Data were collected from January 1, 2018, to June 30, 2018, and processed in the Google Earth Engine.

TABLE 2.4

Summary of Harmonized Landsat 8 OLI and Sentinel-2 products (from NASA/Harmonized Landsat Sentinel-2, 2018).

Product Name	S10	S30	L30	M30
Spatial	10-20-60m	30m	30m	30m
Spectral	As input	OLI-like	OLI-like	Landsat-like
Temporal	As input	As input	As input	5-day (TBC)
BRDF-adj.	No	Yes	Yes	Yes
Projection	UTM	UTM	UTM	UTM
Tiling system	S2 (110*110)	S2 (110*110)	S2 (110*110)	S2 (110*110)

All products are gridded using the same tiling system.

(NPP), BRDF/Albedo Parameter, vegetation continuous fields, water masks, and burned area products. Among these land products, the Surface Reflectance 8-Day L3 Global 500 m (MOD09A1 of Terra and MYD09A1 of Aqua) are appropriate for a large-scale land-cover analysis of the Greater Everglades. A nature color mosaic of the MOD09A1 product during 01/01/2017 to 01/05/2017 with cloud removed in the Greater Everglades is shown in Figure 2.11. In addition, the vegetation index product is useful. Vegetation indices produced on 16-day intervals and at multiple spatial resolutions (250m, 500m, and 1km),

TABLE 2.5

Spectral and spatial specifications of MODIS

Primary Use	Band	Bandwidth (nm)
Land/Cloud/Aerosols Boundaries, 250 m	1	620–670
	2	841–876
Land/Cloud/Aerosols Properties, 500 m	3	459–479
	4	545–565
	5	1230–1250
	6	1628–1652
	7	2105–2155
Ocean Color/Phytoplankton/Biogeochemistry, 1 km	8	405–420
	9	438–448
	10	483–493
	11	526–536
	12	546–556
	13	662–672
	14	673–683
	15	743–753
	16	862–877
Atmospheric Water Vapor, 1 km	17	890–920
	18	931–941
	19	915–965
Surface/Cloud Temperature, 1 km	20	3660–3840
	21	3929–3989
	22	3929–3989
	23	4020–4080
Atmospheric Temperature, 1 km	24	4433–4498
	25	4482–4549
Cirrus Clouds Water Vapor, 1 km	26	1360–1390
	27	6535–6895
	28	7175–7475
Cloud Properties, 1 km	29	8400–8700
Ozone, 1 km	30	9580–9880
Surface/Cloud Temperature, 1 km	31	10780–11280
	32	11770–12270
Cloud Top Altitude, 1 km	33	13185–13485
	34	13485–13785
	35	13785–14085
	36	14085–14385

provide consistent spatial and temporal comparisons of vegetation canopy greenness, a composite property of leaf area, chlorophyll, and canopy structure. Figure 2.12 shows a time series result of Vegetation Indices 16-Day L3 Global 250m product (MOD13Q1) in Everglades National Park. The commonly used MODIS land products and their names are listed in Table 2.6. All these products are available at USGS's Earth Explorer. USGS also provides expedited MODIS (eMODIS) composites, which are built from MODIS Collection 6.0 surface reflectance data. Even though MODIS data are beneficial in vegetation studies,

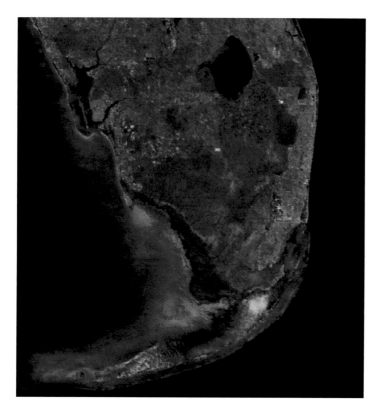

FIGURE 2.11
A screenshot of nature color mosaic of Terra surface reflectance product (MOD09A1 V6) (Bands 1, 4, and 3) with cloud pixels removed in the Everglades. Data were collected during 01/01/2017 to 01/05/2017 and processed in Google Earth Engine.

there have been usability issues encountered with reprojection, file format and subsetting. Therefore, eMODIS was developed to specifically address these issues (USGS/eMODIS, 2018). The products are generated over all areas including the continental US and Alaska and are projected to a regionally specific mapping grid and delivered in a compressed (zipped) GeoTIFF format. Each product delivers acquisition, quality, and NDVI information at 250 m spatial resolution.

2.3 Hyperspectral Remote Sensing

Hyperspectral Remote Sensing (HRS), also known as imaging spectrometry or spectroscopy, is defined as the simultaneous acquisition of images or point spectral data in many relatively narrow, contiguous, and/or noncontiguous spectral bands throughout the ultraviolet, visible, and infrared portions of the electromagnetic spectrum (Jensen, 2015). Goetz (1992) defined HRS as the acquisition of images in hundreds of registered,

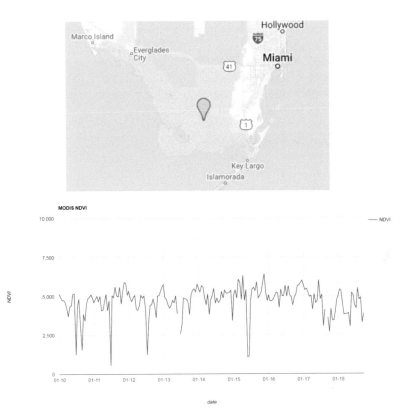

FIGURE 2.12
Time series (01/01/2010 to 01/10/2018) plot of MODIS Normalized Difference Vegetation Index (NDVI) Terra product (Vegetation Indices 16-Day L3 Global 250m, MOD13Q1) at a point (longitude: -80.8°, latitude: 25.37°) in Everglades National Park. The result was produced in Google Earth Engine.

TABLE 2.6

Commonly used land products of MODIS (Land Processes Distributed Active Archive Center of USGS)

Name	Dataset	Product	Spatial Resolution (m)	Temporal Resolution
MOD09A1	Terra MODIS	Reflectance	500	Composites
MOD09CMG	Terra MODIS	Reflectance	5600	Daily
MOD09GA	Terra MODIS	Reflectance	500, 1000	Daily
MOD09GQ	Terra MODIS	Reflectance	250	Daily
MOD09Q1	Terra MODIS	Reflectance	250	Composites
MODOCGA	Terra MODIS	Reflectance	1000	Daily
MODTBGA	Terra MODIS	Reflectance	1000	Daily
MYD09A1	Aqua MODIS	Reflectance	500	Composites
MYD09CMG	Aqua MODIS	Reflectance	5600	Daily

(Continued)

TABLE 2.6 (Cont.)

Name	Dataset	Product	Spatial Resolution (m)	Temporal Resolution
MYD09GA	Aqua MODIS	Reflectance	500, 1000	Daily
MYD09GQ	Aqua MODIS	Reflectance	250	Daily
MYD09Q1	Aqua MODIS	Reflectance	250	Composites
MYD13A1	Aqua MODIS	Vegetation Indices	500	Composites
MYD13A2	Aqua MODIS	Vegetation Indices	1000	Composites
MYD13A3	Aqua MODIS	Vegetation Indices	1000	Monthly
MYD13C1	Aqua MODIS	Vegetation Indices	5600	Composites
MYD13C2	Aqua MODIS	Vegetation Indices	5600	Monthly
MYD13Q1	Aqua MODIS	Vegetation Indices	250	Composites

contiguous spectral bands such that for each picture element of an image, it is possible to derive a complete reflectance spectrum. Some researchers compare HRS to a multiplicity of recording bands (channels) that have relatively narrow bandwidths. Different from multispectral systems such as Landsat, Ikonos, and QuickBird, which usually have several bands (3–10 channels) with broad bandwidths (~100 nm), the HRS indicates much more bands and narrow bandwidth (10–20nm) for each band. For example, NASA's Airborne Visible/Infrared Imaging Spectrometer (AVIRIS) and Earth Observing (EO)-1/Hyperion have more than 200 bands. So far, there is no universal agreement upon the minimum number of bands or bandwidth dimension required for a dataset to be considered hyperspectral. Analysis of HRS data allows extraction of a detailed spectrum for each pixel in the image. Such spectra may allow direct identification of specific materials that appear in the collected scene. Figure 2.13 displays the difference between hyperspectral and multispectral remote sensing data.

2.3.1 Handheld Spectroradiometer to Collect Point Hyperspectral Data

Both point and imaging HRS data can be acquired using hyperspectral sensors. Currently the ASD spectroradiometer is mainly used to collect field and laboratory HRS data. It is specifically designed for environment remote sensing to acquire Visible and Near Infrared (VNIR) and Shortwave Infrared (SWIR) spectra. The ASD spectroradiometer collects data in the range of 350 nm to 2500 nm with different spectral resolution by using different models such as FieldSpec hi-Res and wide-res spectroradiometers. It can measure spectral reflectance, transmittance, absorbance, radiance, and irradiance of the samples. The instrument communicates with a computer that can record the spectral measures using a specifically developed software package RS[3]. The measured spectra can be viewed and processed using the related ViewSpec Pro[TM] package. In addition, the ASD derived spectra binary file can be imported directly into other remote sensing software packages such as ENVI to build a spectral library that can be used as a reference for material identification in remote sensing images. The USGS has built spectral libraries for minerals, rocks, soils, physically constructed as well as mathematically computed mixtures, plants, vegetation communities, microorganisms, and manmade materials (http://speclab.cr.usgs.gov/spectral-lib.html). Other libraries include the Jet Propulsion Laboratory's (JPL's) mineral library, Johns Hopkins University's Spectral Library, IGCP 264's Spectral Library, and University of Texas at El Paso's Vegetation Spectral Library. Our research lab built a coastal sand spectral

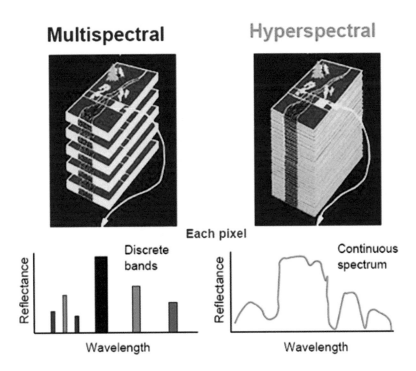

FIGURE 2.13
Comparison of multispectral and hyperspectral data. Multispectral data has several discrete bands; in contrast, hyperspectral data has a more continuous spectral curve, collecting data using many bands, resulting in a 3-D data cube (from NASA, 2019).

library using the ASD spectroradiometer. The collected field spectra can be used to calibrate remote sensing imagery and characterize target materials. Figure 2.14 shows an ASD spectroradiometer and C. Zhang using the instrument to conduct field measures.

2.3.2 Airborne Sensors to Collect Hyperspectral Imagery

Several airborne imaging hyperspectral sensors were developed to collect narrowband hyperspectral imagery for various applications; NASA's Airborne Visible InfraRed Imaging Spectrometer (AVIRIS) data is free to the public. AVIRIS was designed and built by NASA's JPL (NASA/AVIRIS, 2018). It is mounted on four aircraft platforms: NASA's ER-2 jet, Twin Otter International's turboprop, Scaled Composites' Proteus, and NASA's WB-57. The ER-2 flies at approximately 20 km above sea level, at about 730 m/hr. The Twin Otter aircraft flies at 4 km above ground level at 130 km/hr. AVIRIS has flown across the US, Canada, and Europe. The swath and spatial resolution of AVIRIS depend on the flying height and over-ground speed of the aircraft. On the NASA ER-2 jet at 20 km above sea level, AVIRIS covers an 11 km swath at a spatial resolution of 20 m. On the Twin Otter turboprop at 4 km above ground level, AVIRIS covers a 2 km swath at a spatial resolution of 4 m. The sensor records radiance levels at 12-bit radiometric resolution for 224 contiguous spectral bands with wavelengths from 400 nm to 2500 nm. AVIRIS is flown primarily for NASA-funded scientists and researchers. AVIRIS data is normally radiometrically and geometrically

FIGURE 2.14
An ASD spectroradiometer and its application for collecting vegetation spectra.

corrected by the JPL team before it is distributed to the user; however, raw data can be made available. NASA archives the collected AVIRIS imagery and distributes them at no cost currently. Data acquired from 2006 to the present can be browsed and downloaded directly from the AVIRIS Flight Locator Tool (http://aviris.jpl.nasa.gov/alt_locator/). Data acquired from 1992 to 2005 can be located from http://aviris.jpl.nasa.gov/locator_select. php and ordered by filling an AVIRIS Archive Data Request form. Figure 2.15 shows an AVIRIS image collected over Key West, FL, on November 19, 1992.

Hyperspectral imagery is also collected and distributed by NASA Goddard's Lidar, Hyperspectral, and Thermal (G-LiHT) Airborne Imager. G-LiHT is a unique system to simultaneously collect hyperspectral, lidar, and thermal data at a high spatial resolution (~1m) to a better measurement of vegetation structure, foliar spectra, and surface temperature (Cook et al., 2013; NASA/G-LiHT, 2018). G-LiHT collects hyperspectral imagery using a Hyperspec imaging spectrometer that operates in the 400 nm to 1000 nm spectral region. Products of hyperspectral data include at-sensor reflectance and surface reflectance of 114 bands varying from 418 nm to 918 nm with 4.5 nm sampling interval. Data products of G-LiHT are available at https://glihtdata.gsfc.nasa.gov/ or ftp://fusionftp.gsfc.nasa.gov/G-LiHT.

In addition to NASA's airborne hyperspectral imaging sensors, several imaging sensors exist in the market for research applications including Compact Airborne Imaging Spectrometer (CASI), HyMap, and AISA. These sensors are operated by commercial companies, and data are often not free to the public. CASI collects data over the VNIR portion with 288 bands at about 2.5 nm interval. Flown on a standard fixed-wing aircraft configured for aerial surveys, CASI can achieve a spatial resolution in the range of 0.25 m to 1.5 m. HyMap collects data over 450 nm to 2500 nm with 126 bands at a spatial resolution of 3 to 10 m. The

AVIRIS: Key West, Florida 921119

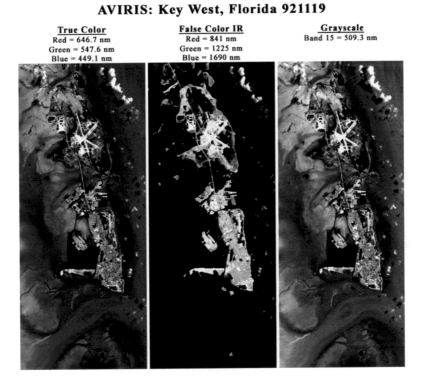

FIGURE 2.15
Comparison of three AVIRIS images: true color composite from bands 646.7nm, 547.6nm, and 449.1nm (left); false color composite from bands 841nm, 1225nm, and 1690nm (center); and grayscale image from band 15 (509.3nm) (right) (from NASA).

sensor is currently operated by the HyVista Corporation. Figure 2.16 shows an image collected by HyMap in the Kissimmee River area. AISA is a hyperspectral sensor collecting high-quality hyperspectral imagery covering VNIR (380–1000 nm), SWIR (1000–2500 nm), and thermal Long-Wave Infrared (LWIR) (7.6–12.4 µm) spectral ranges (www.specim.fi/). The AISA hyperspectral products include AISA Eagle for collecting data covering 400 nm to 1000 nm, AISA Hawk for collecting data covering 970 nm to 2500 nm, AISA Dual for simultaneous acquisition of VNIR and SWIR, and ASIA Owl for collecting thermal hyperspectral imagery. The configuration of these airborne sensors is summarized in Table 2.7.

2.3.3 Spaceborne Sensors to Collect Hyperspectral Imagery

Most spaceborne sensors collect data in broad spectral bands to obtain an adequate signal-to-noise ratio at high spatial resolution. This may make the hyperspectral data collection from space a difficult proposition. However, several efforts have been made. The hyperspectral spaceborne missions started in the early 1990s with NASA's High Resolution Imaging Spectrometer (HRIS), ESA's HRIS, followed by the Australian Resource Information and Environment Satellite (ARIES). Unfortunately, these three missions did not make it out of the design phase into the detailed design and building

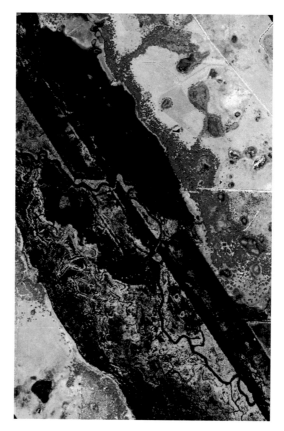

FIGURE 2.16

A true color HyMap image collected in the Kissimmee River floodplain on September 21, 2002, at a spatial resolution of 3.5 m.

phases. Many missions have been initialized, but only two missions were built and successful: NASA's EO-1/Hyperion, and ESA's Compact High Resolution Imaging Spectrometer (CHRIS). With the success of these two missions, more sophisticated sensors such as the Environmental Mapping Program and Hyperspectral Precursor of the Application Mission (PRISMA) have been initialized.

The Hyperion hyperspectral-imaging sensor is onboard NASA's EO-1 spacecraft, which was launched on November 21, 2000 (NASA/EO-1/Hyperion, 2018). It acquires hyperspectral imagery with 220 contiguous bands ranging from 0.40 μm (blue) to 2.5 μm (mid-IR). The EO-1 spacecraft is 1 minute behind Landsat 7, essentially viewing the same atmospheric conditions. The 30 m spatial resolution of Hyperion mimics Landsat's spatial resolution; however, the 7.5 km hyperspectral swath is only a fraction of a 185-km wide Landsat scene. The other hyperspectral sensor Linear Etalon Imaging Spectral Array (LEISA) is also onboard EO-1. This sensor has a 185-km swath at 250 m spatial resolution and collects 246 bands in the mid-IR portion of the spectrum where water vapor absorption is significant. The LEISA data is used to derive atmospheric correction information for the other sensor datasets. Hyperion imagery is archived and distributed by the USGS and placed in the public domain. The archived imagery may be viewed and downloaded at no

TABLE 2.7

Configuration of commonly used airborne hyperspectral sensors

Sensor	Spectral Range	Spectral Resolution (nm)	Bands	Country
AVIRIS	400–2500 nm	9.4–16.0	224	USA/NASA
G–LiHT	418–918 nm	4.5	114	USA/NASA
CASI	430–870 nm	~2.5	288	Canada
HyMap	450–2500 nm	13–17	126	Australian
AISA	400–2500 nm	VNIR: 3.3 SWIR: 12	VNIR: 244 SWIR: 254	Finland

TABLE 2.8

Configuration of spaceborne hyperspectral sensors

Sensors	Launch/End Date	Manufacturer	Spatial Resolution	Bands	Spectral Range
EO-1/Hyperion	2000/2017	USA/NASA	30 m	220	0.40–2.50 μm
HICO	2009/2014	USA/NASA	90 m	128	0.38–0.96 μm
CHRIS	2001–	ESA/UK	17 m/34 m	18/37/6	0.41–1.05 μm

charge from USGS's Earth Explorer (http://earthexplorer.usgs.gov/). The EO-1 mission was decommissioned on February 22, 2017, but the historical record of EO-1 is still available to public users. Products include Level 1Gst and Level 1R. The Level 1Gst is terrain corrected data provided in 16-bit radiance values. The Level 1R product consists of radiometrically corrected images, formatted as HDF files. Metadata are in binary and ASCII formats.

The ESA's CHRIS imaging spectrometer was onboard the Proba-1 satellite launched on October 22, 2001 (ESA/CHRIS, 2018). The Proba-1 mission objectives are primarily for in-orbit demonstration and evaluation of new technologies. The CHRIS instrument is one of these technologies. Data provided from CHRIS is targeted to help with environmental monitoring applications. CHRIS acquires a set of up to five images classified as modes 1 to 5 for each target during each acquisition sequence. It collects imagery at a 17 m spatial resolution in 18 user-selected visible and near-infrared wavelengths. This agile satellite can deliver up to five different viewing angles. Data is limited to registered users.

The Hyperspectral Imager for the Coastal Ocean (HICO) is the first spaceborne imaging spectrometer designed to sample the coastal ocean (http://hico.coas.oregonstate.edu/). HICO samples selected coastal regions at 90 m with full spectral coverage of 380 nm to 960 nm. HICO was developed by the Naval Research Laboratory for the Office of Naval Research as an innovative naval prototype. Support for HICO was then provided by NASA's International Space Station Program. HICO ended its mission in September 2014 due to a severe radiation hit on HICO's computer. HICO demonstrates coastal products including water clarity, bottom types, bathymetry and on-shore vegetation maps. HICO data are publicly available. Note that to access HICO data from the NASA website a registered account is required. Configuration of EO-1/Hyperion, Proba-1/CHRIS, and HICO is shown in Table 2.8.

2.4 Lidar Remote Sensing

Lidar represents light detection and ranging. It is an active remote sensing technique that uses laser light to densely sample the surface of Earth and produces highly accurate x, y, z measurements. Like multispectral and hyperspectral remote sensing, lidar sensors can be mounted on drones, aircraft, and spacecraft. The airborne lidar is primarily used for mapping applications. It is a cost-effective alternative to traditional 3-D surveying techniques such as photogrammetry. A lidar system has three major hardware components: a laser scanner sending laser signals, Global Positioning System (GPS) measuring the location of the aircraft, and Inertial Navigation System (INS) measuring the pitch, roll, and yaw of the lidar system. The laser scanner transmits laser beams toward the targets while moving through the earth's surface. The reflectance of the laser from the target is detected and analyzed by receivers in the lidar system. The range distance between the target and lidar sensor is then calculated using the recorded time of the receivers. Combining the range distance with data collected by GPS and INS, the actual location and elevation of each point reflected from the target can be derived. There are two types of lidar: topographic and bathymetric. Topographic lidar uses a near-infrared laser to map land, while bathymetric lidar uses water-penetrating green light to measure seafloor or riverbed elevation.

2.4.1 Lidar Data Attributes

What lidar collects is known as point cloud data, which is a large collection of 3-D elevation points including horizontal location, elevation, and other attributes. An example of the lidar point cloud over the Shark River Slough in the Everglades National Park is shown in Figure 2.17. In addition to the x, y, z values, each point data has several other attributes delivered by vendors, among which the return number, number of returns, intensity, and classification are also important in lidar applications. A laser pulse travels toward the earth's surface at a certain angle by a rotating mirror in the lidar system. Depending upon the altitude of the lidar instrument and the angle at which the pulse is sent, each pulse illuminates a near-circular area on the ground called the instantaneous laser footprint (e.g., 30 cm in diameter). When the materials within the instantaneous laser footprint have local relief, the laser pulse will be split into several returns to be recorded by the lidar system, resulting in multiple returns of a single output pulse. This often occurs when collecting data in the forests. The laser pulse is reflected off different parts of the forest until it finally hits the ground. The lidar multiple return feature can help identify ground points for DEM generation. The lidar

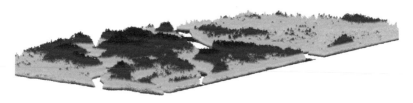

FIGURE 2.17
3-D point cloud data over the Shark River Slough in Everglades National Park. Color is rendered by measured elevation (z). Lidar data was collected by NASA's G-LiHT on December 2, 2017, after Hurricane Irma.

intensity records the amount of energy reflected by the target. This is like the radiation or reflectance recorded by a passive sensor from a target. It can be used to discriminate different targets due to the variation in reflectance of different surfaces. Lidar intensity can be used as a substitute for aerial imagery when none is available. Note that intensity is relative, not quantifiable; the same value off the same target is not expected from flight to flight or from elevation to elevation. Classification of a lidar point defines the type of object that has reflected the laser pulse. This information is derived in the lidar data post-processing procedure. Lidar points can be classified into several categories such as bare earth or ground, top of canopy, and water. The common lidar point attributes in a lidar file are summarized in Table 2.9.

2.4.2 Lidar Data Format

Lidar data has a large data volume. Storing such type of data using ASCII file format is a challenge. A standard of lidar data format, LAS, has been formed for the interchange of lidar data by the American Society for Photogrammetry and Remote Sensing (ASPRS). LAS is a binary format that can largely reduce the file size compared with the ASCII format. ASPRS released LAS 1.0 in 2003; the current version is LAS 1.4, which allows the users to customize the LAS file format to meet their needs. Each LAS file contains metadata of the lidar survey in a header block followed by individual records for each laser pulse recorded. The header block consists of data extents, flight date, flight time,

TABLE 2.9

Description of lidar point attributes

Attributes	Description
Return number	An emitted laser pulse can have up to 5 returns depending on the features it is reflected from and the capabilities of the laser scanner used to collect the data. The first return will be flagged as return number 1, the second as return number 2, and so on.
Number of returns	The number of returns is the total number of returns for a given pulse. For example, a laser data point may be return 2 (return number) within a total number of 5 returns.
Intensity	The return strength of the laser pulse that generated the lidar point
Classification	Every lidar point that is post-processed can have a classification that defines the type of object that has reflected the laser pulse. Lidar points can be classified into several categories including bare earth or ground, top of canopy, and water. The different classes are defined using numeric integer codes in the lidar LAS files.
RGB	Lidar data can be attributed with Red, Green, and Blue (RGB) bands. This attribution often comes from imagery collected at the same time as the lidar survey.
Scan Angle	The scan angle is a value in degrees between -90 and +90. At 0 degrees, the laser pulse is directly below the aircraft at nadir. At -90 degrees, the laser pulse is to the left side of the aircraft, while at +90, the laser pulse is to the right side of the aircraft in the direction of flight. Most lidar systems are currently less than ±30 degrees.
Scan Direction	The scan direction is the direction the laser scanning mirror was traveling at the time of the output laser pulse. A value of 1 is a positive scan direction, and a value of 0 is a negative scan direction. A positive value indicates the scanner is moving from the left side to the right side of the in-track flight direction, and a negative value is the opposite.
Edge of Flight Line	The points will be symbolized based on a value of 0 or 1. Points flagged at the edge of the flight line will be given a value of 1, and all other points will be given a value of 0.
GPS Time	The GPS time stamp at which the laser point was emitted from the aircraft. The time is in GPS seconds of the week.

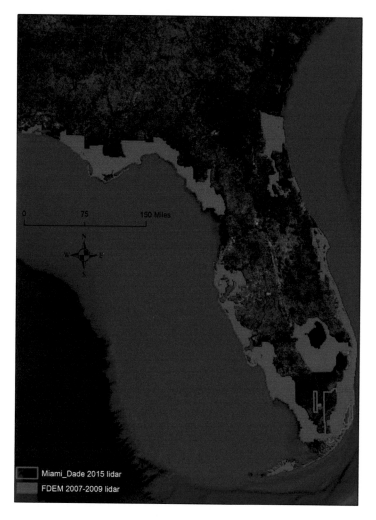

FIGURE 2.18
Lidar data coverage in Florida. 2007–2009 lidar was collected by the Florida Division of Emergency Management (FDEM); 2015 lidar was collected by Miami-Dade County.

number of point records, number of points by return, any applied data offset, and any applied scale factor. Common attributes listed in Table 2.9 are maintained for each lidar point. Lidar software packages such as ArcGIS and FUSION/LDV support LAS format.

2.4.3 Lidar Data Resources in the Everglades

There are two major lidar data sources available for Everglades mapping including 2007–2009 Florida Division of Emergency Management (FDEM) lidar, and 2015 Miami-Dade lidar. The FDEM collected lidar over all Florida coasts to use lidar data as new elevation inputs for updating storm surge models during 2007–2009. Lidar Survey was collected under the guidance of a Professional Mapper/Surveyor. The lidar system acquisition parameters were developed based on a maximum average ground sample distance of 4 feet. The collected lidar point cloud data was classified including bare-earth points (class 2), noise points (class 7), water returns (class 9), and unclassified data (class 1). Class 12 contains lidar points removed from the overlap region between adjacent flight lines. This huge lidar dataset is available at Florida International University's (FIU's) International Hurricane Research Center (http://mapping.ihrc.fiu.edu/). It provides a tile-by-tile download of LAS, DEM, and metadata files from the Florida Coastal lidar project. Currently it has data for the entire FDEM project. Miami-Dade County collected lidar in February (15, 17, 18, 19, 20, 21) and April (2, 3, 11, 12, 13) 2015. This lidar data classifications included 1 (unclassified), 2 (ground), 6 (building), 7 (low noise), 9 (water), 10 (buffered ground around breaklines), 17 (bridge decks), and 18 (high noise). Lidar DEM is also available from this project. This dataset covers 1612 square miles with a reported vertical accuracy of 6.4 cm and horizontal accuracy of 115 cm. This lidar dataset is downloadable at NOAA's digital coast (https://coast.noaa.gov/digitalcoast/). The coverage of 2007 FDEM and 2015 Miami-Dade datasets is also shown in Figure 2.18.

In addition to the above two lidar data sources, various bathymetric lidar datasets along Florida coasts are available from NOAA's digital coast. NASA's G-LiHT lidar data collected in the Everglades is available at https://glihtdata.gsfc.nasa.gov/ or ftp://fusionftp.gsfc.nasa.gov/G-LiHT. FDEM in collaboration with USGS is planning to collect statewide lidar in 2018 and 2019.

References

Claverie, M., Ju, J., Masek, J.G., Dungan, J.L., Vermote, E.F., et al., 2018. The Harmonized Landsat and Sentinel-2 surface reflectance data set. *Remote Sensing of Environment*, 219, 145–161.

Colomina, I., and Molina, P., 2014. Unmanned aerial systems for photogrammetry and remote sensing: A review. *ISPRS Journal of Photogrammetry and Remote Sensing*, 92, 79–97.

Cook, B.D., Corp, L.W., Nelson, R.F. et al., 2013. NASA Goddard's Lidar, Hyperspectral and Thermal (G-LiHT) airborne imager. *Remote Sensing*, 5, 4045–4066.

DigitalGlobe, 2018. www.digitalglobe.com.

Dong, P., and Chen, Q., 2018. *Lidar Remote Sensing and Applications*, CRC Press, Boca Raton, FL.

ESA/CHRIS, 2018. https://earth.esa.int/web/guest/-/proba-chris-level-1a-1488.

ESA/Sentinel Online, 2018. https://earth.esa.int/web/sentinel/.

Goetz, A.F.H., 1992. Imaging spectrometry for earth remote sensing. In: Toselli, F., and Bodechtel, J. (eds.) *Imaging Spectroscopy: Fundamentals and Prospective Applications*, Springer, Dordrecht, Netherlands, 1–20.

González-Jorge, H., Martínez-Sánchez, J., Bueno, M., and Arias, P., 2017. Unmanned aerial systems for civil applications: A review. *Drones*, 1, 2. doi:10.3390/drones1010002

Jensen, J.R., 2007. *Remote Sensing of the Environment: An Earth Resource Perspective*, 2nd Edition, Prentice Hall, Upper Saddle River, NJ.

Jensen, J.R., 2015. *Introductory Digital Image Processing: A Remote Sensing Perspective*, 4th Edition, Prentice Hall, Upper Saddle River, NJ.

Klemas, V.V., 2015. Coastal and environmental remote sensing from unmanned aerial vehicles: An overview. *Journal of Coastal Research*, 31, 1260–1267.

Milas, A.S., Sousa, J.J., Warner, T.A. et al., 2018. Unmanned Aerial Systems (UAS) for environmental applications special issue preface. *International Journal of Remote Sensing*, 39, 4845–4851.

NASA/AVIRIS, 2018. https://aviris.jpl.nasa.gov/. Accessed on 20 July 2018.

NASA/EO-1/Hyperion, 2018. https://cmr.earthdata.nasa.gov/search/concepts/C1379758136-USGS_EROS.html. Accessed on 20 July 2018.

NASA/G-LiHT, 2018. https://gliht.gsfc.nasa.gov/. Accessed on 20 July 2018.

NASA/Harmonized Landsat Sentinel-2, 2018. https://hls.gsfc.nasa.gov/. Accessed on 20 July 2019.

NASA/MODIS, 2018. http://modis.gsfc.nasa.gov/. Accessed on 20 July 2018.

NOAA/National Centers for Coastal Ocean Science, 2018. https://coastalscience.noaa.gov/project/benthic-habitat-mapping-florida-coral-reef-ecosystems/. Accessed on 20 July 2018.

Singh, K.K., and Frazier, A.E., 2018. A meta-analysis and review of unmanned aircraft system (UAS) imagery for terrestrial applications. *International Journal of Remote Sensing*, 39, 5078–5098.

USGS/eMODIS, 2018. https://lta.cr.usgs.gov/emodis. Accessed on 20 July 2018.

USGS/Landsat, 2018. https://landsat.usgs.gov/. Accessed on 20 July 2018.

USGS/National Agriculture Imagery Program (NAIP), 2018. https://lta.cr.usgs.gov/NAIP. Accessed on 20 July 2018.

USGS/National High Altitude Photography (NHAP), 2018. https://lta.cr.usgs.gov/NHAP. Accessed on 20 July 2018.

3

Vegetation Classification Systems in the Everglades

Caiyun Zhang

The distribution, condition, and change of vegetation communities reflect factors that shape the ecosystem of the Florida Everglades including climate, hydrology, and human activities. Changing current vegetation communities is one of the goals in CERP in an effort to restore and preserve the habitat of threatened, endangered, and other wildlife species. Monitoring changes in vegetation communities can measure the progress and effects of restoration on environmental health (Welch et al., 1999; Rutchey et al., 2006). Detailed and accurate spatial data such as vegetation maps are key factors to document change in the Everglades. In CERP, a system-wide program is designed to organize, manage, and provide the highest quality scientific and technical support during the implementation of the restoration program. This program is known as REstoration COordination and VERification (RECOVER, www. evergladesrestoration.gov/ssr/). One component in RECOVER is to document changes in the spatial extent, pattern, and proportion of plant communities within the Everglades landscape. Large-scale remote sensing data have proven effective for vegetation mapping.

For mapping vegetation, it is important to determine a classification system to be used as a standard in the mapping projects. A number of national and statewide classification systems (e.g., Anderson et al., 1976; Cowardin et al., 1979) were available, but they failed to include the desired level of detail and specificity of vegetation classes for Everglades restoration. Thus, a detailed system applicable to the Greater Everglades was developed by Rutchey et al. (2006). This classification system is specifically designed for peninsular south Florida and the Florida Keys, from Lake Okeechobee in the north to Key West in the south. This system is known as Vegetation Classification for South Florida Natural Areas. Other classification systems that were used in the Everglades including Florida Land Use, Cover, and Forms Classification System (FLUCCS), Everglades Vegetation Classification System (Madden et al., 1999), and Kissimmee River Restoration Evaluation Program (KRREP) Baseline Vegetation Classification (Bousquin and Carnal, 2005). These four classification systems are detailed because they are used in projects described in this book for vegetation mapping.

3.1 Vegetation Classification for South Florida Natural Areas

RECOVER required a classification system with enough flexibility and detail to enable the designation of vegetation classes using various identification techniques, from field investigations to remote sensing. The classification system had to be hierarchical and represent distinct ecological communities, individual species, and physical characteristics such as density and height. In addition, it was desirable that the classification system allow exotic species and cattail to be identified using density classes. Rutchey et al. (2006) developed this classification system to meet the needs in CERP, which was used for mapping natural areas in south Florida. Specific areas of interest in this classification system include ENP, Big Cypress National Preserve, Biscayne National Park, Florida Panther National Wildlife Refuge, Loxahatchee National Wildlife Refuge, the WCAs, Holeyland Wildlife Management Area, Rotenberger Wildlife Management Area, J.W. Corbett Wildlife Management Area, Pal-Mar Wildlife Management Area, the Lake Okeechobee Littoral Zones, and additional coastal wetlands of southeastern Miami-Dade County. In addition to being used for mapping of CERP affected areas, the National Park Service-South Florida/Caribbean Network uses the classification for mapping the remaining areas outside the CERP footprint.

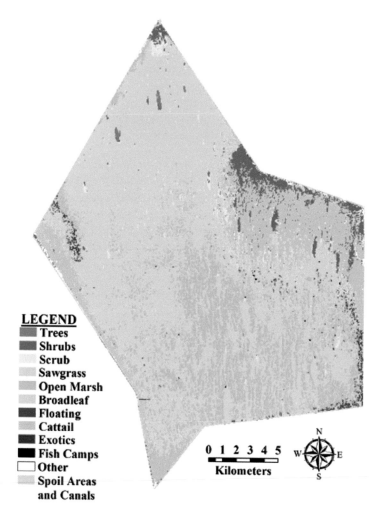

LEGEND
- Trees
- Shrubs
- Scrub
- Sawgrass
- Open Marsh
- Broadleaf
- Floating
- Cattail
- Exotics
- Fish Camps
- Other
- Spoil Areas and Canals

0 1 2 3 4 5
Kilometers

FIGURE 3.1
Vegetation map of WCA-2A produced by Rutchey et al. (2008) (from SFWMD) using the vegetation classification system of "Vegetation Classification for South Florida Natural Areas" and aerial photography collected in 2003.

This classification system is hierarchical, designating up to six levels. Lower levels of the classification (e.g., level 6) are specific classifications nested within higher levels (e.g., level 5). The different levels of this classification system represent distinctions in ecological communities, taxonomy, individual species, and physical characteristics such as density and height. The appropriate level of classification depends on the needs of the project and level of detail discernable from the methods used. Primary vegetation classifications in this system include Forest, Woodland, Shrubland, Scrub, Marsh, Dune, Submerged Aquatic Vegetation, and Exotic. Figure 3.1 shows an application of this classification system for mapping vegetation in WCA-2A from aerial photography collected in 2003.

3.2 FLUCCS

FLUCCS was originally developed by the Florida Department of Transportation (FDOT) in 1976 (FLUCCS Handbook, 1999) and has been modified to meet the requests for specific identification. The current system is the third edition modified in 1999, which is also organized in hierarchical levels (from level I to level IV) with each level containing land information of increasing specificity. The various categories and subcategories reflect the types of data and information that can be extracted from aerial photography of various types (panchromatic, natural color or false color infrared) and scales (large, medium and small) of the current generation of airborne and satellite multispectral imaging systems. Level I is very general in nature, and this type of information can be obtained from satellite imagery (e.g., Landsat) for state-wide or larger scale mapping. Level II is more specific than Level I, and data can be obtained for this level using high altitude imagery supplemented by other materials. Level III can be obtained from medium altitude photography, and the mapping scale typically is 1: 24,000. Level IV is obtained from low altitude photography flown below 10,000 feet. Level IV typically is mapped at the scale of 1: 60,000. Level I includes 8 classes: urban and built-up, agriculture, rangeland, upland forests, water, wetlands, barren land, and transportation/communication/utilities. For the wetland class, level II further breaks into 5 classes including wetland hardwood forest, wetland coniferous forests, wetland forested mixed, vegetated non-forested wetlands, and non-vegetated. Vegetated wetlands are detailed into 23 classes in level III, and level IV gives more details for freshwater marshes (8 classes), saltwater marshes (2 classes), and emergent aquatic vegetation (5 classes).

FLUCCS has been used by the South Florida Water Management District (SFWMD) to produce Land Cover Land Use (LCLU) datasets in south Florida. Archived LCLU datasets from this classification system using manual photo interpretation techniques include LCLU 1988, LCLU 1995, LCLU 1999, LCLU 2004, LCLU 2008, and LCLU 2014–2016. All these LCLU datasets and other vegetation maps in the Greater Everglades are available at no cost in SFWMD. A Level IV classification map of WCA-2A is displayed in Figure 3.2 using LCLU 2004. Figure 3.3 shows the Level I LCLU data in the Greater Everglades combined from LCLU datasets generated by all Water Management Districts in Florida.

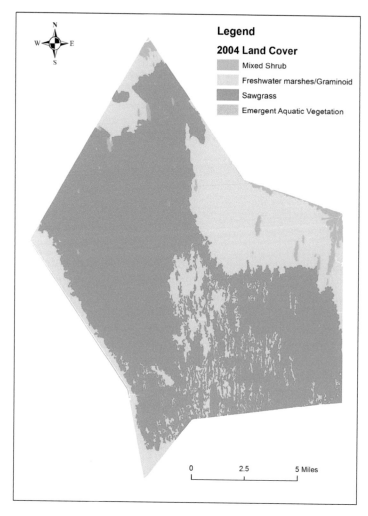

FIGURE 3.2

Vegetation map (Level IV) of WCA-2A in 2004. The dataset used in the map is from LCLU 2004, which was produced by SFWMD using the classification system of FLUCCS and aerial photography collected in 2004.

3.3 Everglades Vegetation Classification System

This classification system is a collaborative effort of the Center for Remote Sensing and Mapping Science at the University of Georgia, the South Florida Natural Resources Center at ENP, and the SFWMD (Madden et al., 1999). The system was developed to map vegetation patterns to the plant-community level within a 12,000 km^2 area including ENP, Big Cypress National Preserve, Biscayne National Park, the Florida Panther National Wildlife Refuge, and WCA 3. The system was developed based on classification systems previously used by researchers mapping ENP and the Big Cypress National Preserve. Classes were also organized hierarchically under 8

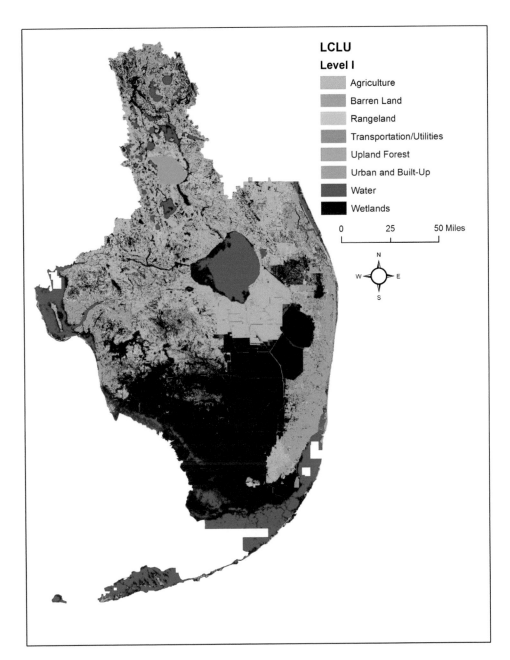

FIGURE 3.3
LCLU Level I map of the Greater Everglades using the classification system of FLUCCS. Data were compiled from LCLU data of all Water Management Districts in Florida.

major vegetation types: forest, scrub, savanna, prairies and marshes, shrublands, exotics, additional class headings, and special modifiers. Each of these major classes is further divided into classes corresponding to plant communities. In cases where individual species can be discerned on the aerial photographs used in the mapping project, a third level of detail was included in the classification system. A three-tiered scheme was also used in the system to accommodate the complex vegetation patterns in ENP. Using this scheme, photo interpreters can annotate each polygon with a dominant vegetation class accounting for more than 50% of the vegetation in the polygon. Secondary and tertiary vegetation classes are then added as required to describe mixed plant communities within the polygon. In addition, one or more of 13 numerical modifiers were attached to each dominant, secondary, and tertiary vegetation label to indicate factors such as human influence, hurricane damage, altered drainage, and extensive off-road vehicles that might influence vegetation growth and distribution. Other modifiers provide information about the vegetation environmental characteristics (e.g., periphyton). Based on this classification, a total of 89 vegetation shape files were produced for the project area from NAPP color infrared aerial photographs of 1994–1995. This vegetation dataset and the classification table are available at Florida Coastal Everglades Long Term Ecological Research (http://fcelter.fiu.edu/). Figure 3.4 displays a vegetation map over Flamingo at ENP using this classification system and datasets produced in the project.

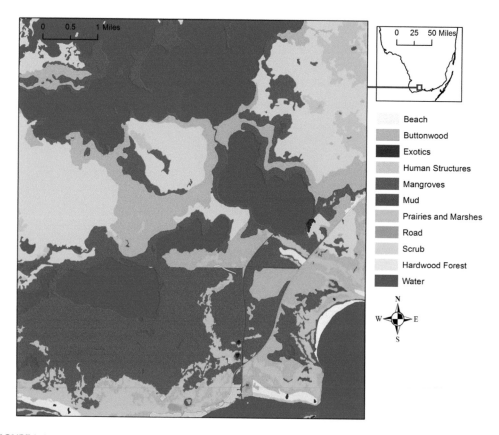

FIGURE 3.4
Vegetation map of Flamingo in ENP. Vegetation datasets were produced by Madden et al. (1999) using the Everglades Vegetation Classification System and aerial photography collected in 1994–1995.

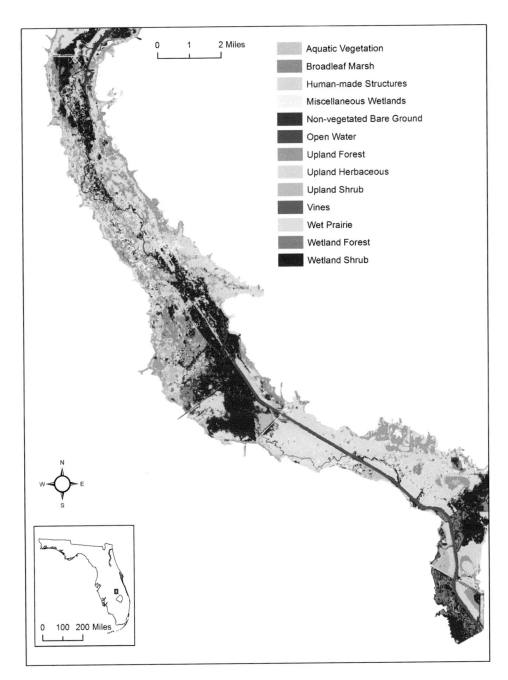

FIGURE 3.5
Vegetation map at the Bcode Group level in the Kissimmee River floodplain. Dataset was produced by SFWMD using KRREP Vegetation Classification System and aerial photography collected in May and June 2011.

3.4 KRREP Baseline Vegetation Classification

This classification system was developed to characterize plant assemblages that occur in the Kissimmee River floodplain in the northern Everglades (north of Lake Okeechobee). It was developed based on previous classification systems that were used to map vegetation in the floodplain prior and immediately following channelization (Bousquin and Carnal, 2005). It provides definitions and decision rules to facilitate consistent, repeatable descriptions of river and floodplain plant assemblages for use by photo interpreters and field data collectors during the KRREP. It is better suited to post-channelization conditions than the previous classification systems. The system also has a hierarchical structure that allows for physiognomic distinctions among plant communities, a basic requirement for linkage with the previous classifications; it allows users flexibility in choice of resolution. The system includes a status category, physiognomic category, Bcode groups, and community type. The status categories group communities by characteristic habitat (e.g., upland, wetland, or aquatic). Physiognomic categories describe the general appearance of communities as forest, shrubland, or herbaceous. Community types (abbreviated Bcodes) are the finest level of classification, capturing communities as distinguished by dominant species. Bcode Groups are groupings of community types into ecologically meaningful categories, at a hierarchical level between community type and physiognomy. Figure 3.5 shows vegetation mapping at the Bcode Group level in the Kissimmee River floodplain using this classification system and aerial photography collected in 2011.

References

Anderson, J.R., Hardy, E.E., Roach, J.T., and Witmer, R.E., 1976. *A Land Use and Land Cover Classification System for Use with Remote Sensor Data: U.S. Geological Survey*, Professional Paper 964, U.S. Government Printing Office, Washington, DC, 28.

Bousquin, S.G., and Carnal, L.L., 2005. Classification of the vegetation of the Kissimmee River and floodplain. Chapter 9, 1-11. In: Bousquin, S.G., Anderson, D.H., Williams, G.E., and Colangelo, D.J. (eds.) *Establishing a Baseline: Pre-restoration Studies of the Channelized Kissimmee River*. South Florida Water Management District, West Palm Beach, FL, Technical Publication ERA #432.

Cowardin, L.M., Carter, V., Golet, F.C., and LaRoe, E.T., 1979. *Classification of Wetlands and Deepwater Habitats of the United States, Fish and Wildlife Service*, U.S. Department of the Interior, Washington, DC, 103.

FLUCCS Handbook, 1999. *Florida Land Use, Cover and Forms Classification System*, 3rd Edition, Florida Department of Transportation, Tallahassee, FL.

Madden, M., Jones, D., and Vilchek, L., 1999. Photointerpretation key for the everglades vegetation classification system. *Photogrammetric Engineering and Remote Sensing*, 65, 171–177.

Rutchey, K., Schall, T., Doren, R., et al., 2006. Vegetation classification for South Florida natural areas: Saint Petersburg, FL, United States geological survey, Open File Report 2006-1240, 142.

Rutchey, K., Schall, T., and Sklar, F., 2008. Development of vegetation maps for assessing Everglades restoration progress. *Wetlands*, 28, 806–816.

Welch, R., Madden, M., and Doren, R., 1999. Mapping the Everglades. *Photogrammetric Engineering and Remote Sensing*, 65, 163–170.

Part II

Multispectral Remote Sensing Applications

4

Applying Aerial Photography to Map Marsh Species in the Wetland of Lake Okeechobee

Caiyun Zhang

4.1 Introduction

In CERP, vegetation information is mainly produced using high spatial resolution aerial photography through manual photo-interpretation techniques to inform the progress and effects of restoration on plant communities as well as others. Examples include Land Cover Land Use (LCLU) datasets produced by the South Florida Water Management District (SFWMD) for the years 1988, 1995, 1999, 2004, 2008, and 2014 to 2016 based on a modified FLUCCS, 2009 vegetation maps produced by the Florida Fish and Wildlife Conservation Commission (FWC) for lake areas in the Kissimmee River floodplain based on FLUCCS, the 1994 to 1995 vegetation dataset in the Everglades National Park (ENP) produced by the University of Georgia using the Everglades Vegetation Classification System, and vegetation maps produced by SFWMD for the Kissimmee River floodplain. Figures 3.1 to 3.5 show the map examples of these products. Applying time series of vegetation maps from aerial photography documents the change of vegetation community in the restoration. Spencer and Bousquin (2014) mapped the change of vegetation communities in the Kissimmee River floodplain using aerial photography to evaluate the effects of Phase I of the Kissimmee River restoration project, which reestablished intermittent inundation of the river's floodplain by backfilling 12 km of the C-38 flood control canal in 2001. The comparison results indicate that the extent of wetland plant communities expanded rapidly within 2 years of the completion of Phase I, and by 2008 wetlands had nearly recovered to pre-channelization levels; hydrology data indicate that the duration and variability of floodplain inundation have not yet achieved restoration targets over the entire Phase I study area, which impact the full recovery of plant communities in the project domain. Aerial photography has also been examined to produce vegetation maps using digital classification procedures. Aerial photography was also applied to map plant species of the wet prairie in ENP (Szantoi et al., 2013, 2015). It has been found that a combination of the texture measurements from the fine spatial resolution aerial photography is valuable for mapping the plant communities.

In this chapter, a framework was presented to apply 1-foot aerial photography for mapping freshwater marsh species in the wetlands of Lake Okeechobee. Lake Okeechobee and its wetlands are at the center of the Greater Everglades. The lake is the heart of the Greater Everglades. The original lake ecosystem was severely modified over the past century by human activities, resulting in many environmental issues in South Florida. A large amount of phosphorus accumulated in lake sediments due to excessive phosphorus loads from the lake's watershed. Large lake freshwater discharges harmed the ecological health of downstream estuaries (Lake Okeechobee Watershed Restoration

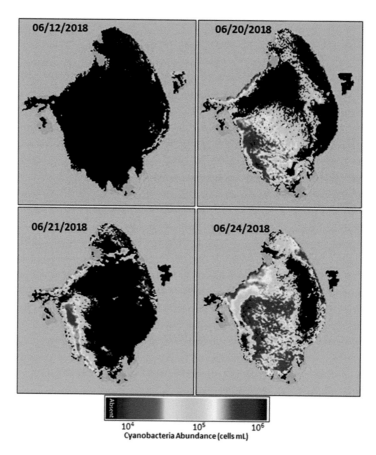

FIGURE 4.1
Satellite image series show the large harmful algae bloom (HAB) on Lake Okeechobee in 2018 (from NOAA).

Project, 2017). In 2018, a large harmful algae bloom (HAB) occurred on Lake Okeechobee, as revealed by the National Oceanic and Atmospheric Administration (NOAA) using a satellite image series (Figure 4.1). The Lake Okeechobee Watershed Restoration Project is currently underway to improve the water quality and quantity, timing of flows, and restoration of the lake's wetlands. This project is part of CERP. For the lake area, detailed marsh maps are needed to assess the effects and progress of the restoration.

Currently, marsh maps in the lake area are mainly generated through manual interpretation of large-scale aerial photography using stereo techniques. In the stereo analysis, a grid is created and superimposed on the georeferenced 3-dimensional stereo imagery first, and then a photo-interpreter manually labels each grid (100 m^2) using a Softcopy Photogrammetry Workstation. Only a certified photogrammetrist is allowed to carry out the manual interpretation, which is time consuming and labor intensive. Inconsistent results can be produced by different mapping personnel. SFWMD has released a series of Lake Okeechobee Littoral Vegetation datasets to the public including years 1972–1973, 1994–1995, 2003, 2006–2007, 2011–2012, and 2015–2016 using the manual procedure. Polygon features were used to define the extent and type of littoral vegetation within the western marsh portion of Lake Okeechobee. With the advance of digital image processing techniques, it is expected that this manual procedure can be semi-automated or automated.

In the literature, numerous efforts have been made to map fresh and salt marshes using remote sensing digital procedures, which can be broadly grouped into three categories based on the application of data sources (Zhang et al., 2018a). The first category is the application of multispectral imagery collected from satellites or airborne sensors (e.g., Belluco et al., 2006; Carle et al., 2014; Kumar and Sinha, 2014; Sun et al., 2016). Results have shown that fine spatial resolution multispectral imagery is useful for mapping marsh species, with an accuracy of 75% for freshwater marsh species classification and an accuracy of over 90% for salt marsh species classification. The second category is the employment of hyperspectral imagery (e.g., Belluco et al., 2006; Judd et al., 2007; Wang et al., 2007; Kumar and Sinha, 2014). A comparison study of five data sources in Belluco et al. (2006) demonstrated that the spatial resolution is more important than the spectral resolution in mapping marsh species; multispectral sensors such as Ikonos and QuickBird can achieve accuracy similar to hyperspectral sensors for mapping saltwater marshes. The third category is the integration of optical imagery with lidar data. Lidar products such as a Digital Elevation Model (DEM) have proven useful for extracting marsh species elevation ranges and distribution (Morris et al., 2005; Sadro et al., 2007) and improving marsh classification (Gilmore et al., 2008; Hladik et al., 2013; Zhang et al., 2018a).

In this chapter, the capability of very high spatial resolution aerial photography was assessed for marsh mapping in the wetlands of Lake Okeechobee because this type of data (1 m or smaller) has been frequently collected over the lake area to support ongoing restoration. The examination of this type of data using digital techniques is limited. Modern Object-based Image Analysis (OBIA), machine learning, and texture analysis techniques were combined in an effort to automate the digital mapping procedure.

4.2 Study Area and Data

The study area is located at the western edge of Lake Okeechobee, a region known as the Moore Haven Marsh (Figure 4.2), the same area as in Zhang et al. (2018a). Note that in Zhang et al. (2018a), the potential of fusing aerial photography with lidar for improving marsh species mapping was investigated. In this chapter, only aerial photography was used. The spatial distribution of marsh species in the Lake Okeechobee area is mainly determined by the hydroperiod (i.e., frequency of inundation, as a percent of days per year) with short hydroperiod regions supporting species such as spikerush and willow, and long hydroperiod regions supporting species such as cattail and sawgrass (Havens and Gawlik, 2005). Freshwater graminoid marsh and floating emergent marsh are two dominant marsh communities. The selected site covers 574 acres (2.32 km^2) and is dominated by six freshwater species to be identified, including willow (*Salix caroliniana*), cordgrass (*Spartina bakeri*), smartweed (*Polygonum spp.*), cattail (*Typha spp.*), common reed (*Phragmites australis*), and sawgrass (*Cladium jamaicense*). These are common species of graminoid marsh over this region. In addition, meadow marsh (graminoid), swamp shrubland, and mud were also present in the study site and mapped. The graminoid marsh is characterized by

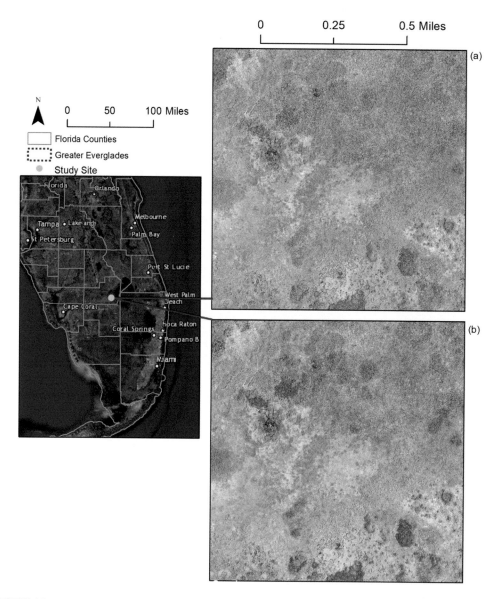

FIGURE 4.2
The Greater Everglades and Lake Okeechobee, and the study site shown as (a) a natural color composite and (b) color infrared composite of the 1-foot aerial photography collected on April 24, 2012.

emergent herbaceous vegetation and is a mosaic of various grasses. If no dominant marsh species can be identified, the region is then labeled graminoid marsh. Swamp shrubland is seasonally to semi-permanently flooded, high-density stands of small trees or shrubs with heights of less than 5 m found throughout Florida. Mud refers to moist, open ground.

As in Zhang et al. (2018a), data used in this chapter include 1-foot (30.48 cm) resolution aerial photography and a reference dataset. The aerial photography was collected on April 24, 2012, using a Microsoft Vexcel UltraCamX Sensor System as part of a project led by the SFWMD to map vegetation in the Lake Okeechobee Moore Haven Marsh. The acquired raw images were geometrically and radiometrically processed and orthorectified into 1-foot (30.48 cm) orthos with four bands (Red, Green, Blue, and Near-Infrared). In total, 223 tiles (5000 feet/1524 m × 5000 feet/1524 m per tile) were produced in the project, but only one tile was selected in this chapter due to the large volume of the aerial photography. This aerial photography product was not available to the public. The author obtained this tile for this study through a collaboration with SFWMD. A natural color and a color infrared composite of the selected tile is displayed in Figure 4.2. The selected site covers the dominant species present in the marsh area. The reference vegetation dataset was from manual interpretation of the aerial photography collected in the project, which collected stereo imagery for using photogrammetric techniques in the interpretation procedure. The stereo imagery was manually interpreted using a stereo plotter to generate a marsh map by the SFWMD. The features were labeled with a classification system developed by Rutchey et al. (2006) for vegetation of south Florida natural areas. The classification system is hierarchical, designating up to six levels with lower levels nested within higher levels. Graminoid and/or herbaceous emergent or floating vegetation in shallow water that stands at or above the ground surface for much of the year is identified as marsh. Marsh is further divided into salt marsh and freshwater marsh (level 2). The freshwater marsh is further detailed to graminoid marsh and floating emergent marsh (level 3). Species labeling belongs to level 4. During manual interpretation, each grid cell (100 m^2) is labeled with the majority vegetation species observed. Level 4 is further detailed by adding more modifications such as density, leading to levels 5 and 6. All levels are labeled manually. The manually interpreted map was provided by the SFWMD and used as a reference dataset in this chapter to assist with training and testing sample selection in the classification.

4.3 Methodology

The methodology flowchart for mapping marsh species using aerial photography is displayed in Figure 4.3. To conduct object-based classification, objects were produced first by applying a segmentation algorithm in the eCognition software package (Trimble, 2014). After segmentation, spatial and spectral features of each object were calculated in the package and exported into ArcGIS for training and testing object selection. In total, 536 image objects were selected as the reference objects by following a spatially stratified data sampling strategy, in which a fixed percentage of samples were selected for each class. The number of samples for each class was roughly estimated with the results of image segmentation and the reference dataset. For minor species (i.e., a species covers a small region), only a limited number of objects were produced. A small number of reference samples would lead to unreliable testing results. Thus, for the minor species present in the study site, a higher percentage was used in the reference sample selection for these species. The reference object samples were then classified using three machine learning classifiers, Support Vector Machine (SVM), Random Forest (RF), and Artificial Neural Network (ANN). The accuracy of each classifier was assessed using the error matrix and Kappa statistical techniques based on a *k*-fold cross-validation technique. If the testing results were acceptable, the whole image's objects were classified, leading to

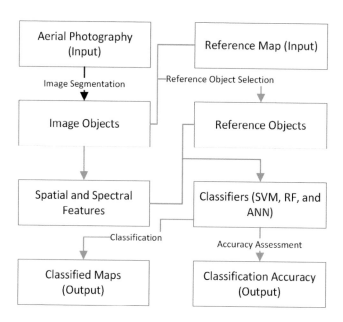

FIGURE 4.3
Methodology flowchart for mapping marsh species using aerial photography.

a final classified marsh map. The major techniques in the mapping procedure include image segmentation, classification, and accuracy assessment, which are detailed in the following subsections.

4.3.1 Image Segmentation to Create Image Objects

OBIA has proven more effective in mapping than pixel-based analyses when fine spatial resolution imagery is applied. As noted in the literature, the pixel-based land cover characterization often ignores the problem that a substantial proportion of signal within a single pixel apparently comes from the land areas surrounding the terrain (Jensen, 2015). OBIA was thus proposed and developed by not only considering the spectral character-istics of a single pixel, but also incorporating the information of the spatial characteristics of the surrounding pixels so that a homogeneous region (an image object) will be analyzed and processed. We have applied OBIA techniques in the Florida Everglades for vegetation mapping (Zhang, 2014; Zhang and Xie, 2012, 2013, 2014; Zhang et al., 2013, 2016, 2017, 2018a), sawgrass biomass modeling and mapping (Zhang et al., 2018b), and mangrove damage and risk analysis from hurricanes (Zhang et al., 2019). All these studies have proven that OBIA is valuable for mapping and monitoring the Florida Everglades. Thus, OBIA was selected in this chapter. This technique was also used in other chapters of this book.

To apply OBIA, image objects are created through an image segmentation proce-dure. A range of image segmentation algorithms have been developed in computer science, which are also applicable for processing remote sensing imagery. eCognition software is one of the popular off-the-shelf packages. It combines several segmenta-tion algorithms including chessboard segmentation, quadtree-based segmentation, contrast split segmentation, multi-resolution segmentation, spectral difference segmentation, multi-threshold segmentation, and contrast filter segmentation (Trimble, 2014). In this chapter, the multi-resolution segmentation algorithm was used to segment the 1-foot aerial imagery. This algorithm starts with single objects of one pixel and repeatedly merges them in several loops in pairs to larger units as long as an upper threshold of homogeneity is not exceeded locally. The homogeneity criterion is defined as a combination of spectral homogeneity and shape homogeneity. Details of the definition and quantification of spectral and shape homogeneity can be found in Jensen (2015) or Trimble (2014). Multi-resolution segmentation is an optimization procedure that, for a given number of image objects, minimizes the average heterogeneity and maximizes respective homogeneity. Running this algorithm in eCognition requires the inputs of several parameters, including scale, color/shape, and smoothness/compactness, among which the scale parameter is the most important input. Efforts have been made to develop approaches for the optimal scale parameter determination (e.g., Grybas et al., 2017). An approach developed by Johnson and Xie (2011) was applied to determine an optimal scale parameter for the aerial imagery. This approach begins with a series of segmentations by setting different scale parameters and then identifies the optimal image segmentation using a method that takes into account global intrasegment and intersegment heterogeneity measures. A global score (*GS*) is calculated in this approach by $GS = V_{norm} + MI_{norm}$, where V_{norm} (normalized weighted variance) measures the global intrasegment goodness, and MI_{norm} (normalized Moran's *I*) measures the global intersegment goodness. More details in computing V_{norm} and MI_{norm} can be found in Johnson and Xie (2011). The *GSs* were used to determine the optimal scale for segmentation. Application of this approach to the aerial photograph revealed that a scale of 75 was optimal. All four spectral bands of the aerial photography were set to equal weights, color/shape weights were set to 0.5/0.5, and smoothness/compactness weights were also set to 0.5/0.5 so as to not favor either compact or non-compact segments. The segmentation results of the aerial photography are displayed in Figure 4.4. As can be seen, relatively homogeneous patches/objects were produced, and each object was considered to have one marsh specie only to be mapped.

Following segmentation, the spectral and texture features of the aerial photography were extracted. Object-based features were extracted for vegetation discrimination. Object-based texture measures from fine spatial resolution imagery have proven valuable for vegetation classification (e.g., Zhang and Xie, 2012, 2013). For the pixel-based analysis, conventional kernel-based texture methods often utilize a fixed-size moving window over which to calculate texture measures for each pixel. It is challenging to determine the appropriate window size. OBIA offers the capability for identifying relatively homogeneous regions of varying shapes and sizes in an image. Texture extraction from image objects is more reasonable (Warner, 2011). Such object-based texture measures were thus extracted and assessed in this chapter. As in Zhang et al. (2018), the first- and second-order texture metrics for each band of the aerial photography were extracted in eCognition and exported into ArcGIS including mean, standard deviation, contrast, dissimilarity, homogeneity, entropy, and angular second moment. The gray level co-occurrence matrix (GLCM) algorithm was used to extract the second-order texture measures by calculating the mean of the texture results in all four directions. Hall-Beyer (2017) suggests choosing one of the contrast measures (contrast, dissimilarity, or homogeneity), one of the orderliness measures (angular second moment, maximum probability, or entropy), and two or

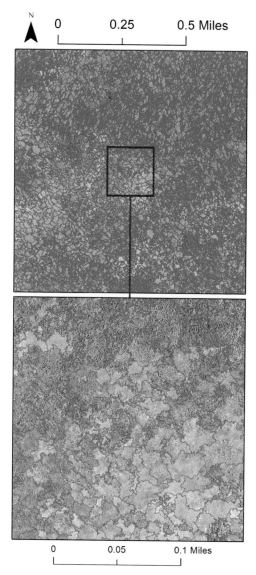

FIGURE 4.4
Segmentation results from the multi-resolution segmentation algorithm.

three descriptive statistics measures (mean, variance, correlation) metrics, because many texture measures are intrinsically similar. Therefore, the first-order mean, standard deviation, and second-order GLCM contrast, entropy, and standard deviation were applied and examined in the classification. Details for the calculation of these metrics at the object level can be found in Trimble (2014). Example maps of the first-order mean and standard deviation of the near infrared band of the aerial photography are displayed in Figure 4.5. In addition, the Normalized Difference Vegetation Index (NDVI) was calculated for each object and used in the classification. NDVI has proven useful for marsh community mapping in ENP (Szantoi et al., 2013).

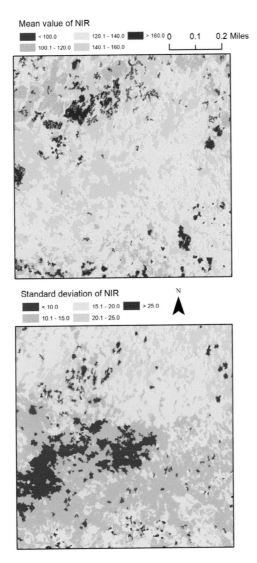

FIGURE 4.5
The first-order texture metrics of mean and standard deviation of near infrared band.

4.3.2 Classification: SVM, RF, and ANN

Several remote sensing image classifiers have been developed, and many of them have been incorporated into off-the-shelf remote sensing image processing software packages such as ENvironment for Visualizing Images (ENVI) and Erdas Imagine. The commonly used supervised classifiers include maximum likelihood, minimum distance, and parallelepiped classification, and unsupervised classifiers include Iterative Self-Organizing Data Analysis Technique (ISODAT) and *k*-means clustering algorithms. Advanced image classifiers such as SVM, RF, and ANN have become the major focus of the literature in image classification (Maxwell et al., 2018). Studies have demonstrated that these classifiers can produce a higher accuracy compared with the traditional classifiers, especially for the input dataset with a high-dimensional feature space. For this study, texture metrics were combined with spectral features of the aerial photography, resulting in high-dimensional input variables. Machine learning algorithms are recognized for their ability to achieve higher accuracy when a high-dimensional dataset is used; thus, this type of algorithms was selected in this study. Previous studies also show that these algorithms are valuable for mapping the Florida Everglades (Zhang, 2014; Zhang and Xie, 2012, 2013, 2014; Zhang et al., 2013, 2016, 2017, 2018a). Here, we assessed three popular machine-learning algorithms, SVM, RF, and ANN.

SVM is a non-parametric supervised classifier with the aim of finding a hyperplane that can separate the input dataset into a discrete predefined number of classes in a fashion consistent with the training samples (Vapnik, 1995). It has been applied in various fields, as reviewed by Mountrakis et al. (2011). Detailed descriptions of SVM algorithms were given by Huang et al. (2002) and Melgani and Bruzzone (2004) in the context of remote sensing. Kernel-based SVMs are commonly used for remote sensing image classification, among which the radial basis function (RBF) and the polynomial kernels are frequently employed. RBF needs to set the kernel width (γ), and polynomial kernel needs to set the degree (p). Both kernels need to define a penalty parameter (C) that controls the degree of acceptable misclassification. The setting of these parameters can be determined by a grid search strategy that tests possible combinations of C and γ in a user-defined range (Hsu et al., 2010). The original SVM algorithm was designed for binary classification. Several strategies including one-against-one and one-against-all have been developed to solve multiclass problems. These solutions divide a multiclass problem into a set of binary problems, making it feasible for multiclass classification. We selected one-against-one strategy, because the one-against-all strategy requires estimating complex discrimination functions (Melgani and Bruzzone, 2004) and thus may lead to unexpected classification results.

RF is a decision tree-based ensemble classifier. To understand this algorithm, it is helpful to first know the decision tree approach. The decision tree splits training samples into smaller subdivisions at "nodes" using decision rules. For each node, tests are performed on the training data to find the most useful variables and variable values for split. The RF consists of a combination of decision trees where each decision tree contributes a single vote for assigning the most frequent class to an input vector. RF increases the diversity of decision trees to make them grow by changing the training set using the bagging aggregating (Breiman, 2001). Bagging creates training data by randomly resampling the original dataset with replacement. Data selected from the input sample for generating the next subset is not deleted. A key feature of RF is that the computational complexity is simplified by reducing the number of input features at each node, which makes it particularly appealing in hyperspectral data classification. Different algorithms can be used to generate the decision trees. The RF often adopts the Gini Index (Breiman, 2001) to measure the best split selection. More descriptions of RF can be found in Breiman (2001) and in remote sensing context in Chan and Paelinckx (2008), Ghimire et al. (2012), and Rodriguez-Galiano et al. (2012).

ANN is also an important technique in image classification. Various ANN algorithms have been developed and applied, as reviewed by Mas and Flores (2008). Here the multilayer perceptron algorithm of ANN was applied. ANN has been motivated by the computational mechanism of the human brain. The brain performs highly complex, nonlinear, and parallel computations using its complex structure and ability to self-organize and build knowledge in storing and learning. An ANN works in a similar way as human's brain to perform a particular task. It has many advantages over traditional computational methods such as nonlinear learning capability, adaption in the learning environment, and ability to deal with incomplete information. The multilayer perceptron algorithm of ANNs is commonly used in various fields. It is a feedforward model based on supervised training. It consists of an input layer, one or more hidden layers, and one output layer. Connections between the layers are made forward on a layer-by-layer basis. The backpropagation algorithm is used in the multilayer perceptron algorithm, which consists of two phases: in the feedforward pass, an input vector is presented to the network and propagated forward to the output; in the backpropagation phase, the network output is compared to the desired output; network weights are then adjusted in accordance with an error-correction rule.

All three classifiers require the inputs of several parameters. For example, SVM needs to set the kernel functions, and RF needs to set the number of selected variables for splitting each node in a tree. In this chapter, each classifier was implemented and tuned in Waikato Environment for Knowledge Analysis (WEKA), a machine learning open-source software package (Hall et al., 2009). WEKA has an experimenter function that can effectively tune an algorithm by changing its required parameters and finding the best model based on the statistical metrics.

4.3.3 Accuracy Assessment

The k-fold cross-validation technique was used in the training and testing procedure. This evaluation method has proven valuable in machine learning classifications (Anguita et al., 2012). It splits the reference data into k-subsets first; then iteratively, some of them are used to train the model, and the others are exploited to assess model performance. The variable k was commonly set to 10 in the literature, which was also used in this chapter. After the iteration, classes are predicted for all input reference objects, which can then be used to calculate the error matrix and Kappa statistics in terms of the true classes of the reference objects. The error matrix can be summarized as an overall accuracy and Kappa value. The overall accuracy is defined as the ratio of the number of validation samples that are classified correctly to the total number of validation samples irrespective of the class. The Kappa value describes the proportion of correctly classified validation samples after random agreement is removed. The McNemar test (Foody, 2004) was applied to evaluate the statistical significance of differences in accuracy between different classifications. In this chapter, the same reference samples were used for accuracy assessment, and thus the samples were not independent, and the McNemar test was selected. The Kappa test assumes the samples used in the calculation are independent, an assumption that was unsatisfied in most studies because the samples were related. The McNemar test is one of alternatives for evaluating the statistical significance of differences in accuracy for related samples. This non-parametric test is based upon a confusion matrix (2 by 2 dimension). The z value of the McNemar test is calculated based upon

$$z = \frac{f_{12} - f_{21}}{\sqrt{f_{12} + f_{21}}} \tag{4.1}$$

where f_{ij} indicates the frequency of testing sites lying in the confusion matrix i, j. The difference in accuracy of a pair of classifications is viewed as statistically significant at a confidence of 95% if the Z-score is larger than 1.96.

4.4 Results and Discussion

4.4.1 Experimental Analyses to Examine the Contribution of Texture Measures

To assess whether the texture measures were useful to improve the marsh species classification, these texture measures were excluded in the classification first, leading to three experimental results (experiments 1–3) from three classifiers, as shown in Table 4.1. An overall accuracy (OA) of less than 70% was produced if texture features were excluded in the classification. SVM produced the highest accuracy with an OA of 67.2% and Kappa value of 0.62. When the spatial texture measures were added, the OA was improved to 81.0% and Kappa value was increased to 0.78 by the ANN classifier, which produced the highest accuracy among three classifiers, shown as experiments 4 to 6 in Table 4.1. The Kappa statistical tests showed that all the classifications from experiments 1 to 6 were significantly better than a random classi-fication. The McNemar tests showed that there was no significant difference among three classifiers using the same dataset, with or without texture measurements. The classifications from an addition of texture measures were significantly better than the classifications without texture measures.

Varying accuracies (59%–99%) have been reported in the literature for marsh species mapping using optical imagery (Belluco et al., 2006; Sadro et al., 2007; Wang et al., 2007). Here a moderate accuracy was obtained if only four bands and NDVI of the aerial photography were used. Three classifiers produced an averaged OA of 66.1% and Kappa value of 0.60. This accuracy is much lower than the accuracy of salt marsh species classification from a similar type of data such as Ikonos or QuickBird in Belluco et al. (2006) who reported an OA of greater than 95%. Salt marsh plant zonation tends to be much stricter along tidal inundation gradients than the zonation observed among freshwater species, allowing for greater species-level classification accuracy (Carle et al., 2014). When the texture measures were added, a higher accuracy was achieved with an averaged OA of 80.3% and Kappa coefficient of 0.77 from three classifiers. This study is comparable to the study from Carle et al. (2014)

TABLE 4.1

Classification accuracies using aerial photography and different classifiers

Experiments	OA (%)	Kappa	z-Score (Kappa)	z-Score (McNemar)
Aerial photography excluding texture				
1. SVM	67.2	0.62	26.16	1.11 (1/2)
2. RF	65.1	0.59	24.77	0.40 (2/3)
3. ANN	65.9	0.60	25.30	0.98 (1/3)
Aerial photography including texture				
4. SVM	80.0	0.77	38.50	0.00, 6.27[a] (4/5, 1/4)
5. RF	80.0	0.77	38.61	0.80, 6.86 [a] (5/6, 2/5)
6. ANN	81.0	0.78	39.79	0.78, 7.30[a] (4/6, 3/6)

SVM: support vector machine; RF: random forest; ANN: artificial neural network; OA: overall accuracy. For the McNemar tests, 1/2, 1/3, 2/3, … 3/6 refer to the test between experiments 1 and 2, 1 and 3, 2 and 3, …, 3 and 6, respectively; [a]Significant with 95% confidence.

who also mapped freshwater marsh species but used WorldView-2 satellite imagery. An OA of 75% and Kappa value of 0.71 was reported in their study for classifying eight freshwater marsh species and three other land cover types. Few studies include texture measures in marsh species mapping. This study revealed an improvement of OA by 14.2% if texture measures were added. This is consistent with the findings from Szantoi et al. (2013, 2015), who also reported the contribution of texture measures derived from fine resolution aerial photography for mapping marsh and wet prairie communities in Everglades National Park. The magnitude of difference in classification accuracies varied across species when texture was incorporated, as shown in Table 4.2 of the per-class accuracy from two datasets with or without texture measures. Some species, such as cattail and cordgrass, showed the greatest improvement in accuracy with the addition of texture as opposed to species such as willow. Further examination showed a high spectral confusion between cattail and smartweed, while an inclusion of texture features improved the discrimination for both species.

As discussed in Zhang et al. (2018a), texture has been of great interest in remote sensing for more than three decades (Warner, 2011). It has been proven that texture measures are useful in image classification but challenging in determining the optimal kernel/window size, an important parameter in texture analysis. Application of OBIA may overcome this challenge by calculating the texture measures in an adaptive window with variable size and shape. OBIA offers the capability to identify regions of varying shapes and sizes in an image, leading to an extraction of texture measures based upon the image objects, rather than a fixed size kernel window. Inclusion of object-based texture measures largely increased the accuracy in marsh species mapping. Other challenges in texture analysis include the specification of texture order, measures, and spectral bands. The setting of these parameters is case and class specific (Warner, 2011). A range of texture features can be extracted. It is a trade-off to use these features in the mapping procedure (Zhang et al., 2018). First, the addition of all of these features will increase the data dimensionality, making the classification more complex. Second, an effective feature selection approach needs to be identified in order to determine the most

TABLE 4.2

Per-class accuracies (%) from different datasets and classifiers

Classes	SVM		RF		ANN	
	D1	D2	D1	D2	D1	D2
1. Willow (*Salix caroliniana*)	94.4	98.1	93.0	96.2	87.7	100.0
2. Cordgrass (*Spartina bakeri*)	57.6	84.8	62.2	78.7	57.1	79.2
3. Smartweed (*Polygonum spp.*)	70.7	72.7	55.6	69.0	50.0	71.7
4. Cattail (*Typha spp.*)	0.0	66.7	28.6	46.7	0.0	50.0
5. Common reed (*Phragmites australis*)	88.5	98.6	93.3	95.9	89.7	94.7
6. Sawgrass (*Cladium jamaicense*)	52.9	64.9	44.8	70.0	54.8	74.0
7. Graminoid	52.3	74.6	46.6	78.1	53.4	75.6
8. Swamp shrubland	58.9	76.4	59.2	77.3	56.8	76.3
9. Mud	100.0	95.8	96.3	96.3	96.2	100.0

D1: dataset of aerial photography excluding texture measures; D2: dataset of aerial photography including texture.

informative features for marsh species classification. Lastly, the effectiveness of these texture features is scale dependent, and selecting an optimal scale for each texture feature is a challenge. In this study, only limited features were used. It is beyond the scope of this study to give a detailed evaluation of each texture feature for marsh species identification.

4.4.2 Object-Based Marsh Species and Spatial Uncertainty Mapping

The experimental analyses showed that the dataset including texture measures produced a high accuracy for marsh species mapping in the lake area. Landis and Koch (1977) suggested that a Kappa value larger than 0.80 indicates a strong agreement or accuracy. All three classifications using the dataset with texture measures produced an OA equal to or more than 80%, and thus marsh species maps were produced using all three classifiers, as shown in Figure 4.6. In general, the classified maps showed a consistent spatial distribution of marshes for the selected study site, which was dominated by graminoid, sawgrass, cordgrass, and swamp shrubland. Cattail and smartweed were sparse; common reed and willow were moderate and shown as bigger patches. An error matrix was also constructed for the ANN classified map, which showed the highest accuracy among three classifiers (OA=81%), as shown in Table 4.3. The user's accuracies ranged from 50.0% (class 4: cattail) to 100.0% (class 1: willow; class 9: mud). The producer's accuracies varied from 33.3% (class 4: cattail) to 100% (class 5: common reed). Cattail, an invasive species in the Greater Everglades, was the most difficult marsh species to classify, while common reed, a giant grass considered a looming threat to the Greater Everglades, was the easiest to identify for the study site.

Though three classifiers showed a similar accuracy in terms of OA and Kappa values, their performance in identifying each individual class was uneven. An ensemble analysis to combine the outputs of three classifications could make the classification more robust (Zhang et al., 2018), but a classification from ensemble

TABLE 4.3

Error matrix of the classified map using the ANN classifier and aerial photography including texture measures

Class	1	2	3	4	5	6	7	8	9	Row Total	PA (%)
1	52							2		54	96.3
2		38		2	1	2	1			44	86.4
3			38	4			7	7		56	67.9
4			8	8		1		7		24	33.3
5					72					72	100.0
6		2				54	6	8		70	77.1
7		2	1			9	59	3		74	79.7
8		6	6	2	2	7	5	87		115	75.7
9					1				26	27	96.3
Col. Total	52	48	53	16	76	73	78	114	26	OA (%): 81.0	
UA (%)	100.0	79.2	71.7	50.0	94.7	74.0	75.6	76.3	100.0	Kappa: 0.78	

PA (%): producer's accuracy; UA (%): user's accuracy; OA (%): overall accuracy. Classification results are displayed in rows, and the reference data are displayed in columns. The name of each class is displayed in Table 4.2.

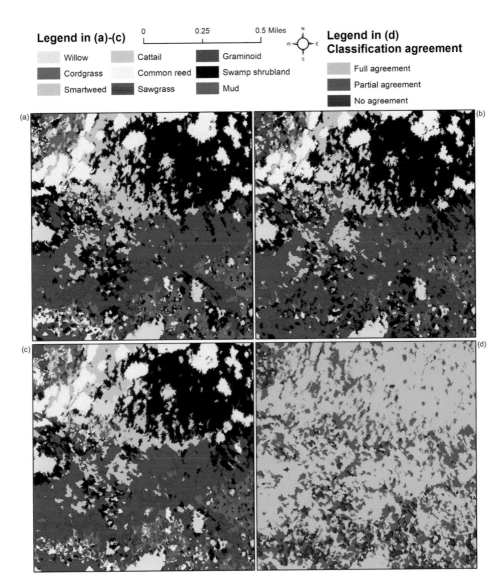

Legend in (a)-(c)

- Willow
- Cordgrass
- Smartweed
- Cattail
- Common reed
- Sawgrass
- Graminoid
- Swamp shrubland
- Mud

0   0.25   0.5 Miles

**Legend in (d)
Classification agreement**

- Full agreement
- Partial agreement
- No agreement

FIGURE 4.6
Classified maps from (a) SVM, (b) RF, (c) ANN, and (d) the uncertainty map.

analysis was not reported in this chapter. Here a spatial uncertainty analysis was conducted based on the outputs of three classifications. If three classifiers vote the same class for an unknown object, a full agreement will be achieved. Conversely, if three votes are completely different, no agreement will be obtained. If two classifiers vote for the same class, a partial agreement will be produced. Consequently, an uncertainty map can be produced in conjunction with the classified maps from the ensemble analysis. The corresponding uncertainty map caused by the classification algorithms is shown in Figure 4.6 (d). It was difficult to visually identify the difference between the classified maps in Figure 4.6(a) to (c), but the uncertainty map successfully revealed consistency and discrepancy between different classifiers. Most regions (64.1%) showed a full agreement of three classifiers (shown in green), indicating a high confidence for being correctly classified. Some areas (29.3%) (shown in blue) were voted by two classifiers, generating a partial agreement in the classification. A few regions (6.6%) displayed a "warning sign" (shown in red), indicating a high probability of being misclassified because there was no agreement from three classifiers for these regions. Overall, the uncertainty map revealed that the final marsh map was robust using the object-based classification procedure with texture measures included. This uncertainty map can help guide the fieldwork for a post-classification analysis.

4.5 Summary and Conclusions

This chapter provides a framework using aerial photography to map marsh species in the wetlands of Lake Okeechobee. The results suggest that the digital procedure to classify very fine spatial resolution aerial photography (1-foot) for marsh species mapping is valuable and can be used as an alternative to the manual procedure. The OBIA, machine learning, and texture analysis techniques are useful in the classification. Major challenges in the digital procedure include the large volume of the dataset of aerial photography, especially with the addition of texture measures, and application of advanced remote sensing digital techniques. The large volume of the dataset makes the developed procedure difficult for mapping a broad area. This issue is expected to be mitigated using modern computational techniques. The OBIA, machine learning, and texture techniques were combined in this chapter to produce an accurate and informative marsh species map. However, most remote sensing users do not have these skill sets. Also note that though a high accuracy was achieved for the selected study site, application of similar techniques and 1-m aerial photography at other regions in the Florida Everglades had a lower accuracy, as demonstrated in Zhang and Xie (2014), Zhang et al. (2013, 2016). A combination of fine spatial resolution aerial photography with lidar or fine spectral resolution hyperspectral imagery can improve the classification compared with the application of aerial photography alone. An example of fusing aerial photography with lidar and hyperspectral data is presented in Chapter 18.

References

Anguita, D., Ghelardoni, L., Ghio, A., et al., 2012. The 'K' in *k*-fold cross validation. *ESANN 2012 Proceedings, European Symposium on Artificial Neural Networks, Computational Intelligence and Machine Learning*, Bruges, Belgium, April 25–27, 441–446.

Belluco, E., Camuffo, M., Ferrari, S., et al., 2006. Mapping salt-marsh vegetation by multispectral and hyperspectral remote sensing. *Remote Sensing of Environment*, 105, 54–67.

Breiman, L., 2001. Random forests. *Machine Learning*, 45, 5–32.

Carle, M.V., Wang, L., and Sasser, C.E., 2014. Mapping freshwater marsh species distributions using WorldView-2 high-resolution multispectral satellite imagery. *International Journal of Remote Sensing*, 35, 4698–4716.

Chan, J.C.-W., and Paelinckx, D., 2008. Evaluation of random forest and Adaboost tree based ensemble classification and spectral band selection for ecotope mapping using airborne hyperspectral imagery. *Remote Sensing of Environment*, 112, 2999–3011.

Foody, G.M., 2004. Thematic map comparison, evaluating the statistical significance of differences in classification accuracy. *Photogrammetric Engineering & Remote Sensing*, 70, 627–633.

Ghimire, B., Rogan, J., Galiano, V., Panday, P., and Neeti, N., 2012. An evaluation of bagging, boosting, and random forests for land-cover classification in Cape Cod, Massachusetts, USA. *GIScience & Remote Sensing*, 49, 623–643.

Gilmore, M.S., Wilson, E.H., Barrett, N., Civco, D.L., Prisloe, S., et al., 2008. Integrating multi-temporal spectral and structural information to map wetland vegetation in a lower Connecticut River tidal marsh. *Remote Sensing of Environment*, 112, 4048–4060.

Grybas, H., Melendy, L., and Congalton, R.G., 2017. A comparison of unsupervised segmentation parameter optimization approaches using moderate- and high-resolution imagery. *GIScience & Remote Sensing*, 54, 515–533.

Hall, M., Frank, E., Holmes, G., Pfahringer, B., Reuteman, P., et al., 2009. The WEKA data mining software, An update. *SIGKDD Explorations*, 11, 10–18.

Hall-Beyer, M., 2017. GLCM Texture, A Tutorial V.3.0. www.fp.ucalgary.ca/mhallbey/tutorial.htm. Accessed on 20 July 2019.

Havens, K.E., and Gawlik, D.E., 2005. Lake Okeechobee conceptual ecological model. *Wetlands*, 25, 908–925.

Hladik, C., Schalles, J., and Alber, M., 2013. Salt marsh elevation and habitat mapping using hyperspectral and lidar data. *Remote Sensing of Environment*, 139, 318–330.

Hsu, C., Chang, C., and Lin, C. 2010. *A Practical Guide to Support Vector Classification*, Final report, National Taiwan University, Taipei City, Taiwan.

Huang, C., Davis, L.S., and Townshend, J.R.G., 2002. An assessment of support vector machines for land cover classification. *International Journal of Remote Sensing*, 23, 725–749.

Jensen, J.R., 2015. *Introductory Digital Image Processing, A Remote Sensing Perspective*, 4th Edition, Prentice Hall, Upper Saddle River, NJ.

Johnson, B., and Xie, Z. 2011, Unsupervised image segmentation evaluation and refinement using a multi-scale approach. *ISPRS Journal of Photogrammetry and Remote Sensing*, 66, 473–483.

Judd, C., Steinberg, S., Shaughnessy, F., and Crawford, G., 2007. Mapping salt marsh vegetation using aerial hyperspectral imagery and linear unmixing in Humboldt Bay, California. *Wetlands*, 27, 1144–1152.

Kumar, L., and Sinha, P., 2014. Mapping salt-marsh land-cover vegetation using high-spatial and hyperspectral satellite data to assist wetland inventory. *GIScience & Remote Sensing*, 51, 483–497.

Lake Okeechobee Watershed Restoration Project, 2017. www.sfwmd.gov/our-work/cerp-project-planning/lowrp. Accessed on 20 July 2019.

Landis, J., and Koch, G.G., 1977. The measurement of observer agreement for categorical data. *Biometics*, 33, 159–174.

Mas, J.F., and Flores, J.J., 2008. The application of artificial neural networks to the analysis of remotely sensed data. *International Journal of Remote Sensing*, 29, 617–663.

Maxwell, A.E., Warner, T.A., and Fang, F., 2018. Implementation of machine learning classification in remote sensing: An applied review. *International Journal of Remote Sensing*, 39, 2784–2817.

Melgani, F., and Bruzzone, L., 2004. Classification of hyperspectral remote sensing images with support vector machines. *IEEE Transactions on Geoscience and Remote Sensing*, 42, 1778–1790.

Morris, J.T., Porter, D., Neet, M., Noble, P.A., Schmidt, L., et al., 2005. Integrating lidar elevation data, multi-spectral imagery and neural network modelling for marsh characterization. *International Journal of Remote Sensing*, 26, 5221–5234.

Mountrakis, G., Im, J., and Ogole, C., 2011. Support vector machines in remote sensing: A review. *ISPRS Journal of Photogrammetry and Remote Sensing*, 66, 247–259.

Rodriguez-Galiano, V.F., Ghimire, B., Rogan, J., Chica-Olmo, M., Rigol-Sanchez, J.P., 2012. An assessment of the effectiveness of a random forest classifier for land-cover classification. *ISPRS Journal of Photogrammetry and Remote Sensing*, 67, 93–104.

Rutchey, K., Schall, T.N., Doren, R.F., et al., 2006. Vegetation classification for South Florida natural areas. Saint Petersburg, Florida, United States Geological Survey, Open-File Report 2006-1240, 142.

Sadro, S., Gastil-Buhl, M., and Melack, J., 2007. Characterizing patterns of plant distribution in a southern California salt marsh using remotely sensed topographic and hyperspectral data and local tidal fluctuations. *Remote Sensing of Environment*, 110, 226–239.

Spencer, L.J., and Bousquin, D.G. 2014. Interim responses of floodplain wetland vegetation to phase I of the Kissimmee River restoration project: Comparisons of vegetation maps from five periods in the river's history. *Restoration Ecology*, 22, 397–408.

Sun, C., Liu, Y., Zhao, S., Zhou, M., Yang, Y., et al., 2016. Classification mapping and species identification of salt marshes based on a short-time interval NDVI Time-series from HJ-1 optical imagery. *International Journal of Applied Earth Observation and Geoinformation*, 45, 27–41.

Szantoi, Z., Escobedo, F., Abd-Elrahman, A., Smith, S., and Pearlstine, L., 2013. Analyzing fine-scale wetland composition using high resolution imagery and texture features. *International Journal of Applied Earth Observation and Geoinformation*, 23, 204–212.

Szantoi, Z., Escobedo, F., Abd-Elrahman, A., Pearlstine, L., Dewitt, B., and Smith, S., 2015. Classifying spatially heterogeneous wetland communities using machine learning algorithms and spectral and textural features. *Environmental Monitoring and Assessment*, 187, 262.

Trimble, 2014. *eCognition Developer 9.0.1 Reference Book*. Trimble Germany GmbH, Arnulfstrasse 126, D-80636, Munich.

Vapnik, V.N., 1995. *The Nature of Statistical Learning Theory*, Springer-Verlag, New York.

Wang, C., Menenti, M., Stoll, M., Belluco, E., and Marani, M., 2007. Mapping mixed vegetation communities in salt marshes using airborne spectral data. *Remote Sensing of Environment*, 107, 559–570.

Warner, T., 2011. Kernel-based texture in remote sensing image classification. *Geography Compass*, 5, 781–798.

Zhang, C., 2014. Combining hyperspectral and lidar data for vegetation mapping in the Florida Everglades. *Photogrammetric Engineering & Remote Sensing*, 80, 733–743.

Zhang, C., Denka, S., and Mishra, D.R., 2018a. Mapping freshwater marsh species in the wetlands of Lake Okeechobee using very high-resolution aerial photography and lidar data. *International Journal of Remote Sensing*, 39, 5600–5618.

Zhang, C., Denka, S., Cooper, H., and Mishra, D.R., 2018b. Quantification of sawgrass marsh aboveground biomass in the coastal Everglades using object-based ensemble analysis and Landsat Data. *Remote Sensing of Environment*, 204, 366–379.

Zhang, C., Durgan, S., and Lagomasino, D., 2019. Modeling risk of mangroves to tropical cyclones: A case study of Hurricane Irma. *Estuarine, Coastal, and Shelf Science*, 224, 108–116.

Zhang, C., Selch, D., and Cooper, H., 2016. A framework to combine three remotely sensed data sources for vegetation mapping in the central Florida Everglades. *Wetlands*, 36, 201–213.

Zhang, C., Smith, M., Lv, J., and Fang, F., 2017. Applying time series Landsat data for vegetation change analysis in the Florida Everglades water conservation area 2a during 1996–2016. *International Journal of Applied Earth Observations and Geoinformation*, 57, 214–223.

Zhang, C., and Xie, X., 2012. Combining object-based texture measures with a neural network for vegetation mapping in the Everglades from hyperspectral imagery. *Remote Sensing of Environment*, 124, 310–320.

Zhang, C., and Xie, Z., 2013. Object-based vegetation mapping in the Kissimmee River watershed using HyMAP data and machine learning techniques. *Wetlands*, 33, 233–244.

Zhang, C., and Xie, Z., 2014. Data fusion and classifier ensemble techniques for vegetation mapping in the Everglades. *Geocarto International*, 29, 228–243.

Zhang, C., Xie, Z., and Selch, D., 2013. Fusing lidar and digital aerial photography for object-based forest mapping in the Florida Everglades. *GIScience & Remote Sensing*, 50, 562–573.

5

Unmanned Aircraft System (UAS) for Wetland Species Mapping

Sara Denka Durgan and Caiyun Zhang

5.1 Introduction

The REstoration COordination and VERification (RECOVER) program utilizes vegetation maps to measure the success of Everglades restoration activities carried out by the Comprehensive Everglades Restoration Plan (CERP) (Doren et al., 1999; Rutchey et al., 2008). Previous vegetation mapping efforts in the Everglades have consisted of aerial photointerpretation techniques (Welch et al., 1999; Rutchey et al., 2008), digital image analysis of manned airborne and spaceborne multispectral imagery (Rutchey and Vilcheck, 1994; Jensen et al., 1995; Szantoi et al., 2013), hyperspectral imagery (Hirano et al., 2003; Zhang and Xie, 2012, 2013), or a fusion of multiple sensors (Zhang et al., 2016, 2018). While these techniques have proven successful for mapping vegetation at a regional level, the limited spatial and spectral resolutions of manned airborne and spaceborne images presents a challenge in mapping wetlands at a local-scale with small vegetation units (Adam et al., 2010). For manned airborne sensors, data availability is often limited due to the high cost of data collection (Grenzdörffer et al., 2008; Nex and Remondino, 2014). For spaceborne sensors, frequent cloud cover over the Everglades presents an issue in optimal image selection for mapping purposes (Jones, 2015). Emerging technologies in Unmanned Aircraft System (UAS) remote sensing present opportunities to address these challenges in Everglades vegetation mapping.

UAS has several advantages over satellite and manned airborne remote sensing platforms including a relatively low cost of data acquisition, ability for rapid deployment, and high resolution of output imagery (Colomina and Molina, 2014; Nex and Remondino, 2014). In the field of coastal and environmental remote sensing, UAS has shown immense potential for obtaining data in dynamic, complex, and difficult-to-access environments (Klemas, 2015). Wetland ecosystems in particular are often remote and contain dense vegetation, making field surveying difficult (Davis and Fitzgerald, 2004; Mahdavi et al., 2018). These environments also undergo frequent change due to inundation; therefore, repeated surveys are necessary (Gallant, 2015). As a result, the field of wetland remote sensing is expected to benefit greatly from UAS remote sensing techniques (Kalacska et al., 2017). In this chapter, we examine the development of UAS as a remote sensing platform and the utilization of UAS-based remote sensing data products for detailed, species-level vegetation mapping in a restored wetland in the Greater Everglades. The following subsections present more details of UAS and relevant techniques in this emerging field.

5.1.1 Background of UAS

Klemas (2015) defines Unmanned Aerial Vehicles (UAV) as "powered aircraft operated remotely or autonomously with preprogrammed flight planning," referring to what is considered modern unmanned aircraft. The term UAS encompasses both the aircraft itself (UAV) and the payload, operators, ground control, and communications systems. UAVs can be rotary, fixed-wing, or a combination of both. The strength of rotary platforms lies in their ability to hover, which permits data collection at several scales (Klemas, 2015). Fixed-wing aircraft are more aerodynamic and thus support a greater flight range. Hybrid UAVs are emerging on the market in an attempt to combine the strengths of both configurations (Cetinsoy et al., 2012). These hybrid UAVs often support vertical takeoff and landing characteristic of a rotary system but have the flight range of a fixed-wing aircraft. Given the wide array of UAVs on the market, selecting the appropriate UAV for a project is vital and highly dependent on the survey area, environmental conditions, and expertise of the operator.

The availability of different payloads or sensors for UAS has increased in recent years due to a trend in miniaturization of these systems. Off-the-shelf Red-Green-Blue (RGB) cameras are the standard payloads of most low-cost UASs and are sufficient for most photogrammetry applications. RGB cameras can also be modified to support image collection in the Near-Infrared (NIR) spectrum for vegetation studies. More complex multispectral systems also exist for improved spectral resolution to support more rigorous remote sensing applications. Sensors such as the Micasense RedEdge or the Parrot Sequoia are often used in precision agriculture (Duan et al., 2017; Potgieter et al., 2017; Johansen et al., 2018). More customizable multispectral sensors, such as the Tetracam Micro-MCA and the Mapir Kernel, promote interchangeable filters to select the ideal wavelengths for different applications (Deng et al., 2018; Turner et al., 2018). Hyperspectral imagers and light detection and ranging (lidar) sensors outfitted for UAS payloads are also becoming increasingly available (Klemas, 2015; Madden et al., 2015). Yet, the higher cost, larger size, and generally coarser resolution of these sensors as compared with RGB cameras have restricted the widespread adoption of these technologies (Wallace et al., 2012; Zhou et al., 2012; Pande-Chhetri et al., 2017). RGB or modified-RGB cameras have proven to be sufficient for many applications and have the added benefit of producing simultaneous 3-D information using contemporary photogrammetry.

5.1.2 Structure from Motion Photogrammetry

Advances in the field of computer vision have produced an alternative technique for generating 3-D data using UAS imagery collected at a high rate of overlap. This photogrammetric approach, known as Structure from Motion (SfM)-Multi View Stereopsis (MVS), can be achieved using the data captured by instruments that come standard with most UAS, including images, global positioning system (GPS) position, and altitude information (Vallet et al., 2011). SfM was introduced as a photogrammetric technique for generating 3-D objects from 2-D information in the beginning of the 1980s (Ullman, 1979). SfM algorithms identify several matching features within an assemblage of digital images while resolving camera location and orientation using the positions of the matched features to produce a 3-D point cloud of the surveyed area (Carrivick et al., 2016). The use of SfM alone produces a relatively sparse point cloud, limiting its application in many cases. Therefore, MVS emerged as an additional step to apply the resolved camera parameters to generate a completed, dense point cloud. A merger of these two processes is referred to as SfM-MVS, and this innovation has had a drastic impact on the use of photogrammetry in modern remote sensing.

The main advantage of SfM-MVS is the ability to generate high-resolution 3-D information via relatively unstructured image acquisition, unlike traditional photogrammetric methods that rely on parallel flight lines (Fonstad et al., 2013). Snavely et al. (2008) demonstrated this capability by generating a 3-D scene reconstruction of several destinations around the world using images from the internet with no GPS information. This aspect of SfM-MVS allows for the expansion of imagery available for generating 3-D information, including higher resolution images captured via low-cost UAS platforms. Westoby et al. (2012) details the immense potential for SfM-MVS in geoscience applications and presents a general outline of how 3-D models with comparable, if not higher, accuracy than those generated by lidar can be obtained. SfM-MVS has been applied in numerous applications in remote sensing, yet its full capability as a 3-D modeling approach has not been fully determined in any one field.

Several software packages employ the SfM-MVS workflow including Agisoft Photoscan, Pix4D, Trimble Business Center, and Drone Deploy. An overview of various commercial and open-source SfM-MVS software can be found in Carrivick et al. (2016). The SfM-MVS workflow can produce point clouds, Digital Surface Models (DSMs), bare-earth Digital Elevation Models (DEMs), and orthoimagery from the same set of images. The ability to generate these data products from the same set of images provides more opportunities for analysis and minimizes concerns over fusing data from different sources or collected at different times.

5.1.3 UAS Data Collection

While UAS presents a promising solution to data collection limitations associated with manned airborne and satellite platforms, the field is still in its infancy and numerous difficulties exist. The accuracy of UAS imagery mainly depends on the precision of the GPS system, capability of the sensor, stability of the sensor mounted on the aircraft, and the flight characteristics of the mission (Kalacska et al., 2017). GPS systems provided with most commercially available UAS often do not have the required precision for a UAS survey. It is a common practice to use high accuracy GPS measures taken using a survey-grade GPS instrument, referred to as Ground Control Points (GCPs), to increase geometric accuracy (Harwin and Lucieer, 2012; Carrivick et al., 2016; Kalacska et al., 2017). Standard practices for the design and deployment of GCPs are still being developed but it is generally agreed that GCPs should be adequately sized and colored in a way that contrasts with the survey environment (James et al., 2017) and deployed to be evenly spaced throughout the study area (Harwin and Lucieer, 2012). Recently, the American Society for Photogrammetry and Remote Sensing (ASPRS) updated their positional accuracy standards to include requirements for GCPs used for aerial triangulation. These requirements provide guidelines for GCP accuracy, density, and distribution as well as the use of checkpoints for accuracy estimation of derived products (ASPRS, 2014).

Alternative techniques for georeferencing UAS data are constantly emerging. One such technique is direct georeferencing, which utilizes high accuracy positional and GPS data onboard the aircraft and can produce accuracies higher than or comparable to using GCPs (Turner et al., 2014). UASs with this capability are starting to become available but at a high cost, therefore their use in published works is limited. Furthermore, UASs with direct georeferencing capability still require the use of a few GCPs for accuracy estimation.

The flight plan for data acquisition also has a large impact on the accuracy of UAS survey products. In general, UAS image acquisition missions should be flown at solar noon in order to minimize the effect of shadows (Laliberte et al., 2010). For SfM-MVS photogrammetry, high rates of overlap (80%) and side lap (60%) are generally required to compensate for aircraft instability (Colomina and Molina, 2014). Dandois et al. (2015) assessed the accuracy of SfM-MVS 3-D models from images collected with varying levels of lighting, altitude, and side lap in order to determine optimal parameters for delineating forest canopy metrics. They found that missions flown under clear skies and with a maximum photographic overlap were able to best penetrate the canopy and produce point clouds with the highest densities. Due to the above factors, UAS data accuracy is highly dependent upon the conditions in which they were collected; therefore, care must be taken to optimize the flight plan, GCP deployment, and lighting conditions for specific needs.

5.1.4 UAS for Vegetation Mapping

As a result of increasing UAS technologies and advances in the SfM-MVS algorithm, applications of UAS-derived datasets for vegetation mapping are growing. Vegetation classification using UAS imagery has mostly been employed in precision agriculture (Laliberte et al., 2010; Torres-Sánchez et al., 2014; Tamouridou et al., 2017) and forestry (Gini et al., 2014; Nevalainen et al., 2017). Vegetation classification for ecological applications is less prominent, but several studies demonstrate promising results for species-level mapping using UAS with limited spectral resolution. Lu and He (2017) mapped grass species composition in a grassland in Ontario, Canada, using a modified-RGB camera. They achieved overall accuracies (OAs) ranging from 82-86% for images acquired on different dates using object-based image analysis (OBIA) and the random forest (RF) classification model. Prošek and Šímová (2019) examined the fusion of a UAS-derived multispectral orthoimage with a canopy height model (CHM), both derived from SfM-MVS photogrammetry, for mapping shrubland plant species. They found that the fused dataset produced significantly higher accuracies compared with the spectral data alone. These studies indicate that the ultra-high spatial resolution UAS data and simultaneous 3-D information can generate species classification models with high quality despite limited spectral resolution of the images.

Applying UAS data for species-level vegetation mapping in wetland environments remains rare in published literature. Li et al. (2017) and Cao et al. (2018) demonstrated the capability of UAS-derived hyperspectral imagery fused with a photogrammetric DSM for wetland species mapping in China. It should be noted that unlike lidar, UAS photogrammetry does not penetrate a dense vegetation canopy to capture bare earth terrain (Kalacska et al., 2017; Meng et al., 2017). Because of this limitation, both studies only considered relative height differences between plants rather than canopy height above the ground. In both studies, the inclusion of the photogrammetric DSM improved species classification accuracy.

Wetland vegetation mapping in the Everglades using UAS data is even more limited in published literature, and only two comparable studies are presently available. Zweig et al. (2015) applied RGB imagery from a UAS to map plant communities in Water Conservation Area (WCA) 3A South in the Everglades. They achieved an OA of 69% and a Kappa value of 0.65 for nine community classes. One major finding was that classifications performed with image resolutions higher that 0.5 m resulted in low accuracies; therefore, the UAS imagery was resampled to 0.5 m for the final classification. This result argues against the feasibility of ultra-high resolution UAS data for vegetation classification because the higher resolution data is too complex and produces error. Pande-Chhetri et al. (2017) used UAS-derived RGB imagery to classify freshwater wetland vegetation near Lake Okeechobee, FL. They compared object-based and pixel-based classification procedures using different classifiers and achieved maximum OA of 70.8% using the support vector machine (SVM) and OBIA approach. This study is the most similar to the case study presented in this chapter except they did not include vertical variables derived from SfM-MVS or a spectral band in the NIR, which has been shown to increase classification accuracy for wetland species (Li et al., 2017; Cao et al., 2018).

Previous studies have demonstrated that UAS-based wetland vegetation classification is a viable method in a few cases. However, further research is necessary to investigate the full potential of UAS applications in wetland species mapping. To date, there has been no examination of UAS-based species mapping for a high diversity, restored wetland ecosystem in the coastal Everglades. Furthermore, whether a fusion of UAS data products including multispectral imagery and a DSM from the SfM-MVS technique can improve wetland species mapping remains unknown. To this end, the main objective of this chapter is to explore the potential of fusing a UAS-derived orthoimage (spectral data) and DSM (vertical data) for ultra-high resolution species mapping in a restored coastal wetland with a high diversity of plant species.

5.2 Study Site and Data Collection

5.2.1 Study Site

The study site is a restored wetland in the coastal Greater Everglades with a size of 7.5 acres in a region known as North Cutler Wetlands in Cutler Bay, FL (Figure 5.1). The area is a coastal saltwater marsh that is landward of a large mangrove forest adjacent to Biscayne Bay. Historically, the area was an extensive freshwater wetland with a short hydroperiod and a relatively narrow mangrove band. The entire area was developed, which caused habitat destruction, reduction of freshwater flow, fragmentation, and the spread of invasive species. In 2013, a partnership of the Institute for Regional Conservation, National Park Service, Tropical Audubon Society of Florida, Palmetto Bay Village Center, South Florida Water Management District (SFWMD), Fairchild Tropical Botanic Garden, and the US Fish and Wildlife Service Atlantic Coast Joint Venture obtained funding from the US Fish and Wildlife Division of Bird Habitat to begin the North American Wetland Conservation Act (NAWCA) Coastal Palmetto Bay and Cutler Bay Habitat Restoration project. The goal of this project was to restore 370 acres of pine rockland and freshwater and saltwater marsh for migratory bird habitat (Martin, 2015). As a result of this project, North Cutler Wetlands is actively undergoing restoration in the form of invasive species removal, controlled burns, and native species plantings.

The project area also lies within the CERP denoted as Biscayne Bay Coastal Wetlands. The US Army Corps of Engineers (USACE) has implemented Phase 1 of the project to construct pump stations, culverts, and spreader canals and to fill mosquito control ditches in order to rehydrate the wetlands and reduce discharges

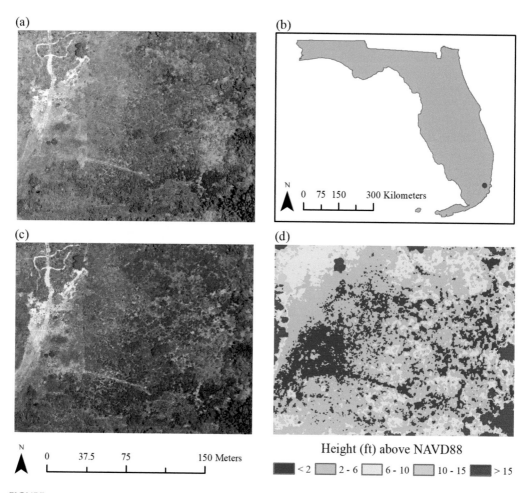

FIGURE 5.1
Maps of the study area within North Cutler Wetlands in Cutler Bay, FL, (a) shown as a true-color composite of the collected UAS imagery, (b) the location of the study area, (c) a color-infrared composite of the collected UAS imagery, and (d) DSM generated from the UAS imagery.

into Biscayne Bay. At the local level, the restoration of North Cutler Wetlands is an integral component of an initiative to preserve natural areas and green space in the Town of Cutler Bay. This research provides essential information and data regarding the current success of restoration activities in North Cutler Wetlands and demonstrates a novel approach for continued monitoring of the area over the lifetime of these projects.

5.2.2 UAS Data Collection

We conducted both UAS and in-situ vegetation surveys in the study area between June 22 and 27, 2018. We used a 3DR Solo quadcopter for the UAS survey. The 3DR Solo is a 1.5 kg quadcopter with an approximate flight time of 14 minutes per battery, maximum flight speed of 15 m/s, 0.8 km range, and wind speed limitation of 11 m/s. The 3DR Solo uses a Pixhawk 2.0 flight controller system and ArduPilot copter software. The 3DR Solo payload consists of four Mapir Survey 3 modified-RGB cameras equipped with Mapir Survey 2 lenses to capture images in Red (650 nm), Green (548 nm), Blue (450 nm), and NIR (850 nm) wavelengths. The Mapir Survey 3 uses a Sony Exmor R IMX117 12MP sensor (4000 x 3000 px) with a rolling shutter. The Mapir Survey 3 can capture images in 12-bit RAW or 8-bit JPG and supports full-motion video. The Mapir Survey 2 lenses are 23 mm with 84° FOV. These four cameras are mounted to the 3DR Solo using a quad static mount equipped with rigid and flexible dampening balls to absorb excess vibration during flight. The cameras are connected to a Survey 3 Standard GPS receiver that uses a ublox UBX-G7020-KT chip, which provides GPS tags in the EXIF files of the images.

The UAS images used in this study were collected during three flights on June 26, 2018, at 12:00 pm with each flight of approximately 9 minutes of flight time. The UAS was flown autonomously using the flight planning application Tower for Android mobile devices. The flights were conducted at a planned altitude of 100 m and speed of 10 m/s. The flight paths were planned in Tower using a grid pattern at a calculated frontal overlap and side lap of 85%. All images contained GPS tags in the WGS 84 (egm96) datum.

In total, 15 GCPs were surveyed using a Leica Viva GS14 GNSS Real-time Kinematic (RTK) with a reported horizontal accuracy of 8 mm and vertical accuracy of 15 mm in Network RTK mode. GCP position was recorded in the NAD83 FL East FIPS 0901 (feet) horizontal datum and the NAVD88 (feet) vertical datum. The GCPs were constructed using 2 feet × 2 feet plywood boards painted with a black and white checkerboard pattern. The GCPs were evenly distributed throughout the study area in open areas free of dense vegetation. A total of 3,437 images were collected over the 3 flights by the 4 cameras (approximately 850 images each).

5.2.3 In-situ Data Collection

To perform the species classification and accuracy evaluation, ground truth data was collected through an in-situ vegetation survey. The survey was planned using a stratified random sampling scheme where the study area was divided into 20 m × 20 m grids and the centroids of each grid were calculated. During the survey, the centroid of a grid was marked using a PVC pipe, and random compass headings and distances were generated using a random number generator. At each survey location, horizontal and vertical position was recorded using the RTK, and dominant species, minor species, and maximum canopy height were measured within a 1 foot × 1 foot transect. A total of 211 survey points were generated. Species identification was carried out with help from two field biologists from the SFWMD. Images and samples were also collected during the survey to aid in species identification. Figure 5.2 shows the in-situ data collection at the study site.

The field survey identified 28 species present at the study site. For this analysis, only 17 species are considered due to limited observations of several species. The most abundant species found during the survey were saltbush (*baccharis halimifolia*), leather fern (*acrostichum danaeifolium*), and ragweed (*ambrosia artemisiifolia*). Due to difficulties with identification, sand cordgrass (*spartina bakeri*) and muhly grass (*muhlenbergia capillaris*) are grouped together under the designation of native grasses. Muhly grass and sand cordgrass are visually very similar and often can only be distinguished when they are flowering. In total, 184 survey points were utilized for this study, and the number of observations for each species is listed in Table 5.1.

FIGURE 5.2
In-situ data collection at the study site. Images taken on June 22nd (a, b) and June 24th (c), 2018.

TABLE 5.1

Number of species observations from the vegetation survey and reference objects used in the classification. A * denotes that a large community was found and noted in the survey

Common Name	Scientific Name	Number of Observations	Number of Reference Objects
Saltbush	*Baccharis halimifolia*	30	115
Leather fern	*Acrostichum danaeifolium*	19*	108
Ragweed	*Ambrosia artemisiifolia*	16	21
Spanish needles	*Bidens alba*	15*	68
Dog fennel	*Eupatorium capillifolium*	15	15
Carolina willow	*Salix carolina*	14	54
Nealley's sprangletop	*Leptochloa nealleyi*	13	25
Late boneset	*Eupatorium serotinum*	13	24
Native grasses	*Spartina bakeri, muhlenbergia capillaris*	12	59
Napier grass	*Pennisetum purpureum*	7	35
Saltmarsh fleabane	*Pluchea odorata*	7	23
Mullein nightshade	*Solanum donianum*	6	33
Vasey's grass	*Paspalum urvillei*	5	22
Lead tree	*Leucaena leucocephala*	4	10
Green buttonwood	*Conocarpus erectus*	3	14
White mangrove	*Laguncularia racemosa*	3	14
Railroad vine	*Ipomoea pes-caprae*	2*	41

5.3 Methodology for Species Mapping

A methodology framework for the UAS data processing and classification procedure is shown in Figure 5.3. The UAS images were prepared and processed in SfM-MVS photogrammetry software to obtain the orthomosaic and DSM. The orthomosaic was radiometrically corrected, scaled, and then segmented into image objects. Spectral and textural image features were extracted and fused with vertical data from the DSM. The image objects were spatially matched with the in-situ species observations to create training and testing data for classification model development. RF classification was performed using the training data and classification accuracy was assessed in terms of the error matrix, Kappa statistic, and k-fold cross-validation approaches. Finally, the RF classification model was applied to all image objects to obtain a species map of the study area. The details of these steps are given in the following subsections.

5.3.1 UAS Image Pre-processing

The UAS images were first converted from 16-bit RAW files to 16-bit TIFFs using Mapir Camera Control (MCC) software. The images were simultaneously converted from triple-

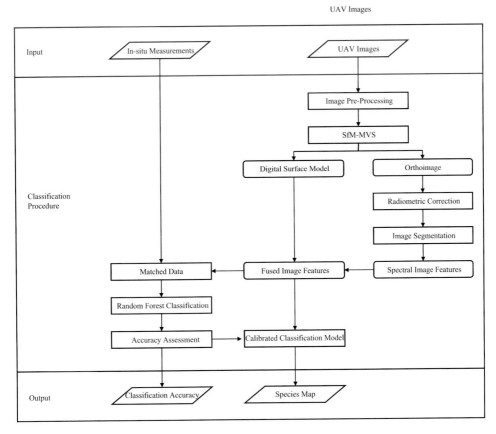

FIGURE 5.3
The framework to fuse UAS-derived spectral and vertical data for wetland species mapping.

band RGB images to single band grayscale images during this step. Manual filtering of the output TIFFs was performed in order to remove images taken on the ground or during takeoff and landing. SfM-MVS photogrammetry of the pre-processed TIFFs was performed using Pix4D Mapper Pro to obtain a 3-D point cloud, orthomosaic, and DSM. The single band greyscale images from each of the four cameras were processed using separate camera models, which were then merged into one project in Pix4D in order to align the output datasets. The initial processing step of Pix4D computes the key points on the images, calculates key point matches, resolves the camera position and orientation, and performs Bundle Block Adjustment. The GCP horizontal and vertical coordinates measured by RTK-GPS were loaded into Pix4D and marked in the images to georeference the point cloud. Geometric accuracy of the UAS data products was estimated by utilizing 8 of the 15 GCPs as checkpoints in Pix4D. Unlike GCPs, checkpoints are not utilized in the georeferencing process; they therefore provide an unbiased estimate of positional accuracy of the output data products. The average geometric accuracy in terms of Root Mean Square Error (RMSE) for the 8 checkpoints is 0.031 m in X error, 0.027 m in Y error, and 0.084 m in Z error. To ensure the highest possible positional accuracy, all 15 GCPs were utilized to generate the final datasets.

The second phase of processing in Pix4D is the densification of the point cloud, otherwise known as MVS. MVS applies the resolved camera parameters to the model in order to generate a completed, dense point cloud (Carrivick et al., 2016). Finally, the third processing step in Pix4D is the generation of a DSM and orthomosaic. Each camera model generated a separate orthomosaic and point cloud, but a single DSM was obtained as the average of the 4 camera models. The output resolution of the DSM and each orthomosaic was 4.23 cm/pixel, and the average point density of the point clouds was 54.76 points/m^2. The final UAS-derived products, the orthoimagery, and DSM are shown in Figure 5.1.

5.3.2 UAS Orthoimage Radiometric Correction and Segmentation

Radiometric calibration was performed on each orthomosaic using the Mapir camera reflectance calibration ground target and the MCC post-processing software. Following the image acquisition mission, images of the reflectance calibration target were obtained from the ground. The reflectance calibration target contains 4 targets of varying brightness that have been measured using a spectrometer (Figure 5.4). MCC is an open-source software package that utilizes a modified empirical line method to convert the radiance recorded by the sensor into surface reflectance through regression analysis using the known target reflectance values (Smith and Milton, 1999). Each single-band orthomosaic and the reflectance target images from each camera were loaded into MCC simultaneously to perform radiometric calibration and scaling of the pixel values. The calibrated orthomosaic was then stacked together in ENVI 5 software to generate a 4-band orthoimage.

Image objects were generated from the calibrated UAS orthomosaic using the multiresolution segmentation algorithm in eCognition Developer 9.0 (Trimble, 2014). The color/shape and smoothness/compactness parameters were set to equal weights (0.5/0.5) to favor color, texture, compact, and non-compact segments equally. Several scale factors were examined for this study in terms of how well the segments captured the diverse vegetation communities. A scale factor of 100 was chosen as the largest scale that accurately identified differing plant communities. During segmentation, mean and standard deviation values from each band were extracted to use as input variables for

FIGURE 5.4
A photo of the Mapir camera reflectance calibration target.

the classification procedure. An additional textural variable was derived using the grey level co-occurrence matrix (GLCM) algorithm to calculate mean texture for each object in all directions. Additional information on these metrics and their calculation can be found in Trimble (2014). Finally, descriptive statistics (maximum, minimum, mean, and standard deviation) of the DSM were calculated for each object using the Zonal Statistics tool in ArcGIS Desktop software.

5.3.3 Data Matching and Manual Interpretation

Due to the high diversity of plant species at the study site, additional ground truth samples were necessary for training and accuracy assessment of the classification procedure. We identified more reference samples through a manual interpretation of the high-resolution UAS orthoimage based upon our field knowledge of the study site, leading to 681 reference samples for model calibration and validation. The number of reference objects for each species is shown in Table 5.1.

5.3.4 Species Classification

Several machine-learning classifiers were examined for the classification procedure, but RF produced optimal results and thus was selected and presented in this chapter. RF is a machine-learning classifier combining an ensemble of decision-trees (Breiman, 2001). RF is the ideal classification algorithm because it has shown to be robust to parameter settings, small training samples, and uncertain data quality (Maxwell et al., 2018). RF has also proven successful for species classification using UAS imagery in previous studies (Feng et al., 2015; Lu and He, 2017). Parameter tuning and RF implementation were carried out in the free statistical software tool R (www.r-project.org/). Specifically, the caret package was used within R, which provides functions for machine learning classification and regression (Kuhn et al., 2016).

5.3.5 Accuracy Assessment

Accuracy assessment was performed using the k-fold cross-validation technique. This approach is a widely accepted method for estimating classification accuracy for machine learning applications (Anguita et al., 2012). k-fold cross validation separates the dataset into k subsets (i.e., folds) of approximately equal size then uses some to train and others to test model performance (Kohavi, 1995). This calculation is performed iteratively in order to cross-validate the training and testing folds and calculate accuracy across all reference samples. The variable k was set to 4 in this study due to a relatively small number of reference samples for several classes. Error matrix and Kappa values were produced in the k-fold cross-validation procedure.

5.4 Experimental Analysis and Results

We conducted four experimental analyses to investigate whether a fusion of the UAS products would be more effective for the species classification in the study area. We first considered all 17 species and identified them by using the spectral and textural variables only (referred to as spectral) and using a fusion of spectral, textural, and vertical variables (referred to as spectral and vertical). We then only considered the 10 most abundant species and classified them with spectral variables alone and the spectral and vertical fusion. The resulting accuracies of the 4 experimental analyses are given in Table 5.2.

As expected, identifying all species in the classification produced a lower accuracy than the identification of major species. Classification of all species using spectral data only produced an OA of 67.6% and Kappa value of 0.64, while discrimination of 10 major species achieved an OA of 79.7% and Kappa value of 0.69. The inclusion of more species increased the confusion among species in spectral and vertical structure, leading to decreased accuracy. Several minor species had a very low accuracy in the classification. This might be attributed to the limited number of training samples for these species. An exclusion of these minor species achieved a higher

TABLE 5.2

Classification accuracies of four experimental analyses

Classification	Overall Accuracy (%)		Kappa Coefficient	
	Spectral	Spectral and Vertical	Spectral	Spectral and Vertical
All Species	67.6	72.0	0.64	0.69
Major Species	79.7	84.7	0.76	0.82

TABLE 5.3

Per-class accuracies for the major species classification

Species	User's Accuracy (%)		Producer's Accuracy (%)	
	Spectral	Spectral and Vertical	Spectral	Spectral and Vertical
1. Native Grasses	64.4	73.8	79.7	76.3
2. Nealley's Sprangletop	50.0	55.0	28.0	44.0
3. Carolina Willow	74.4	83.3	59.3	74.1
4. Railroad Vine	84.6	84.8	80.5	95.1
5. Leather Fern	96.2	98.1	92.6	93.5
6. Mullein Nightshade	84.8	87.9	84.8	87.9
7. Saltmarsh Fleabane	70.4	83.3	82.6	87.0
8. Spanish Needles	82.1	87.9	80.9	85.3
9. Vasey's Grass	59.1	77.3	59.1	77.3
10. Saltbush	81.7	83.2	89.6	90.4

accuracy by considering only the major species in the classification. The results also revealed that an inclusion of the vertical variables from the DSM increased the species classification for both the major species and all species analyses. An addition of DSM-derived vertical structure increased the OA to 72.0% and Kappa value to 0.69 for identifying all species and the OA to 84.7% and Kappa value to 0.82 for classifying the 10 major species.

The per-class accuracies for the major species classification are shown in Table 5.3. With the inclusion of the vertical variables, the user's and producer's accuracies increased for all major species except the producer's accuracy for native grasses. The user's accuracies (UAs) varied between 50.0% and 96.2%, and the producer's accuracies (PAs) were within 28.0% and 92.6% when the spectral data was used alone. When the vertical data was included, the UAs varied from 55.0% to 98.1%, and PAs were in the range of 44.0% and 95.1%.

The experimental analyses showed that a combination of spectral and vertical data from UAS was effective for species classification, and we thus produced species maps using the fused dataset. To generate the species maps, we first masked out non-vegetation covers in the study area including GCPs and the bare ground. We performed a classification using RF to delineate the area into three classes: vegetation, bare ground, and GCPs; we achieved an OA of 99.4% and a Kappa value of 0.98. The bare ground and GCP classes were then removed so that only vegetation objects remained. Then the all species classification scheme including the vertical variables (OA: 72%, Kappa: 0.69) was applied to the vegetation objects to obtain a species map of the study area. Figure 5.5 shows the final species map for all 17 species considered in the study. Figure 5.6 shows the 10 species considered in the major species classification scheme. The maps reveal that plant species in the selected study area have a high spatial heterogeneity. Leather fern (light blue) and saltbush (red) are two dominant species.

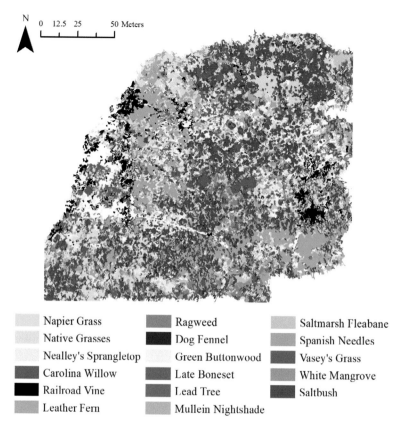

FIGURE 5.5
Species map of all species considered in the study.

5.5 Discussion

This research demonstrates that accurate species classification can be achieved for several species by fusing spectral, textural, and vertical data collected with a small UAS in a high diversity, restored coastal wetland. The inclusion of the vertical data increased overall classification accuracy and Kappa value. These findings demonstrate that the vertical data provided from SfM-MVS as applied to UAS-imagery is beneficial for species mapping in a coastal wetland ecosystem. The major species classification performed for 10 species (OA: 84.7%, Kappa: 0.82) is an improvement over the 70.8% in OA achieved by Pande-Chhetri et al. (2017) for 8 species classifications in the freshwater wetlands of Lake Okeechobee. This improvement in accuracy is likely due to the inclusion of an additional spectral band of NIR as well as including the vertical variables from the DSM. Furthermore, Pande-Chhetri et al. (2017) indicated that much of the error in their classification was likely due to radiometric differences amongst the images. The application of a target-based

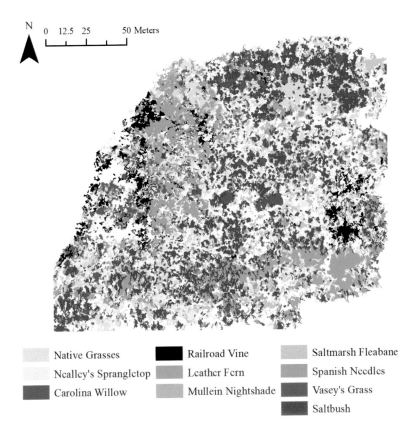

FIGURE 5.6
Species map of the major species identified in the study.

radiometric calibration on the UAS-derived orthomosaic used in this study may have also led to higher classification accuracies.

Poor accuracy was produced in the identification of several species when all 17 species were considered. In particular, dog fennel, lead tree, green buttonwood, and white mangrove all had 0% accuracy in the classification. Dog fennel (*Eupatorium capillifolium*) was prevalent at the site but only as an individual plant instead of a larger community; therefore, it was not captured well following image segmentation. To address this issue, a multi-scale segmentation approach comparable to the method developed by Pande-Chhetri et al. (2017) should be investigated in further work. Lead tree (*Leucaena leucocephala*), green buttonwood (*Conocarpus erectus*), and white mangrove (*Lagunclaria racemosa*) all had very low numbers of samples, likely leading to the poor classification results. Lead tree, green buttonwood, and white mangrove species were also found as both juvenile and mature plants at the site leading to additional errors in the classification.

Napier grass (UA: 51.4%, PA: 50.0%), late boneset (UA: 37.5%, PA: 42.9%), and ragweed (UA: 52.4%, PA: 84.6%) also generated relatively poor results in the classification despite a suitable number of reference samples. Napier grass producer's and user's accuracies actually decreased from 53.7% to 51.4% and 62.9% to 51.4%, respectively, when the vertical variables were included in the classification. Napier grass (*Pennisetum purpureum*) is a common invasive plant species found throughout Florida and the Everglades; therefore, it is an important species to map using this methodology. With the inclusion of the vertical variables, napier grass was increasingly misclassified with mullein nightshade (*Solanum donianum*). Methods to increase the classification accuracy of napier grass should be examined for future studies.

Classification of the 10 major species produced higher accuracies by removing the above species that highly increased the rate of misclassification. The species with the highest accuracies include saltbush, leather fern, mullein nightshade, Spanish needles, and railroad vine. Saltbush (*Baccharis halimifolia*) is a native deciduous shrub that grows in tidal and brackish marshes in Florida and is an important habitat for small birds (Miller and Skaradek, 2006). Saltbush was the most prevalent species found at the site and was found in dense clusters, which likely aided in the achievement of high accuracy in the classification. Leather fern (*Acrostichum danaeifolium*) is a common marsh species that is often found in the saline substrates landward of the mangrove band (Adams and Tomlinson, 1979). The vegetation map shows a large community of leather fern in the southeastern corner of the study area, which represents one of the lowest elevations at the site. The unique color and texture of this species, coupled with the high number of reference samples, are likely the factors responsible for the high classification accuracy.

Due to their small height compared to the other species in the study, Spanish needles (*Bidens alba*) and railroad vine (*Ipomoea pes-caprae*) were both expected to perform well in the classification when vertical variables were included. However, only the PA for railroad vine significantly increased when including the vertical variables from 80.5% to 95.1%, due to decreased misclassification with mullein nightshade and native grasses. Similarly, Spanish needles and railroad vine were found extensively at the site and are native to the area.

The high classification accuracies achieved by the species mentioned above testify to the argument that for machine learning classifiers the number of reference samples affects the accuracy of the classification (Prošek and Šímová, 2019). However, the species mullein nightshade (*Solanum donianum*) generated one of the highest accuracies in the major species classification (UA: 87.9%, PA: 87.9%) despite having a similar number of samples to napier grass, which performed poorly. This result shows that classification accuracy is in fact dependent on the spectral, textural, and vertical characteristics of the considered species and that while very limited sample sizes can lead to low accuracies, a high number of reference samples is not the main determinant of accuracy. This finding furthers the assessment that this UAS-based species mapping procedure is a viable method for mapping several coastal wetland species in the Everglades. The very high spatial resolution UAS data is a promising data source for mapping detailed vegetation types down to the species level for regions with a high diversity of species types and a high spatial heterogeneity.

5.6 Conclusion

We combined UAS-derived spectral and vertical data to improve wetland species classification in a high diversity, restored coastal wetland. It is found that UAS is a promising approach for wetland species mapping. Additionally, this study demonstrates a major advantage of UAS photogrammetry, the ability to generate both spectral and vertical datasets using the same set of images. This data acquisition and classification procedure can inform future wetland management strategies by providing maps of vegetation change over time. In particular, the successful monitoring of several important native species including saltbush, leather fern, and mullein nightshade can be carried out using the developed UAS procedure.

References

Adam, E., Mutanga, O., and Rugege, D., 2010. Multispectral and hyperspectral remote sensing for identification and mapping of wetland vegetation: A review. *Wetlands Ecology and Management*, 18, 281–296.

Adams, D.C., and Tomlinson, P.B., 1979. Acrostichum in Florida. *American Fern Journal*, 69, 42–46.

American Society for Photogrammetry and Remote Sensing, 2014. ASPRS positional accuracy standards for geospatial data. *Photogrammetric Engineering & Remote Sensing*, 3, A1–A26.

Anguita, D., Ghelardoni, L., Ghio, A., et al., 2012. The 'K' in k-fold cross validation. *ESANN 2012 Proceedings, European Symposium on Artificial Neural Networks, Computational Intelligence and Machine Learning*, Bruges, Belgium, April 25–27, 441–446.

Breiman, L., 2001. Random forests. *Machine Learning*, 45, 5–32.

Cao, J., Leng, W., Liu, K., Liu, L., He, Z., et al., 2018. Object-based mangrove species classification using unmanned aerial vehicle hyperspectral images and digital surface models. *Remote Sensing*, 10, 89.

Carrivick, J.L., Smith, M.W., and Quincey, D.J. 2016. *Structure from Motion in the Geosciences*, 1st Edition, John Wiley & Sons, Hoboken, NJ.

Cetinsoy, E., Dikyar, S., Hancer, C., Oner, K.T., Sirimoglu, E., et al., 2012. Design and construction of a novel quad tilt-wing UAV. *Mechatronics*, 22, 723–745.

Colomina, I., and Molina, P., 2014. Unmanned aerial systems for photogrammetry and remote sensing: A review. *ISPRS Journal of Photogrammetry and Remote Sensing*, 92, 79–97.

Dandois, J.P., Olano, M., and Ellis, E.C., 2015. Optimal altitude, overlap, and weather conditions for computer vision UAV estimates of forest structure. *Remote Sensing*, 7, 13895–13920.

Davis, R.A., and Fitzgerald, D. 2004. Coastal Wetlands. In: *Beaches and Coasts*, John Wiley & Sons Hoboken, NJ, 264–277.

Deng, L., Mao, Z., Li, X., Hu, Z., Duan, F., et al., 2018. UAV-based multispectral remote sensing for precision agriculture: A comparison between different cameras. *ISPRS Journal of Photogrammetry and Remote Sensing*, 146, 124–136.

Doren, R.F., Rutchey, K., and Welch, R., 1999. The Everglades: A perspective on the requirements and applications for vegetation map and database products. *Photogrammetric Engineering and Remote Sensing*, 65, 155–161.

Duan, T., Chapman, S.C., Guo, Y., and Zheng, B., 2017. Dynamic monitoring of NDVI in wheat agronomy and breeding trials using an unmanned aerial vehicle. *Field Crops Research*, 210, 71–80.

Feng, Q., Liu, J., and Gong, J., 2015. UAV remote sensing for urban vegetation mapping using random forest and texture analysis. *Remote Sensing*, 7, 1074–1094.

Fonstad, M.A., Dietrich, J.T., Courville, B.C., and Jensen, J., 2013. Topographic structure from motion: A new development in photogrammetric measurement. *Earth Surface Processes and Landforms*, 38, 421–430.

Gallant, A.L., 2015. The challenges of remote monitoring of wetlands. *Remote Sensing*, 7, 10938–10950.

Gini, R., Passoni, D., Pinto, L., and Sona, G., 2014. Use of Unmanned Aerial Systems for multispectral survey and tree classification: A test in a park area of northern Italy. *European Journal of Remote Sensing*, 47, 251–269.

Grenzdörffer, G.J., Engel, A., and Teichert, B., 2008. The photogrammetric potential of low-cost UAVs in forestry and agriculture. The International Archives of the Photogrammetry. *Remote Sensing, and Spatial Information Sciences*, 31 (B3), 1207–1214.

Harwin, S., and Lucieer, A., 2012. Assessing the accuracy of georeferenced point clouds produced via multi-view stereopsis from unmanned aerial vehicle (UAV) imagery. *Remote Sensing*, 4, 1573–1599.

Hirano, A., Madden, M., and Welch, R., 2003. Hyperspectral image data for mapping wetland vegetation. *Wetlands*, 23, 436–448.

James, M.R., Robson, S., d'Oleire-Oltmanns, S., and Niethammer, U., 2017. Optimising UAV topographic surveys processed with structure-from-motion: Ground control quality, quantity and bundle adjustment. *Geomorphology*, 280, 51–66.

Jensen, J.R., Rutchey, K., Kock, M.S., and Narumalani, S., 1995. Inland wetland change detection in the Everglades Water Conservation area 2A using a time series of normalized remotely sensed data. *Photogrammetric Engineering and Remote Sensing*, 61, 199–209.

Johansen, K., Raharjo, T., and McCabe, M.F., 2018. Using multi-spectral UAV imagery to extract tree crop structural properties and assess pruning effects. *Remote Sensing*, 10, 854.

Jones, J., 2015. Efficient wetland surface water detection and monitoring via Landsat: Comparison with in situ data from the everglades depth estimation network. *Remote Sensing*, 7, 12503–12538.

Kalacska, M., Chmura, G.L., Lucanus, O., Bérubéc, D., and Arroyo-Morad, J.P., 2017. Structure from motion will revolutionize analyses of tidal wetland landscapes. *Remote Sensing of Environment*, 199, 14–24.

Klemas, V.V., 2015. Coastal and environmental remote sensing from unmanned aerial vehicles: An overview. *Journal of Coastal Research*, 31, 1260–1267.

Kohavi, R., 1995. A study of cross-validation and bootstrap for accuracy estimation and model selection. *International Joint Conference on Artificial Intelligence*, 14 (2).

Kuhn, M., Wing, J., Weston, S., Williams, A., Keefer, C., et al., 2016. Caret: Classification and regression training. *R package version 6.0-73*. https://cran.r-project.org/web/packages/caret/index.html.

Laliberte, A.S., Herrick, J.E., Rango, A., and Winters, C., 2010. Acquisition, orthorectification, and object-based classification of unmanned aerial vehicle (UAV) imagery for rangeland monitoring. *Photogrammetric Engineering & Remote Sensing*, 76, 661–672.

Li, Q.S., Wong, F.K.K., and Fung, T., 2017. Assessing the utility of UAV-borne hyperspectral image and photogrammetry derived 3D data for wetland species distribution quick mapping. *The International Archives of the Photogrammetry, Remote Sensing and Spatial Information Sciences*, XLII-2/W6, 209–215.

Lu, B., and He, Y., 2017. Species classification using Unmanned Aerial Vehicle (UAV)-acquired high spatial resolution imagery in a heterogeneous grassland. *ISPRS Journal of Photogrammetry and Remote Sensing*, 128, 73–85.

Madden, M., Jordan, T., Cotten, D., O'Hare, N., and Pasqua, A., 2015. The future of unmanned aerial systems (UAS) for monitoring natural and cultural resources. *Proceedings of Photogrammetric Week*, 15, 369–384.

Mahdavi, S., Salehi, B., Granger, J., Amani, M., 2018. Remote sensing for wetland classification: A comprehensive review. *GIScience & Remote Sensing*, 55, 623–658.

Martin, S., 2015. The coastal Palmetto Bay and Cutler Bay habitat restoration project. https://regional conservation.org/ircs/NAWCA.pdf.

Maxwell, A.E., Warner, T.A., and Fang, F., 2018. Implementation of machine-learning classification in remote sensing: An applied review. *International Journal of Remote Sensing*, 39, 2784–2817.

Meng, X., Shang, N., Zhang, X., and Li, C., 2017. Photogrammetric UAV mapping of terrain under dense coastal vegetation: An object-oriented classification ensemble algorithm for classification and terrain correction. *Remote Sensing*, 9, 1187.

Miller, C., and Skaradek, W., 2006. Plant fact sheet: Eastern baccharis. *United States Department of Agriculture Natural Resources Conservation Service*. https://plants.usda.gov/factsheet/pdf/fs_baha.pdf.

Nevalainen, O., Honkavaara, E., Tuominen, S., and Viljanen, N., 2017. Individual tree detection and classification with UAV-based photogrammetric point clouds and hyperspectral imaging. *Remote Sensing*, 9, 185.

Nex, F., and Remondino, F., 2014. UAV for 3D mapping applications: A review. *Applied Geomatics*, 6, 1–15.

Pande-Chhetri, R., Abd-Elrahman, A., Liu, T., Morton, J., and Wilhelm, V.L., etal., 2017. Object-based classification of wetland vegetation using very high-resolution unmanned air system imagery. *European Journal of Remote Sensing*, 50, 564–576.

Potgieter, A.B., George-Jaeggli, B., Chapman, S.C., and Laws, K., 2017. Multi-spectral imaging from an unmanned aerial vehicle enables the assessment of seasonal leaf area dynamics of sorghum breeding lines. *Frontiers in Plant Science*, 8, 1–11.

Prošek, J., and Šímová, P., 2019. UAV for mapping shrubland vegetation: Does fusion of spectral and vertical information derived from a single sensor increase the classification accuracy? *International Journal of Applied Earth Observation and Geoinformation*, 75, 151–162.

Rutchey, K., Schall, T., and Sklar, F., 2008. Development of vegetation maps for assessing Everglades restoration progress. *Wetlands*, 28, 806–816.

Rutchey, K., and Vilcheck, L., 1994. Development of an Everglades vegetation map using a SPOT image and the global positioning system. *Photogrammetric Engineering and Remote Sensing*, 60, 767–775.

Smith, G.M., and Milton, E.J., 1999. The use of the empirical line method to calibrate remotely sensed data to reflectance. *International Journal of Remote Sensing*, 20, 2653–2662.

Snavely, N., Seitz, S.M., and Szeliski, R., 2008. Scene reconstruction and visualization from internet photo collections. *International Journal of Computer Vision*, 80, 189–210.

Szantoi, Z., Escobedo, F., Abd-Elrahman, A., Smith, S., and Pearlstine, L., 2013. Analyzing fine-scale wetland composition using high resolution imagery and texture features. *International Journal of Applied Earth Observation and Geoinformation*, 23, 204–212.

Tamouridou, A.A., Alexandridis, T.K., Pantazi, X.E., and Lagopodi, A., 2017. Evaluation of UAV imagery for mapping Silybum marianum weed patches. *International Journal of Remote Sensing*, 38, 2246–2259.

Torres-Sánchez, J., Peña, J.M., de Castro, A.I., and López-Granados, F., 2014. Multi-temporal mapping of the vegetation fraction in early-season wheat fields using images from UAV. *Computers and Electronics in Agriculture*, 103, 104–113.

Trimble, 2014. *eCognition Developer 9.0.1 Reference Book*. Trimble Germany GmbH, Arnulfstrasse 126, D-80636, Munich.

Turner, D., Lucieer, A., Malenovský, Z., and Kang, D., 2018. Assessment of Antarctic moss health from multi-sensor UAS imagery with random forest modeling. *International Journal of Applied Earth Observation and Geoinformation*, 68, 168–179.

Turner, D., Lucieer, A., and Wallace, L., 2014. Direct georeferencing of Ultrahigh-Resolution UAV Imagery. *IEEE Transactions on Geoscience and Remote Sensing*, 52, 2738–2745.

Ullman, S., 1979. The interpretation of structure from motion. *Proceedings of the Royal Society of London. Series B. Biological Sciences*, 203 (1153), 405–426.

Vallet, J., Panissod, F., Strecha, C., and Tracol, M., 2011. Photogrammetric performance of an ultra light weight swinglet UAV. *International Archives of the Photogrammetry, Remote Sensing and Spatial Information Sciences*, XXXVIII (C22), 253–258.

Wallace, L., Lucieer, A., Watson, C., and Turner, D., 2012. Development of a UAV-lidar system with application to forest inventory. *Remote Sensing*, 4, 1519–1543.

Welch, R., Madden, M., and Doren, R., 1999. Mapping the Everglades. *Photogrammetric Engineering and Remote Sensing*, 65, 163–170.

Westoby, M.J., Brasington, J., Glasser, N.F., Hambrey, M.J., and Reynolds, J. M., 2012. 'Structure-from-Motion' photogrammetry: A low-cost, effective tool for geoscience applications. *Geomorphology*, 179, 300–314.

Zhang, C., Denka, S., and Mishra, D.R., 2018. Mapping freshwater marsh species in the wetlands of Lake Okeechobee using very high-resolution aerial photography and lidar data. *International Journal of Remote Sensing*, 39, 5600–5618.

Zhang, C., Selch, D., and Cooper, H., 2016. A framework to combine three remotely sensed data sources for vegetation mapping in the central Florida Everglades. *Wetlands*, 36, 201–213.

Zhang, C., and Xie, X., 2012. Combining object-based texture measures with a neural network for vegetation mapping in the Everglades from hyperspectral imagery. *Remote Sensing of Environment*, 124, 310–320.

Zhang, C., and Xie, X., 2013. Data fusion and classifier ensemble techniques for vegetation mapping in the coastal Everglades. *Geocarto International*, 29, 228–243.

Zhou, G., Yang, J., Li, X., and Yang, X., 2012. Advances of flash lidar development onboard UAV. *International Archives of the Photogrammetry, Remote Sensing and Spatial Information Sciences*, 39, B3.

Zweig, C.L., Burgess, M.A., Percival, H.F., and Kitchens, W.M., 2015. Use of unmanned aircraft systems to delineate fine-scale wetland vegetation communities. *Wetlands*, 35, 303–309.

6

Spaceborne Multispectral Sensors for Vegetation Mapping and Change Analysis

Caiyun Zhang

6.1 Introduction

In CERP, vegetation maps or LCLU data are mainly produced using aerial photography through a manual interpretation procedure. Efforts have been made to apply satellite imagery for vegetation mapping via digital image classification techniques. Examples include application of 20-m satellite imagery for mapping vegetation in WCA-1 (Richardson et al., 1990) and WCA-2A (Rutchey and Vilchek, 1994, 1999), respectively, as well as the use of time series satellite imagery for vegetation change analysis in WCA-2A (Jensen et al., 1995). Jensen et al. (1995) were the first to attempt to document vegetation changes by processing time series satellite images. In this work, a 1991 SPOT classified image created by Rutchey and Vilchek (1994) was used as a base from which to analyze the historical trends of cattail coverage in WCA-2A. Landsat multispectral scanner (MSS) data (1973, 1976, and 1982) and SPOT data (1987) were normalized to the base year 1991 SPOT imagery. An unsupervised clustering approach was applied to generate the vegetation map for each individual year. A post-classification comparison change detection approach was used to reveal the trend of cattail coverage in the years 1973 to 1991. Cattail is an invasive species in the Everglades, and its population growth is threatening to throw the ecosystem out of balance. The abovementioned studies have demonstrated that the application of medium spatial resolution satellite imagery (20–30 meters) has a potential to map and inventory dominant plant communities in WCAs, even though this type of data might not be useful to map other regions with a high degree of heterogeneity in the Greater Everglades. In this chapter, we introduce an application of Landsat time series data to reveal the vegetation change during the period 1996 to 2016 in WCA-2A by developing object-based change analysis techniques and producing more recent vegetation maps based on the classification system of Vegetation Classification for South Florida Natural Areas (Rutchey et al., 2006).

6.2 Data

Landsat images used in this application were collected during 1996 to 2016. Landsat Collection Level-1 Tier 1 Products of relevant cloud-free images were downloaded and clipped to WCA-2A. These images were collected by three Landsat sensors/missions, including Landsat-5 TM, Landsat-7 ETM+, and Landsat-8 OLI. A natural color composite from the 2003 Landsat-7 ETM+ data for the WCA-2A is shown in Figure 6.1. Six

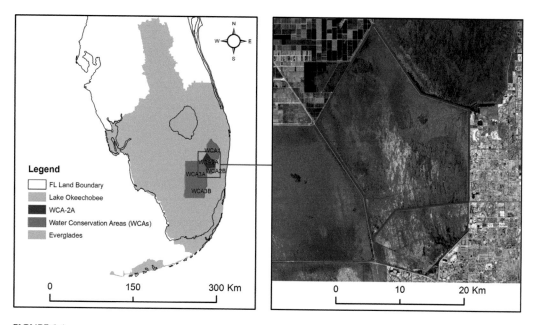

FIGURE 6.1
Applying Landsat imagery to map vegetation in WCA-2A that is shown as a natural color composite from a Landsat 7 ETM+ image collected on February 13, 2003 (bands 3, 2, and 1).

spectral bands (three visible, one near infrared, and two shortwave infrared) with a spatial resolution of 30 meters were used and the thermal channel was excluded. A 2003 vegetation map created by Rutchey et al. (2008) was used as a reference map to assist with training/testing sample selection from the Landsat imagery (Figure 3.1). This map was produced by stereoscopic analysis of 1:24,000 scale color infrared aerial photography flown in January and February of 2003. It reveals that the dominant plant communities in WCA-2A are sawgrass (65%), open marsh (14%), cattail (14%), and shrubs (3%). Other communities (e.g., trees and scrubs) as well as non-vegetation land covers (e.g., spoil areas and canals) span 1% or less in WCA-2A.

6.3 Methodology

To map vegetation changes during the years 1996 to 2016, multiple data processing steps were conducted, as shown in Figure 6.2. First, the 2003 image is selected as the base image since the reference data is only available for this year (i.e., the 2003 vegetation map). The radiometric normalization is then conducted for other images (1996, 2007, 2011, and 2016). The normalized images and the 2003 base image are then segmented for object-based classification and change analysis. A hybrid pixel/object-based change detection approach is applied to collect training/testing samples for the normalized images (Zhang et al., 2017). A machine learning classifier support vector machine (SVM) (Vapnik, 1995) is then applied to classify the 2003 base image and normalized images,

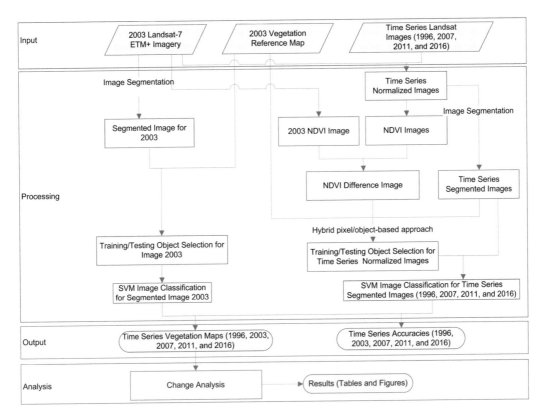

FIGURE 6.2
Methodology flowchart to map vegetation change in WCA-2A during 1996–2016 using Landsat time series data (from Zhang et al., 2017).

resulting in a time series of vegetation maps from 1996 to 2016. Accuracy assessment approaches are applied to evaluate the generated maps. Change analysis is finally conducted based upon an object-based post-classification change analysis procedure. The major techniques include image normalization, image segmentation, training sample selection, image classification and accuracy assessment, and object-based post-classification change analysis. These techniques are detailed below.

6.3.1 Time Series Image Normalization

Image normalization is a common step using time series remote sensing images for change analysis. It can reduce the variation of pixel brightness values (BVs) caused by factors in the image acquisition, such as sun angle, Earth/sun distance, atmospheric conditions, and detector calibration differences between sensors (Jensen et al., 1995). Multi-date image normalization involves selecting a base image and then transforming all other images obtained on different dates/years to have approximately the same radiometric scale as the base image (Jensen, 2015). Here the February 13, 2003, Landsat-7 ETM+ image was selected as the base image to which the 1996, 2007, 2011, and 2016 scenes were normalized. Note that the radiometric scale used in the multi-date image normalization should be the simple brightness values (BVs) rather than scaled surface reflectance (i.e., Landsat Level-1) that is distributed as scaled, and calibrated digital numbers should be used, rather than Landsat Level-2 or Landsat Analysis Ready Product (ARD). Regression equations were applied to the images to be normalized; this could predict what a pixel BV would be if it had been acquired under the same conditions as the 2003 base image. To develop the regression equations, pseudo-invariant features (PIFs) were selected first. PIFs are features or targets that are assumed constant on both the scene being normalized and the base scene.. Deep clean water bodies, bare soil, large rooftops, or other homogeneous features are candidates of PIFs. Here, 20 to 25 PIFs were selected for each scene, including unchanged water bodies, manmade homogeneous constructions, and road intersections. Based on the pixel BVs of the PIFs, regression equations were established for the correlated bands of two scenes. An example of the relationship between the PIFs in the 2007 image and the 2003 base image is displayed in Figure 6.3. These equations were then applied to compute a normalized 2007 image. Regression equations developed for years 1996, 2007, 2011, and 2016 and the coefficient of determination (R^2) are listed in Table 6.1. Note that the 2016 image from Landsat-8 OLI was acquired as 16-bit data with a BV varying from 1 to 65,535, while other images were acquired as 8-bit with a BV ranging from 1 to 255. This resulted in a different scale in slope and intercept for the 2016 image. The final normalized images and the base year 2003 image shown as a color infrared composite are displayed in Figure 6.4.

6.3.2 Image Segmentation

To conduct object-based image classification and change analysis, image objects need to be produced first. The multi-resolution segmentation algorithm in eCognition Developer 9.0 (Trimble, 2014) was applied to generate image objects for each individual image. The segmentation algorithm starts with one-pixel image segment and merges neighboring segments together until a heterogeneity threshold is reached. The heterogeneity threshold is determined by a user-defined scale parameter, as well as the color/shape and smoothness/compactness weights. The image segmentation is scale-dependent, and the quality of segmentation and classification depends on the scale of the segmentation. To find an optimal scale for image segmentation, an unsupervised image segmentation evaluation approach (Johnson and Xie, 2011) was used. This approach begins with a series of segmentations using different scale parameters and then identifies the optimal image segmentation using an unsupervised evaluation method that considers global intrasegment and intersegment heterogeneity measures. All six spectral channels of the images were set to equal weights. Color/shape weights were set to 0.9/1.0 so that spectral information would be considered most heavily for segmentation. Smoothness/compactness weights were set to 0.5/0.5 so as to not favor either compact or non-compact segments. Figure 6.5 shows examples of segmentation results for normalized images of years 2007 and 2016. Following segmentation, object-based features were extracted (mean and standard deviation), resulting in 12 values available for each object in further analysis.

6.3.3 Training/Testing Sample Selection in the Classification

Training samples are required for all supervised classifiers. Training samples can be collected either from field surveys or on-screen selection. It is impossible to go back in time and collect in-situ data as the training samples, thus the on-screen selection approach is frequently used for historical images. On-screen selection of training data

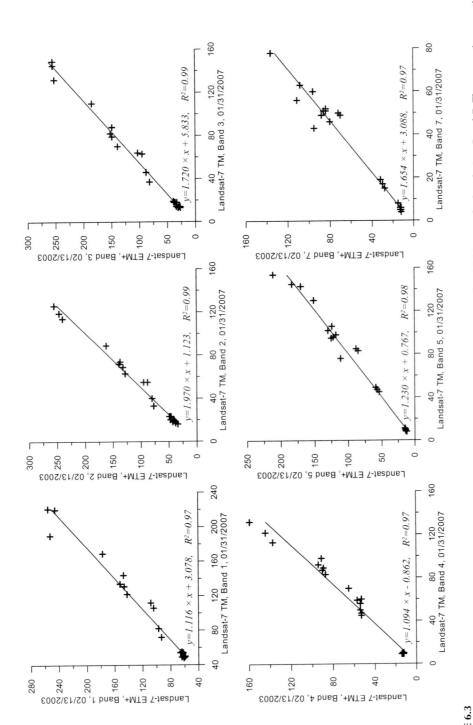

FIGURE 6.3

Relationship between PIFs found on February 13, 2003, Landsat ETM+ and January 31, 2007, Landsat TM images for bands 1 to 5 and 7. The regression equations were used to normalize the January 31, 2007, imagery (from Zhang et al., 2017).

TABLE 6.1

Equations used to normalize the historical images
with the 2003 image as the base

Date	Equations and R^2	
February 18, 1996	$1.144 \times b1 + 4.464$,	$R^2=0.92$
	$2.202 \times b2–1.720$,	$R^2=0.91$
	$1.931 \times b3 + 1.642$,	$R^2=0.93$
	$1.058 \times b4 + 3.454$,	$R^2=0.96$
	$1.225 \times b5 + 3.169$,	$R^2=0.98$
	$1.627 \times b7 + 5.658$,	$R^2=0.94$
January 31, 2007	$1.116 \times b1 + 3.078$,	$R^2=0.97$
	$1.970 \times b2 + 1.123$,	$R^2=0.99$
	$1.720 \times b3 + 5.833$,	$R^2=0.99$
	$1.094 \times b4–0.862$,	$R^2=0.97$
	$1.230 \times b5 + 0.767$,	$R^2=0.98$
	$1.654 \times b7 + 3.088$,	$R^2=0.97$
November 10, 2011	$0.833 \times b1 + 21.381$,	$R^2=0.98$
	$1.499 \times b2 + 14.15$,	$R^2=0.96$
	$1.515 \times b3 + 9.712$,	$R^2=0.98$
	$0.842 \times b4 + 7.598$,	$R^2=0.95$
	$1.113 \times b5 + 5.271$,	$R^2=0.97$
	$1.663 \times b7 + 4.179$,	$R^2=0.97$
January 24, 2016	$0.018 \times b2–78.265$,	$R^2=0.99$
	$0.016 \times b3–66.076$,	$R^2=0.99$
	$0.017 \times b4–76.661$,	$R^2=0.98$
	$0.006 \times b5–20.227$,	$R^2=0.99$
	$0.012 \times b6–50.000$,	$R^2=0.99$
	$0.013 \times b7–55.290$,	$R^2=0.97$

Note: all regression equations were significant at the
0.001 level; b1-b7 represents bands 1 to 7 of each
Landsat dataset.

is a difficult task because it requires expert knowledge in both remote sensing and
plants commonly found in the Everglades. Selecting training samples for the 2003
base year is straightforward due to the availability of the 2003 vegetation reference
map. To effectively collect training samples for other images, a hybrid pixel/object-
based training sample selection scheme was applied. This approach was developed
based on the idea of the binary change detection approach (Zhang et al., 2017).

The binary change detection approach is commonly used to extract quantitative
binary "changed/unchanged" information from two dates of imagery, using raster
image algebra such as the band differencing/ratioing logic or comparing a vegetation
index for a better detection of vegetation changes (Jensen, 2015). Again, it must be noted
that the vegetation reference map is only available for the year 2003. If the "changed/
unchanged" information between the base image (2003) and the other image (such as
1996) can be determined, this information in conjunction with the 2003 vegetation
reference map can then assist with the training sample selection for the 1996 image.
This was achieved by the following steps with the 1996 image as an example. First, the
Normalized Difference Vegetation Index (NDVI) was calculated for the 2003 base image
and 1996 image, respectively. Second, the NDVI differencing between 1996 and 2003

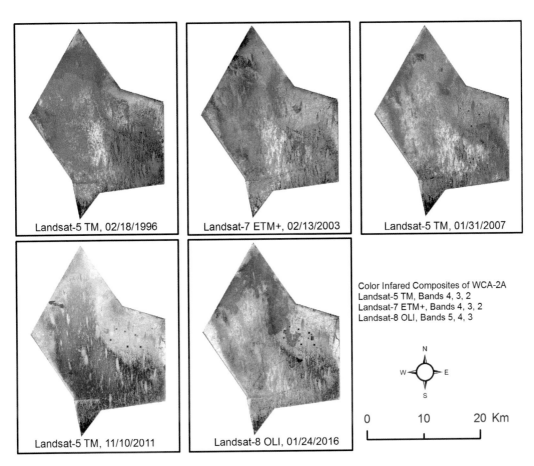

FIGURE 6.4
Color infrared composites of Landsat images collected during 1996 to 2016 in WCA-2A. Images for years 1996, 2007, 2011, and 2016 were normalized to the base year image 2003 (from Zhang et al., 2017). These images were used for vegetation change analysis.

was calculated. Third, the average of NDVI differencing for each 1996 image object was calculated. The 1996 image objects should be converted to vector polygons first in this step. Fourth, "changed/unchanged" image objects of 1996 were identified by setting a threshold to the NDVI differencing, resulting in a binary "changed/unchanged" map between 1996 and 2003. The threshold setting was determined by evaluating the histogram of the NDVI differencing. A relatively narrow range around the mean was used because the purpose of this change detection procedure was to assist with training sample selection, rather than to detect all possible change/unchanged regions. The robust unchanged regions should hover closer about the mean (Jensen, 2015). Last, overlaying the "changed/unchanged" map to the 2003 vegetation reference map and the 1996 segmented imagery, on-screen training sample selection was conducted by randomly selecting the unchanged image objects on the 1996 imagery. These selected image objects can be used as reference samples of 1996. Steps 1 and 2 were conducted at the pixel level, while other steps were conducted at the object level. Thus, the training sample selection scheme was referred to as a hybrid pixel/object-based

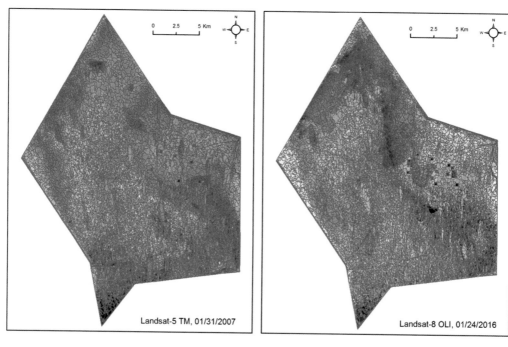

FIGURE 6.5
Examples showing segmentation results of normalized images for years 2007 and 2016.

approach (Zhang et al., 2017). By applying this approach to the segmented images 2007, 2011, and 2016, respectively, training/testing samples were effectively selected for each individual date. The total number of reference image objects for each date is listed in Table 6.2. These reference object samples were split into two halves with one for calibration/training and the other for validation/testing in the classification.

6.3.4 Image Classification and Accuracy Assessment

The SVM was applied to classify the time series segmented images using the training samples selected from the hybrid pixel/object-based approach. SVM is a non-parametric supervised classifier. The aim of SVM is to find a hyperplane that can separate the input dataset into a discrete predefined number of classes in a fashion consistent with the training samples (Vapnik, 1995). SVM research in remote sensing has increased in the past decade, as evidenced by a review in Mountrakis et al. (2011). Detailed descriptions of SVM algorithms were given by Huang et al. (2002) in the context of remote sensing. Kernel based SVMs are commonly used in the classification, among which the radial basis function (RBF) and the polynomial kernels are frequently employed. This classifier is available in remote sensing software package ENVI. It is also included in open source packages such as WEKA and R. Here we tested both kernels to find the best model for each image. The accuracy of each classified map was assessed using the traditional error matrix and Kappa statistic approaches, which have served as the standard in accuracy assessment (Congalton and Green, 2009). The overall accuracy, Kappa value, and producer's and user's accuracy were calculated for each map.

TABLE 6.2

Classification accuracies (%) of each plant community for each date in WCA-2A

	Class	PA (%)	UA (%)	Overall Accuracy (%)	Kappa	Total References
February 18, 1996	Sawgrass	92.4	90.7	92.5	0.89	323
	Cattail	79.3	88.5			
	Shrub	100	100			
	Open Marsh	97.0	94.1			
February 13, 2003	Sawgrass	95.9	85.5	93.1	0.90	320
	Cattail	93.5	95.6			
	Shrub	93.8	93.8			
	Open Marsh	89.8	100			
January 31, 2007	Sawgrass	93.7	85.8	88.4	0.82	606
	Cattail	88.9	90.1			
	Shrub	81.3	100			
	Open Marsh	79.5	90.6			
November 10, 2011	Sawgrass	99.2	77.7	87.5	0.80	591
	Cattail	83.6	98.3			
	Shrub	68.7	100			
	Open Marsh	77.1	98.5			
January 24, 2016	Sawgrass	95.8	86.5	89.5	0.83	494
	Cattail	85.3	95.1			
	Shrub	37.5	100			
	Open Marsh	88.2	90.0			

PA: Producer's Accuracy; UA: User's Accuracy.

6.3.5 Object-based Post-classification Change Analysis

For object-based post-classification change analysis, researchers commonly convert the object-based classifications into the pixel-based raster images first and then create the "from-to" information in further analysis. Directly comparing the varying shape and size of objects from two dates of imagery is difficult. For the Everglades restoration, scientists are more interested in the geographic expansion/reduction of a specific plant community, such as cattail encroachment. Here we applied an object-based quantification approach in change analysis to assess the expansion/reduction of a specific class based on the object-based classifications of two dates of imagery. For example, to quantify the cattail expansion during 1996–2016, the approach starts with the conversion of the 1996 object-based map into a pixel-based raster map, and then calculates how many 1996 cattail pixels are within a 2016 cattail object. Consequently, the percentage of 1996 cattail for a 2016 cattail object is derived through dividing the total number of 1996 cattail pixels by the total number of pixels within this 2016 cattail object. In this way, the expansion of cattail for each individual 2016 cattail object is quantified and mapped. In contrast, the reduction of cattail can be derived by calculating the percentage of 2016 cattail pixels within a 1996 cattail object.

6.4 Results and Discussion

6.4.1 Time Series Vegetation Maps in WCA-2A

Vegetation maps in the Everglades produced from aerial photography using the manual photo-interpretation procedure are vector shape file format, while those from satellite data using digital image analysis are commonly raster format because a pixel-based analysis procedure is used in the classification. Here, we applied the OBIA technique and produced vegetation maps in a vector format, as shown in Figure 6.6. The maps reveal a generally consistent geographic pattern of four dominant communities. Sawgrass encompassed the greatest area and was mainly distributed in the northern portion of WCA-2A. Open marsh, a mixture of sparse graminoids, herbaceous, and/or emergent freshwater vegetation, was the second largest community in WCA-2A. It was found mainly over the middle of the study area north to south. Cattail was mainly observed in the eastern portion, western edge, and the northernmost region. It was the third largest community in WCA-2A. Small patches of shrubs were mainly distributed along the

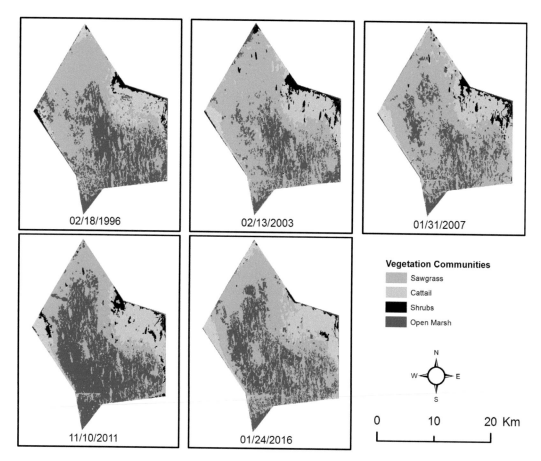

FIGURE 6.6
Vegetation maps produced from Landsat images for WCA-2A during 1996–2016 (from Zhang et al., 2017).

eastern edge. Shrubs are high-density stands of small trees and/or shrubs with heights less than five meters. In WCAs, shrubs include willow, primrose willow, and wax myrtle. Willow and primrose willow primarily occurred as a co-dominant species within cattail areas, and for that reason, shrubs are adjacent to the cattail patches. Spatial patterns of these four dominant communities were also consistent with the 2003 vegetation reference map shown in Figure 3.1.

The accuracy of each map was assessed using the selected reference samples. Several trials were conducted using both polynomial and RBF kernels with varying specifications of required parameters to identify the best SVM algorithm for each individual image. The classification accuracies of each image from the best SVM test are listed in Table 6.2. The 2003 base image produced the best result with an overall accuracy of 93.1% and Kappa value of 0.90, while the 2011 image generated the lowest accuracy with an overall accuracy of 87.5% and Kappa value of 0.80. This was expected because robust training samples could be selected from the 2003 vegetation reference map. The 2011 imagery was acquired in the wet season of the Everglades (May-November), while other images were acquired in the dry season (December-April). Unfortunately, there is no cloud-free imagery available for dry season in this year. Mapping plant communities in the wet season is more difficult than the dry season in WCAs due to a relatively higher water level that causes emergence of other aquatic plants. The producer's accuracies varied from 37.5% in 2016 to 100% in 1996 for shrub identification. For cattail discrimination, the producer's accuracies ranged from 79.3% in 1996 to 93.5% in 2003. Classifications of sawgrass and open marsh were also encouraging with the producer's accuracies in the range of 77.7%-99.2%. The user's accuracies were also acceptable with the range of 77.7%-100% for all the communities.

Promising time series vegetation maps were achieved by combining OBIA, a machine learning classifier, and the training sample selection approach. However, delineation of small or narrow patches of communities is still a challenge using the 30-meter Landsat imagery. Note that in this application, the classification was conducted in a non-exhaustive manner (i.e., only four dominant communities (sawgrass, cattail, open marsh, and shrubs) were specified in the training and mapping procedure, and other minor communities such as floating plants and scrub shown in Figure 3.1 were excluded. The non-exhaustive training inevitably brings errors into the classification because a pixel/object representing an area of an untrained class must be allocated, erroneously, to one of specified classes in the training stage (Foody, 2002). Thus, the accuracies listed in Table 6.2 might not be "true accuracy." Inclusion/exclusion of the minor communities in the training is a trade-off in the mapping procedure. For WCA-2A restoration, scientists are more concerned about the spatial and temporal variation of dominant plant communities; classes may be excluded from the training stage deliberately if they are not of interest. In this study, training sample selection for the minor communities in WCA-2A, such as scrub, trees, and other exotics was a difficult task, especially for the historical images. Exclusion of these minor communities in the training stage was actually an advantage in reducing the potential for spectral confusion (Foody, 2002). The study illustrates that non-exhaustive training and mapping is reasonable and effective to map the dominant communities in WCA-2A.

6.4.2 Vegetation Change Analysis Results

To reveal the vegetation change in WCA-2A, the statistics of the classification maps are summarized in Table 6.3. Sawgrass was the largest community with varying areas from

TABLE 6.3

Hectares of dominant plant communities in WCA-2A based upon the analysis of five dates of Landsat images collected during 1996–2016

Date	Sawgrass	Cattail	Open Marsh	Shrub	Total Hectares of WCA-2A
February 18, 1996	21503.2	5938.4	13792.0	1091.0	
February 13, 2003	25469.7	6549.9	8375.0	1930.0	
January 31, 2007	24937.7	7301.5	8341.2	1719.5	42353.0
November 10, 2011	14937.4	7015.4	18652.3	1708.4	
January 24, 2016	21715.6	7074.9	13098.1	419.9	

25469.7 ha in 2003 to 14937.4 ha in 2011. Sawgrass easily converts into open marsh when a high water level is maintained in the wet season, as revealed by the mapping results (Figure 6.6). Again, note that the 2011 imagery was acquired in the wet season, during which more water was reserved in WCA-2A, resulting in an expansion of open marsh community in the region.

Cattail has continued spreading with patches expanding throughout the eastern portion of the impoundment and along the southwestern boundary from 1996 to 2007 (Figure 6.6). This is consistent with findings from Rutchey et al. (2008). Manual interpretation of aerial photography revealed that the total area of monotypic cattail and cattail dominant mix was 5590.3 ha in 1995 (Rutchey and Vilchek, 1999) and 6039 ha in 2003 (Rutchey et al., 2008), respectively. Based on the classified imagery, the estimated coverages of cattail were 5938.4 ha in 1996 and 6549.9 ha in 2003, respectively (Table 6.3). Both displayed a continued expansion of cattail during 1995/1996-2003, but the semi-automated procedure overestimated the coverage of cattail compared to the manual interpretation. This was expected because here a non-exhaustive training was used in the classification, and other minor communities could be classified as cattail if they had a high degree of spectral confusion with cattail. The coverage of cattail reached a maximum of 7301.5 ha in 2007, then decreased to 7015.4 ha in 2011, and was relatively stable with 7074.9 ha in 2016. This finding was not reported before. For the restoration of WCA-2A, several projects have been conducted since 1996 to constrain cattail expansion by reducing the concentrations and loadings of phosphorus in WCA-2A (Newman et al., 2006; Rutchey et al., 2008). In one of the projects, several cattail treatment openings/ plots (6.25 ha) were created in phosphorus impacted regions with dense cattail, as shown in dark spots on images 2007, 2011, and 2016 (Figure 6.4). These openings are able to experience greater nutrient fluxes and constrain cattail expansion. The analyses of images 2007, 2011, and 2016 demonstrated that cattail expansion has been constrained in the eastern portion where cattail treatment openings were built and the dominance of sawgrass has been enhanced. Statistics of shrubs confirmed that this community comprised ~1%-4% in WCA-2A. The total area of each dominant community might be overestimated due to the exclusion of other minor communities in the training and classification.

In general, based on the Landsat images collected during the years 1996 to 2016 and the digital analysis procedure, the coverage of monotypic cattail or cattail dominant mix was increased from 1996 to 2007 and remained relatively stable since 2007 based on the statistics of the classified maps. The expansion of cattail in the eastern portion has been constrained with the creation of openings, while expansion trend was

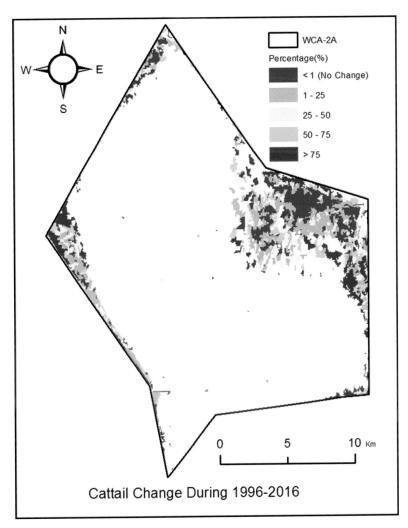

FIGURE 6.7
Map of cattail expansion in percentage in WCA-2A during 1996 to 2016. Blank areas are non-cattail regions. Unchanged regions are shown in blue, and large expansions are shown in red (from Zhang et al., 2017).

observed in the northern portion. The map of cattail expansion from 1996 to 2016 in a quantitative category using the object-based change analysis approach is displayed in Figure 6.7. The unchanged areas were mainly observed in the middle of cattail-dominated region, while expansions were found at the edge of this community. Transitional patches occurred between the unchanged and changed regions. A further observation revealed that the cattail expansion was heterogeneous. Such object-based change analysis not only can reveal the expansion/reduction of a specific class in a quantitative manner, but can also indicate the direction of the expansion/reduction.

References

Congalton, R., and Green, K., 2009. *Assessing the Accuracy of Remotely Sensed Data, Principles and Practices*, 2nd Edition, CRC Taylor & Francis, Boca Raton, FL.

Foody, G.M., 2002. Hard and soft classifications by a neural network with a non-exhaustively defined set of classes. *International Journal of Remote Sensing*, 23, 3853–3864.

Huang, C., Davis, L.S., and Townshend, J.R.G., 2002. An assessment of support vector machines for land cover classification. *International Journal of Remote Sensing*, 23, 725–749.

Jensen, J.R., 2015. *Introductory Digital Image Processing. A Remote Sensing Perspective*, 4th Edition, Prentice Hall, Upper Saddle River, NJ.

Jensen, J.R., Rutchey, K., Koch, M., and Narumalani, S., 1995. Inland wetland change detection in the Everglades Water Conservation Area 2A using a time series of normalized remotely sensed data. *Photogrammetric Engineering and Remote Sensing*, 61, 199–209.

Johnson, B., and Xie, Z., 2011. Unsupervised image segmentation evaluation and refinement using a multi-scale approach. *ISPRS Journal of Photogrammetry and Remote Sensing*, 66, 473–483.

Mountrakis, G., Im, J., and Ogole, C., 2011. Support vector machines in remote sensing: A review. *ISPRS Journal of Photogrammetry and Remote Sensing*, 66, 247–259.

Newman, S., Hagerthey, S.E., and Cook, M.I., 2006. Cattail habitat improvement project. In: *Options for Accelerating Recovery of Phosphorus Impacted Areas of the Florida Everglades*, South Florida Water Management District, West Palm Beach, FL.

Richardson, J.R., Bryant, W.L., Kitchens, W.M., and Mattson, J.E., 1990. *An Evaluation of Refuge Habitats and Relationships to Water Quality, Quantity, and Hydroperiod (A Synthesis Report Prepared for the Arthur R. Marshall Loxahatchee National Wildlife Refuge, Boynton Beach, Florida)*, Florida Cooperative Fish and Wildlife Research Unit, Gainesville, FL.

Rutchey, K., Schall, T., Doren, R., et al., 2006. Vegetation Classification for South Florida Natural Areas. Saint Petersburg, Florida, United States Geological Survey, Open–File Report 2006–1240, 142.

Rutchey, K., Schall, T., and Sklar, F., 2008. Development of vegetation maps for assessing Everglades restoration progress. *Wetlands*, 28, 806–816.

Rutchey, K., and Vilchek, L., 1994. Development of an Everglades vegetation map using a SPOT image and the global positioning system. *Photogrammetric Engineering and Remote Sensing*, 60, 767–775.

Rutchey, K., and Vilchek, L., 1999. Air photointerpretation and satellite imagery analysis techniques for mapping cattail coverage in a northern Everglades impoundment. *Photogrammetric Engineering and Remote Sensing*, 65, 185–191.

Trimble, 2014. *eCognition Developer 9.0.1 Reference Book*. Trimble Germany GmbH, Arnulfstrasse 126, D-80636 Munich.

Vapnik, V.N., 1995. *The Nature of Statistical Learning Theory*, Springer–Verlag, New York.

Zhang, C., Smith, M., Lv, J., and Fang, C., 2017. Applying time series Landsat data for vegetation change analysis in the Florida Everglades Water Conservation Area 2A during 1996-2016. *International Journal of Applied Earth Observations and Geoinformation*, 57, 214–223.

7

Water Quality Modeling and Mapping using Landsat Data

Caiyun Zhang

7.1 Introduction

In CERP, we want to know what effects hydrologic modifications have on water quality in the Everglades. The quality of Everglades water pertains to its ability to support life and sustain the natural Everglades ecosystem. Water quality data are mainly collected through fixed stations and field surveys, which is time consuming and labor intensive. In addition, it is difficult to generate a reasonable spatial pattern of a water quality parameter if limited field data are available. Remote sensing has proven valuable in assessing water quality, as reviewed by Gholizadeh et al. (2016). In the Everglades, Landsat and MODIS have been applied to estimate water quality parameters with an aim to develop remote sensing models to map water quality on a large spatial scale. For example, Chang et al. (2012) applied MODIS to estimate Chlorophyll-a in Lake Okeechobee. Chebud et al. (2012) applied Landsat data in the Kissimmee River floodplain to estimate phosphorus, turbidity, and chlorophyll-a. Lagomasino et al. (2014) applied Landsat reflectance of mangroves in Everglades National Park to indirectly indicate water chemistry in the mangrove environment. Barnes et al. (2014) used time series Landsat data to track historical water quality changes in the Florida Keys marine environment. Gholizadeh and Melesse (2017) applied Landsat data to examine the spatial and temporal variation of water quality parameters (chlorophyll-a, phosphate, turbidity, and nitrogen) in Florida Bay.

Our research group has also developed Landsat-based models to assess water salinity in Florida Bay (Zhang et al., 2012; Xie et al., 2013). Salinity is a fundamental characteristic of the physical conditions of the Everglades. Salinity affects water quality, plant associations, and the spatial distribution of vegetative communities. Effective salinity monitoring is critical in CERP, especially with sea level change. To assist in salinity monitoring, the USGS developed a boat-mounted measuring system conducting bimonthly salinity surveys in Florida Bay from 1994 to 2001. Surveys typically take 3 to 5 days to collect data along several predesigned transects that research boats can access. Little or no salinity data can be collected for the shallow regions inaccessible to research boats. The surveyed datasets, although supplemented with station-collected salinity, are still inadequate for effective salinity monitoring because of the high degree of spatial and temporal heterogeneity of the bay. Thus, salinity simulation and forecast models were constructed to examine the spatial and temporal salinity change in the bay, including statistical models and mechanistic models (Marshall and Nuttle, 2008). Statistical models depend on accurately describing observed salinity variations and correlative relationships with other parameters (Marshall et al., 2009). Mechanistic models include mass-balance models (Nuttle et al., 2000) and hydrological models (Hamrick and Moustafa, 2003), which rely on accurately accounting for the physical processes that drive changes in salinity. The accuracy of these models is limited by the data available to describe patterns of salinity and their driving processes, and their applications are still in preliminary stages (Marshall and Nuttle, 2008).

The literature has demonstrated that remote sensing has the capability to assess water salinity. The lower microwave frequency has been identified as the ideal spectral channel to directly sense water salinity (Lagerloef et al., 1995). Several research experiments with the NOAA's scanning low-frequency microwave radiometer (SLFMR) demonstrated the proof-of-concept and operational capability of airborne salinity data acquisition. A reasonable salinity pattern was obtained in Florida Bay with the use of the SLFMR in a pilot effort (D'Sa et al., 2002). The adoption of airborne SLFMR for salinity monitoring, however, is unpractical due to the high cost in data collection. In 2011, a space-borne microwave instrument, Aquarius, was launched as the first satellite platform to measure sea surface salinity. This instrument was designed to provide global salinity maps on a monthly basis with a spatial resolution of 150 km. Such a spatial resolution limits its application in coastal regions. Water salinity can be indirectly assessed from concentrations of detritus and colored dissolved organic material (CDOM) in water. In coastal regions, a large concentration of CDOM is terrestrial in origin and thus associated with fresh water (Opsahl and Benner, 1997). An inverse relationship is expected between salinity and CDOM. The CDOM is commonly estimated by two satellite remote sensors: Sea-viewing Wide Field-of-view Sensors (Sea-WiFS) and MODerate resolution Imaging Spectroradiometer (MODIS). SeaWiFS and MODIS have a suitable temporal resolution of 1 day in data acquisition, which makes them attractive for salinity monitoring. However, at a spatial resolution of 1.13 km for regional-scale applications, SeaWiFS data are of little use in shallow and small water bodies (Liu et al., 2003). As far as MODIS is concerned, Bands 8 to 16 were specifically designed to estimate ocean color, phytoplankton, and biogeochemistry at a spatial resolution of 1 km. Again, the resolution reduces its applications in Florida Bay. Researchers also made efforts to estimate water quality using the Advanced Very High Resolution Radiometer (AVHRR), but its spatial resolution of 1.09 km, again, limits its application in Florida Bay. Florida Bay is divided into numerous discrete basins by a series of interconnected carbonate mudbanks, which function as barriers to water circulation, thus leading to marked spatial differences in water salinity (Hall et al., 2007). Landsat has a suitable resolution to monitor shallow water quality (Gholizadeh et al., 2016). Compared to other satellite missions, the major benefits of Landsat TM for water quality monitoring include its mission continuity, cost-free data products, and absolute calibration. In this chapter, we introduce several approaches for applying Landsat data for salinity modeling and mapping in Florida Bay.

7.2 Study Area

The capability of Landsat for water quality modeling is tested in Florida Bay, which is a marine lagoon at the southern end of Florida with an average depth of less than 1 meter. It is bounded on the east and south by the Florida Keys and on the west by the Gulf of Mexico. Its northern boundary represents the primary interface between the bay and the upgradient ecosystems of the Everglades. Fresh water from the southern Everglades enters the bay and mixes with the saltwater from the Gulf of Mexico, resulting in a salinity gradient pattern in the bay. The variation of salinity is highly dependent on local rainfall and evaporation. Cells of hypersaline water are common during the dry season.

7.3 Developing a Linear Model to Map Water Salinity in Northeast Florida Bay

7.3.1 Data

The capability of Landsat for salinity modeling and mapping was first examined in northeast Florida Bay, as shown in Figure 7.1. The northeastern bay area comprises the discharge locations of the wide C-111 canal and Taylor Slough carrying a large volume of fresh water entering the bay, which makes its water mass different from that of its surroundings. Data include field-surveyed salinity data and Landsat TM images collected over the northeastern Florida Bay. The USGS developed a project "Monitoring and Assessment Plan" (MAP) as the primary tool for assessing the system-wide performance of CERP over the Everglades. MAP has been operating and maintaining monitoring stations and performs salinity surveys based on boat-mounted systems along the southwest coast of Everglades National Park, the Everglades wetlands, coastlines of northeastern Florida Bay and northwest Barnes Sound. For the boat-mounted systems, salinity is measured along several predesigned transects using a YSI water quality monitor. Position is determined using a GPS unit that interfaces with the YSI monitor. Data collection occurs every 5 seconds and is stored in the YSI 650 data acquisition system. All salinity meters are checked in known conductivity standards prior to and following all surveys. The surveyed data are posted on the website of SOuth Florida Information Access (SOFIA, http://sofia.usgs.gov/). The collected salinity data are available for Water Year 2004–2006 during which 12 boat-based surveys were conducted. The USGS defines Water Year as the 12-month period from October 1 of one year to September 30 of the following year and designates it by the calendar year in which it ends.

Landsat 5 TM acquired imagery during the temporal window of field data collection. Salinity surveys during Water Year 2004–2006 were not conducted concurrently at the time of Landsat satellite overpass. The surveyed data can only be matched to their closest temporary window during which TM images are available. If the TM data were contaminated by cloud, then the matched datasets were dropped. The resulting 6 matched datasets to be used for model development and application are listed in Table 7.1. Figure 7.1 shows the spatially and temporally matched datasets in Water Year 2006. The non-synchronization between the field survey data and TM data is not a problem for salinity modeling at the seasonal scale because of the uniqueness of Florida Bay. Rather than the large, open system it appears to be on a map, Florida Bay is made up of many shallow basins that are separated by an intricate network of mudbanks. These mudbanks extend throughout the bay and function as barriers that severely restrict water circulation. The mudbanks along the western margin are especially broad, several miles wide, and can effectively prevent free mixing of bay water with the Gulf of Mexico, even though these two water bodies share an open water boundary (Zhang et al., 2012). As a consequence, water is held in Florida Bay for

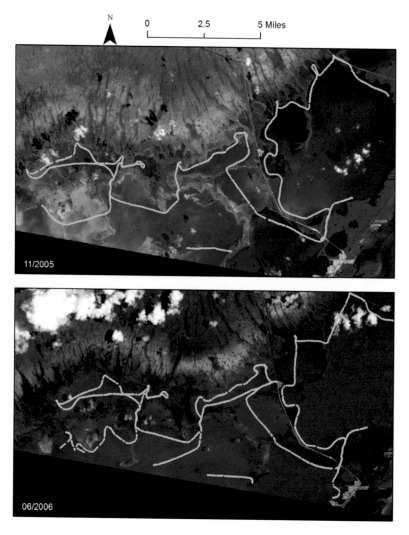

FIGURE 7.1
Field sampling locations overlaid on two temporally matched Landsat images collected in November 2005 and June 2006, respectively.

a long period of time. Some of the inner basins, such as the northeastern area, take as long as a year for water to be completely flushed by tides or wind. This effectively dampens tidal effects on water salinity estimation (D'Sa et al., 2002).

Other factors also reduce tidal effects on water salinity estimates. Salinity in northeast Florida Bay is mainly controlled by rainfall, evaporation, and runoff from C-111 canal and Taylor Slough. These factors are highly dependent on seasonal variations in the Everglades. A seasonal mapping of salinity is thus possible even though the data was not concurrently collected. The Everglades has two seasons: dry and wet. The dry season is from about November through April and the wet season from May until October. The matched datasets were divided into two groups representing the dry season (11/12/2003, 10/03/2005, and 20/11/2005) and the wet season (02/06/2004,

TABLE 7.1

Spatially and temporally matched field data and Landsat 5 TM images to be used for salinity model development and mapping (Zhang et al., 2012)

Water Year	Field Survey Date	Matched Landsat Scenes	Number of Field Samples
2004	12/11/2003[*]	12/06/2003[*]	8358[*]
	06/02/2004[**]	05/30/2004[**]	3391[**]
2005	10/14/2004[**]	10/21/2004[**]	3353[**]
	03/10/2005[*]	03/14/2005[*]	3779[*]
2006	11/10/2005[*]	11/09/2005[*]	2129[*]
	06/28/2006[**]	06/21/2006[**]	2410[**]

[*] denotes data collected in dry season; [**] denotes data collected in wet season.

14/10/2004, and 28/06/2006) with an aim to delineate seasonal variations in salinity. The surveyed salinity sample locations in latitude/longitude coordinates were transformed into the Universal Transverse Mercator, Zone 17N to be consistent with the coordinate system of the TM images. An atmospheric correction of the TM data was conducted, using the Fast Line-of-sight Atmospheric Analysis of Spectral Hypercubes (FLAASH) module in the ENVI software package. The TM reflectance values for each sample location were then extracted to generate spatially matched data with the surveyed salinity records. To remove the potential noise in the TM data and errors in the sample point locations, the matched dataset was spatially resampled into a resolution of 90 meters by using a moving window (3×3 pixels), which output the average reflection of the 3 x 3 window in the central kernel of the window. This resulted in 2059 and 3177 samples for dry season and wet season respectively to be used for model development. The thermal band was not used. Datasets for Water Years 2004 and 2005 were used to calibrate the models, and the dataset for Water Year 2006 was employed to validate the models.

7.3.2 Methodology and Results

To effectively assess salinity from Landsat 5 TM data, exploration of the original data is important. Scatter plots revealed the nonlinear character of the relationship between salinity and each TM band, suggesting that a nonlinear transformation of the datasets was necessary. The simple logit transformation for the TM records was found to best typify the data in this case. The correlation matrix of the data (Table 7.2) illustrates that three visible bands were highly correlated (Bands 1, 2, and 3). Similarly, two mid-infrared bands (Bands 5 and 7) were also highly correlated. The near-infrared band (Band 4) was more correlated to the mid-infrared bands than the visible bands. Among the three visible bands, Band 1 generated the highest correlation to the salinity ($r=0.55$). Band 4 also presented a higher correlation with salinity ($r=0.50$). As far as the two mid-infrared bands are concerned, Band 5 presented a relative higher correlation to the salinity ($r=0.52$). Results of the correlation matrix for the dry season and wet season were consistent, although differences were observed for the derived correlation coefficients (r) between two seasons.

TABLE 7.2

Data correlation matrix for the dry/wet seasons (Zhang et al., 2012)

	Landsat 5 TM Bands					
r	1	2	3	4	5	7
Salinity	0.55/0.45	0.36/0.30	0.18/0.18	0.50/0.43	0.52/0.43	0.48/0.42
Band 1		0.96/0.97	0.88/0.92	0.72/0.57	0.60/0.54	0.55/0.54
Band 2			0.94/0.95	0.60/0.47	0.46/0.45	0.43/0.45
Band 3				0.67/0.56	0.49/0.51	0.46/0.51
Band 4					0.87/0.87	0.79/0.82
Band 5						0.91/0.94

Univariate regression and multivariate regression analyses were performed with the observed salinity as the dependent variable and one TM band or a combination of TM bands as independent variables in order to determine the optimal bands for establishing the empirical models. The results are presented in Table 7.3. Univariate regression results illustrated that all bands were significantly predictive of salinity for both dry season and wet season. Bands 1 and 5 explained the highest variance among them. However, adoption of a single band is inadequate for salinity estimation because no single band explained more than 30% of the variation in salinity. Using all bands as independent variables in a multivariate regression model greatly increased its accuracy. A combination of all Landsat bands could explain a variance of 83% for the dry season and 70% for the wet season, respectively. The estimated coefficient for Band 7, however, was statistically insignificant because of the multi-collinearity among these independent variables. After dropping this band, all estimated coefficients were statistically significant with a p-value less than 0.05. The R^2 was not changed for the dry season and was slightly decreased for the wet season (Table 7.3).

To further refine the model, a stepwise regression analysis was conducted. The result suggested that Bands 1, 3, and 4 are the most effective variables in salinity prediction. To validate this selection, a partial F-test was carried out. The partial F-test examines whether the difference is statistically significant between a full model and a reduced model. The results were consistent with the stepwise regression outcomes.

TABLE 7.3

Results from univariate and multivariate regression models for dry/wet season (Zhang et al., 2012)

Univariate model with each band as the independent variable

	1	2	3	4	5	7
R^2	0.30/0.23	0.12/0.09	0.04/0.02	0.23/0.22	0.30/0.23	0.24/0.20

Multivariate model with all bands as independent variables (1–5 and 7)

R^2	0.83/0.70

Multivariate model with reduced bands (1–5) as independent variables

R^2	0.83/0.70

Multivariate model with reduced bands (1, 3, and 4) as independent variables

R^2	0.82/0.70

The preferred empirical regression models for quantitative assessment of salinity in northeast Florida Bay are (Zhang et al., 2012):

Dry season:

$$salinity = -173.3 + 78.5 \times ln \ (Band \ 1) - 52.1 \times ln \ (Band \ 3) + 11.7 \times ln \ (Band \ 4) \quad (7.1)$$

Wet season (N=3177):

$$salinity = -156.4 + 76.5 \times ln \ (Band \ 1) - 54.2 \times ln \ (Band \ 3) + 13.0 \times ln \ (Band \ 4) \quad (7.2)$$

The models were validated using the field surveyed salinity data collected in Water Year 2006. The root mean squared error (RMSE) can be used to evaluate the accuracy of estimations. Salinity values were estimated for the sample locations using TM data extracted from two TM scenes collected on 11/09/2005 (dry season) and 06/21/2006 (wet season). The estimations were compared with the field surveyed data obtained on 11/10/2005 (dry season) and 06/28/2006 (wet season), respectively. A RMSE of 5.8 parts per thousand (PPT) was generated for the dry season, and a lower RMSE of 4.8 PPT was produced for the wet season. The scatter plots of the TM estimations and surveyed data revealed dozens of outliers in both dry and wet seasons. A geographical projection of the locations of these outliers on the map presented a clustered pattern, suggesting that systematic errors may be occurring during the surveys. Omission of these outliers could improve the model results.

A map of salinity for northeast Florida Bay can be generated from a TM scene using the empirical models. The water body over the study area needs to be identified first. This was achieved using the near-infrared band because of the strong absorption of water over this spectral region. Salinity values were then calculated for the cloud-free water pixels using equation 7.1 or 7.2 based on the acquisition date of the TM data. The generated salinity maps for the selected TM scenes during Water Years 2004–2006 are shown in Figure 7.2. For comparison purposes, the salinity maps derived from the surveyed data are also presented in Figure 7.2. The distribution of salinity over this region showed a gradient pattern, with lower salinity (in blue) being observed along the coastline of Everglades National Park, from where the salinity increased southward to the Florida Keys. Lower salinity values along the coast are expected due to the fresh-water inflow from the Taylor Slough and C-111 canal. The TM estimated salinity pattern was in general agreement with the salinity maps derived from the field-surveyed data in this area. A large number of cells of hypersaline water with salinity more than 40 PPT (in red) were observed on 05/30/2004. Lower salinity values for the entire northeastern bay were observed on 12/11/2003 and 11/09/2005.

A qualitative assessment of salinity was also examined using remote sensing classification techniques. The minimum distance classification method and a neural network approach were applied. The field samples were grouped into several classes representing low, medium, high, and hyper salinity. The sample data for each class were then randomly divided into two parts with one as training data and the other as testing data. The minimum distance classifier was examined first using all TM bands. The conventional error matrix approach was employed to evaluate the classification results (Jensen, 2014). The total accuracy was calculated from the number of correctly discriminated salinity classes against the total number of valida-tion samples. The Kappa value was also calculated to quantify the classification accuracy. An average of the total accuracy of 58.9% and Kappa coefficient of 0.43 was obtained for the dry season after running the algorithm 50 times with different training data and testing data. Correspondingly, an average total accuracy of 46.3% and Kappa coefficient of 0.23 were generated for the wet season. The minimum distance approach assumes that each class has one spectral signature that is the spectral mean vector of the training data for its class. Poor outcomes will be

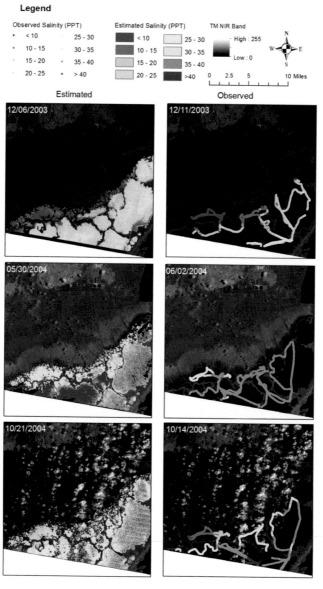

FIGURE 7.2
Produced salinity maps for Water Year 2004–2006 using the developed regression models and their comparison with field-surveyed data.

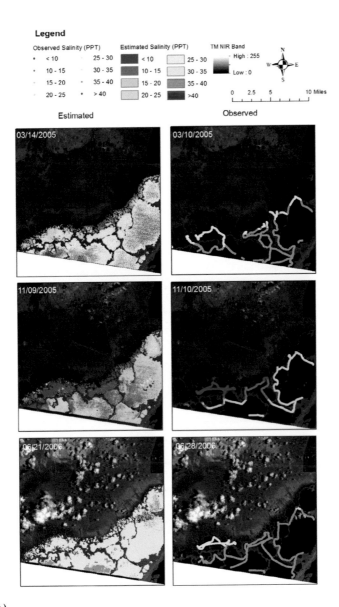

FIGURE 7.2 (Cont.)

generated if multiple spectral signatures exist for each class. A supervised neural network developed by Qiu and Jensen (2004) was tested in an attempt to obtain a higher accuracy for the qualitative assessment of salinity. One of the advantages of this neural network is its capability to model the multiple spectral signatures within a class and catch the spectral difference between classes. Similarly, after running the neural network algorithm 50 times, an average of 61.5% for the total accuracy and an average of 0.47 for the Kappa coefficient were obtained for the dry season. Poor results were generated for the wet season with an average of total accuracy of 56.6% and an average of Kappa coefficient of 0.37. This confirms the findings in the empirical algorithms that it is more difficult to assess salinity in the wet season. Analysis of the error matrix revealed that it is easier to identify water pixels with low salinity or hyper salinity than water bodies with medium or high salinity. The neural network approach, although generating better results than the minimum distance method, was inferior to the empirical regression algorithms. A salinity map using the neural network classifier is shown in Figure 7.3.

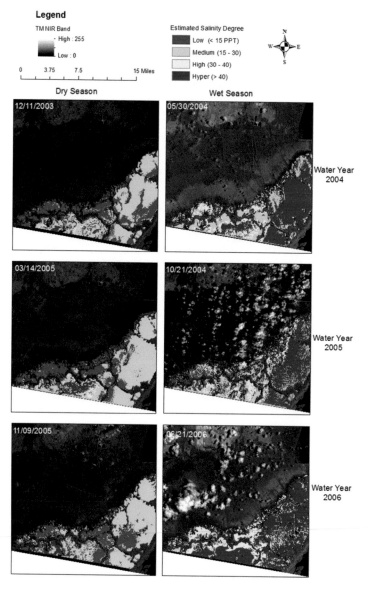

FIGURE 7.3

Applying a neural network classifier to map salinity using Landsat 5 TM data.

7.4 Applying Geographically Weighted Regression (GWR) to Map Water Salinity in Florida Bay

The linear regression models achieved an encouraging result in water salinity modeling and mapping using Landsat data in northeast Florida Bay. However, extrapolating these two empirical models to map salinity in the entire bay had a poor performance. The commonly used regression models have limitations in geospatial modeling due to the nonlinear and non-stationary property of data, especially when the large scale remote sensing data are used in the model (Foody, 2003). For a broad area modeling such as surface salinity assessment in Florida Bay, the salinity can be considered a non-stationary process due to the complex interactions of the land, ocean, and the relatively shallow water depth (Xie et al., 2013). For such cases, the geographically weighted regression (GWR) modeling techniques might work. GWR has been designed to incorporate the spatial dependency into (non-stationary) regression models. To achieve this goal, a local regression model is constructed at every location, and different relationships (regression coefficients) are possible at different points (Brunsdon et al., 1996; Fotheringham et al., 2002). GWR is therefore regarded as a truly local approach capable of capturing non-uniform spatial dependency or local spatial variation (O'Sullivan and Unwin, 2010). In this part, we examined GWR for salinity modeling and mapping in Florida Bay.

7.4.1 Data

USGS also conducted boat-mounted salinity survey in the entire bay from November 1994 to December 2001, and the data are posted on USGS' SOFIA website. Here we downloaded field salinity data collected in December 1998 and February 1999 and temporally matched them to Landsat 5 TM products, which were atmospherically corrected using FLAASH in ENVI. Figure 7.4 shows the matched field data and Landsat imagery displayed in a color infrared composite.

7.4.2 Methodology and Results

The model was constructed using the 1999 matched dataset and then applied to hindcast salinity in 1998 with the 1998 Landsat as input. For the surface salinity modeling in Florida Bay, the GWR model is described as:

$$S_i = b_{0i} + b_{1i}X_{1i} + b_{2i}X_{2i} + \ldots + b_{ni} + \varepsilon_i, \tag{7.3}$$

where S_i is the salinity at location i, X_{ji} is the independent variables for location i (i.e., spectral values of Landsat band j), and b_{ji} is the regression coefficient for location i. Note that now coefficient b varies in space rather than being constant for the entire study domain.

To construct the GWR model, it is critical to adequately specify relevant parameters. GWR relies on weighted linear regression to construct a local model for each location. For each local model, certain field samples are included and spatially weighted, depending on their proximity to the location. The weighting needs to be specified using either a Gaussian or a kernel function (Fotheringham et al., 2002), for which the bandwidth is usually more important (O'Sullivan and Unwin, 2010). The kernel bandwidth can be fixed or adaptive (to accommodate variable sample density). Here an adaptive kernel with Akaike

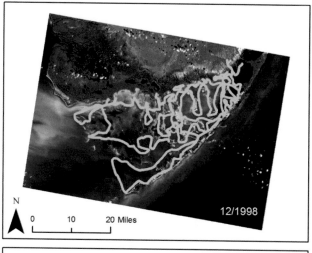

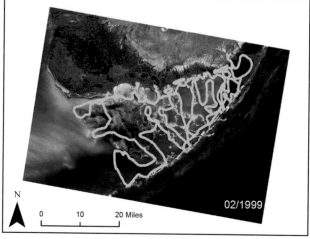

FIGURE 7.4
Field sampling locations overlaid on two temporally matched Landsat images (Bands 4, 3, 2 composite) collected in December 1998 and February 1999, respectively.

information criterion (AICc) bandwidth method was used. For successfully using GWR, independent variables should not be severely correlated. However, Landsat bands have a high correlation, as demonstrated in Table 7.2. To reduce the high correlation between each band and maximize the Landsat information, the principle component analysis (PCA) was used first. PCA is a technique that transforms the original remotely sensed dataset into a substantially smaller and easier-to-interpret set of uncorrelated variables that represents most of the information present in the original dataset.

For the 1999 Landsat imagery, the first three PCA components account for a total variance of 94%, and they were adequate to be used as independent variables for developing the GWR model for salinity mapping. In model validation, to examine the impact of spatial autocorrelation of input data, different scenarios of selecting training and testing data were tested, including: (1) the first out of every 3 survey points was used for GWR model training, and the remaining two-thirds were used for validation; (2) the first and fifth of every 10 for training and testing, respectively; (3) the first and the 13th of every 25 for training and validation, respectively; (4) the first and the 25th of every 50 for training and validation, respectively; (5) the first and the 50th of every 100 for training and validation, respectively; (6) the first and the 100th of every 200 for training and validation, respectively; and (7) the first and the 150th of every 300 for training and validation, respectively.

The results of different scenarios showed a clear decrease of model performance when less data was used in training. When there were adequate training samples (e.g., more than 100 for the study domain), GWR could produce a R^2 of more than 0.9. A traditional linear regression model only produced a R^2 of 0.05. To hindcast salinity in 1998, all the field data collected in February 1999 were used to train the GWR model that was then used to estimate salinity in 1998. Similarly, the first three PCA components of 1998 Landsat imagery were used as the inputs. The estimations were validated using all of the field data collected in December 1998. A higher RMSE (3.5 ppt) was produced compared with the 1999 GWR models for 1999 estimations. Further examination of the field-surveyed data revealed that large errors appeared in a cluster where lower salinity was found in a small bay at the north. Elimination of this cluster could reduce the RMSE to 2.5 ppt. The final salinity map for February 1999 and December 1998 from the GWR model and their comparison to the field surveyed data are shown in Figure 7.5. As expected, the salinity maps are similar in pattern because there is a short period between December 1998 and February 1999 and water circulation in the bay area is small. The model had a poor performance in estimating the lower salinity in the small bay observed in December 1998. The results demonstrate that GWR is effective for salinity estimation using Landsat data for the entire bay because of its capability to catch the non-stationary character of salinity in Florida Bay.

7.5 Exploring an Object-Based Machine-Learning Approach to Assessing Water Salinity

GWR achieved an encouraging result for modeling salinity in the entire bay by considering the non-stationary feature of salinity in the bay. However, for regions with limited field data or no field data, the estimated regression coefficients tend to be unreliable, leading to more uncertainties over these regions. Data exploration in the northeast Florida Bay area has shown a nonlinear relationship between salinity and Landsat spectral data, which indicates that machine-learning regression techniques may have a better performance in salinity modeling and mapping. Machine-learning regression techniques have proven valuable for modeling the nonlinear relationship (Zhang et al., 2018). Here we explored the potential of the Random Forest regression technique for modeling and mapping salinity in Florida Bay. So far, the salinity modeling and mapping using the linear regression model in northeast Florida Bay and GWR in the entire bay were conducted at the pixel level (i.e., matching field samples with the pixels of Landsat imagery). In practice, information of salinity over a patch/region might be more useful. OBIA offers such a potential for modeling and mapping salinity at the object level. Here we explored the object-based machine-learning modeling and mapping salinity in Florida Bay.

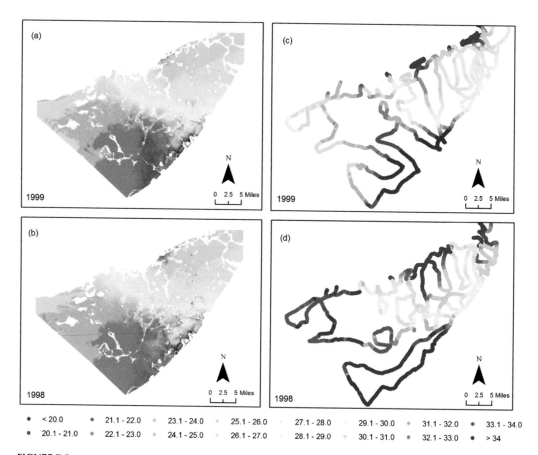

| ● | < 20.0 | ● | 21.1 - 22.0 | ● | 23.1 - 24.0 | ● | 25.1 - 26.0 | ● | 27.1 - 28.0 | | 29.1 - 30.0 | ● | 31.1 - 32.0 | ● | 33.1 - 34.0 |
| ● | 20.1 - 21.0 | ● | 22.1 - 23.0 | ● | 24.1 - 25.0 | | 26.1 - 27.0 | | 28.1 - 29.0 | | 30.1 - 31.0 | ● | 32.1 - 33.0 | ● | > 34 |

FIGURE 7.5
Produced salinity (ppt) maps for (a) February 1999, (b) December 1998 using the developed GWR model and field surveyed salinity maps for (c) February 1999, and (d) December 1998 for comparison purposes.

7.5.1 Data

The field salinity datasets collected in December 1998 and February 1999 with the GWR approach were used again, but here we applied the USGS newly released ARD Landsat products; thus atmospheric correction was not needed.

7.5.2 Methodology and Results

To conduct object-based modeling and mapping, image objects were produced first using eCognition Developer software package for each image. After segmentation, the spectral mean and standard deviation of an object for all bands were exported to be used for model development and salinity mapping. The objects were spatially and temporally matched with the field surveyed salinity data. To produce a robust training and testing dataset, only image objects with at least 10 field samples within an object were selected and the salinity of this object was calculated by taking the mean of salinity of all field samples within it. Two matched datasets were produced including

December 1998 with 544 samples and February 1999 with 519 samples, which were both used in model calibration and validation. Landsat imagery was also used to delineate water bodies of Florida Bay by using the Normalized Difference Water Index (NDWI):

$$NDWI = \frac{B_{green} - B_{nir}}{B_{green} + B_{nir}}, \tag{7.4}$$

where B_{green} and B_{nir} represent the spectral value of green (Band 2) and near infrared (Band 4) of Landsat 5 TM data. Image objects with a value of NDWI more than 0.5 were considered water objects, and non-water objects were dropped for further analysis.

The Random Forest regression approach was applied for model development. Random Forest is a decision tree based ensemble approach that constructs numerous small regression trees contributing to the predictions. It has been widely used in image classification and barely applied in remote sensing modeling. The performance of this

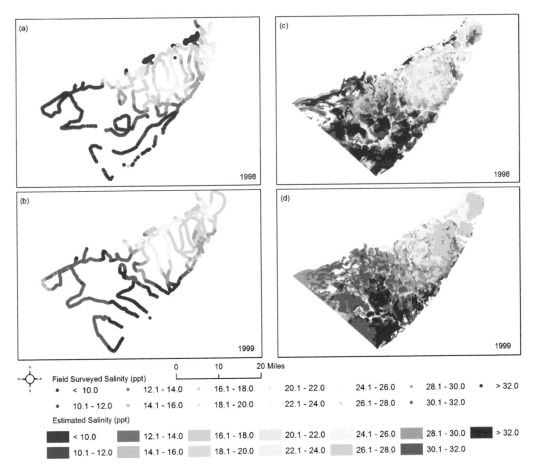

FIGURE 7.6
Field surveyed salinity maps for (a) December 1998 and (b) February 1999, and produced salinity (ppt) maps for (c) December 1998, and (d) February 1999 using the object-based Random Forest model.

machine-learning approach was evaluated using the k-fold cross-validation technique. This method splits the matched dataset into k divisions first and then, iteratively, some of them are used to train the model, and the others are exploited to assess the model performance. RMSE and R^2 could be produced after iteration.

Random forest produced a R^2 of 0.56 and 0.50 for the 1998 and 1999 matched dataset, respectively, based on the 10-fold cross-validation. In contrast, an application of a multiple linear regression approach to the matched datasets only generated a R^2 of 0.12 and 0.01 for 1998 and 1999, respectively. This, again, showed that traditional linear regression did not work for the entire bay area. Compared with the GWR model, a lower R^2 was produced; however, the final estimated salinity maps for the selected two years from the machine-learning model were more reasonable in terms of the field surveyed patterns, as shown in Figure 7.6. The lower salinity along the coast in 1998 was delineated from the Random Forest estimation, while GWR failed the prediction. The nonlinear machine-learning approach better delineated the heterogeneity of salinity in the bay than the GWR model, and the object-based mapping was more informative. In addition, the object-based salinity map products are in vector format, while the GWR outputs are in raster format. Validation results of the machine-learning model by either applying the 1998 model to predict 1999 salinity or applying the 1999 model to predict 1998 salinity showed a reduced R^2. This was considered mainly from the uncertainties in ARD products that were atmospherically corrected. However, difference in reflectance value was found for the same unchanged objects such as manmade features. To improve the cross-scene prediction, image normalization should be conducted to reduce the uncertainties from atmospheric correction applied by USGS.

In summary, Landsat shows a great potential in salinity modeling and mapping in Florida Bay. However, cloud cover and time delays in Landsat data acquisition may reduce the potential of the Landsat to serve as an independent salinity-monitoring tool in spite of its predictive capability. Therefore, it may provide a supplementary tool for reducing the cost of overall salinity monitoring programs in CERP.

References

Barnes, B.B., Hu, C., Holekamp, K.L., et al., 2014. Use of Landsat data to track historical water quality changes in Florida Keys marine environments. *Remote Sensing of Environment*, 140, 485–496.

Brunsdon, C.F., Fotheringham, A.S., and Charlton, M.E., 1996. Geographically weighted regression: A method for exploring spatial non-stationarity. *Geographical Analysis*, 28, 281–298.

Chang, N.-B., Yang, Y.J., Daranpob, A., et al., 2012. Spatiotemporal pattern validation of chlorophyll-a concentrations in Lake Okeechobee, Florida, using a comparative MODIS image mining approach. *International Journal of Remote Sensing*, 33, 2233–2260.

Chebud, Y., Naja, G.M., Rivero, R.G., and Melesse, A.G., 2012. Water quality monitoring using remote sensing and an artificial neural network. *Water Air and Soil Pollution*, 223, 4875–4887.

D'Sa, E.J., Zaitzeff, J.B., Yentsch, C.S., et al., 2002. Rapid remote assessments of salinity and ocean color in Florida Bay. In: Porter, J.W., and Porter, K.G. (eds.), *The Everglades, Florida Bay, and Coral Reefs of the Florida Keys: An Ecosystem Sourcebook*, CRC Press, Boca Raton, Florida, Chapter 15, 451–459.

Foody, G.M., 2003. Geographical weighting as a further refinement to regression modelling: An example focused on the NDVI–Rainfall relationship. *Remote Sensing of Environment*, 88, 283–293.

Fotheringham, A.S., Brunsdon, C., and Charlton, M., 2002. *Geographically Weighted Regression: The Analysis of Spatially Varying Relationships*, Wiley, New York, 269.

Gholizadeh, H.G., and Melesse, A.M., 2017. Study on Spatiotemporal variability of water quality parameters in Florida Bay using remote sensing. *Journal of Remote Sensing & GIS*, 6, 3.

Gholizadeh, H.G., Melesse, A.M., and Reddi, L., 2016. A comprehensive review on water quality parameters estimation using remote sensing techniques. *Sensors*, 16, 1298. doi:10.3390/s16081298.

Hall, M.O., Madley, K., Durako, M.J., et al., 2007. Florida Bay. In seagrass status and trends in the northern gulf of Mexico: 1940–2002. In: Handley, L., Altsman, D., and DeMay, R. (eds.), *US Geological Survey Scientific Investigations Report 2006-5287 and US Environmental Protection Agency 855-R-04-003*, US Geological Survey, Reston, Virginia, USA, 243–254.

Hamrick, J.H., and Moustafa, M.Z., 2003. Florida Bay hydrodynamic and salinity model analysis. *Conference abstract from Joint Conference on the Science and Restoration of the Greater Everglades and Florida Bay Ecosystem.*

Jensen, J.R., 2015. *Introductory Digital Image Processing, A Remote Sensing Perspective*, 4th Edition, Prentice Hall, Upper Saddle River, NJ.

Lagerloef, G.S.E., Swift, C., and Le Vine, D., 1995. Sea surface salinity: The next remote sensing challenge. *Oceanography*, 8, 44–50.

Lagomasino, D., Price, R.M., Whitman, D., et al., 2014. Estimating major ion and nutrient concentrations in mangrove estuaries in Everglades National Park using leaf and satellite reflectance. *Remote Sensing of Environment*, 154, 202–218.

Liu, Y., Islam, M.A., and Gao, J., 2003. Quantification of shallow water quality parameters by means of remote sensing. *Progress in Physical Geographer*, 27, 24–43.

Marshall, F.E., and Nuttle, W.K., 2008. *Task 7: Simulating and Forecasting Salinity in Florida Bay: A Review of Models. Critical Ecosystems Studies Initiative Project Task Report for Everglades National Park.* Cetacean Logic Foundation, New Smyrna Beach, FL. http://sofia.usgs.gov/publications/reports/salinity_flbay/index.html. Accessed on February 2, 2012.

Marshall, F.E., Wingard, G.L., and Pitts, P., 2009. A simulation of historic hydrology and salinity in Everglades National Park: Coupling paleoecologic assemblage data with regression models. *Estuaries and Coasts*, 32, 37–53.

Nuttle, W.K., Fourqurean, J.W., Cosby, B.J., et al., 2000. The influence of net freshwater supply on salinity in Florida Bay. *Water Resources Research*, 36, 805–1822.

Opsahl, S., and Benner, R., 1997. Distribution and cycling of terrigenous dissolved organic matter in the ocean. *Nature*, 386, 480–482.

O'Sullivan, D., and Unwin, D.J., 2010. *Geographic Information Analysis*, 2nd Edition, John Wiley and Sons, New York, 405.

Qiu, F., and Jensen, J.R., 2004. Opening the black box of neural networks for remote sensing image classification. *International Journal of Remote Sensing*, 25, 1749–1768.

SOFIA, http://sofia.usgs.gov/, last accessed on October 2, 2019

Xie, Z., Zhang, C., and Berry, L., 2013. Geographically weighted modeling of surface salinity in Florida Bay using Landsat TM data. *Remote Sensing Letters*, 4, 76–84.

Zhang, C., Denka, S., Cooper, H., and Mishra, D.R., 2018. Quantification of sawgrass marsh aboveground biomass in the coastal Everglades using object-based ensemble analysis and Landsat data. *Remote Sensing of Environment*, 204, 366–379.

Zhang, C., Xie, Z., Roberts, C., et al., 2012. Salinity assessment in northeastern Florida Bay using Landsat TM data. *Southeastern Geographer*, 52, 267–281.

8

Mapping Sawgrass Aboveground Biomass using Landsat Data

Caiyun Zhang

8.1 Introduction

Sawgrass marsh occupies about 70% of the Everglades and distributes from the north to south Everglades. The Everglades is a peatland with significant carbon storage coming from on-site plant production. The emergent plant marshes such as sawgrass are particularly productive. Plant biomass is renewable energy coming from plants. It contains stored energy from the sun by absorbing the sun's energy in the photosynthesis process. Estimation of plant biomass not only benefits the understanding of the carbon and energy cycles but also provides information on plant health conditions and standing fuel load for fire management in the Everglades (Lauck and Benscoter, 2015). Some sawgrass communities may be considered fire climaxes because they have been burned annually during periods of drought to improve habitat for wildlife and reduce large accumulations of fuels (Loveless, 1959).

Plant biomass data are mainly collected through field and laboratory procedures using destructive or non-destructive methods (Childers et al., 2006; Lauck and Benscoter, 2015), which is time-consuming and labor-intensive. The Florida Coastal Everglades Long Term Ecological Research (http://fcelter.fiu.edu/) has been collecting aboveground biomass data of a few plots of sawgrass and mangroves since 2000, but this dataset cannot delineate the spatial and temporal variation of plant biomass in the Everglades. Remote sensing has been widely used to estimate plant biomass, as reviewed by Klemas (2013) for coastal wetland biomass estimation and Lu et al. (2014) for forest aboveground biomass estimation. Kellndorfer et al. (2013) have developed a National Biomass and Carbon Dataset (NBCD) for the year 2000. This dataset provides a 30-meter baseline estimate of basal area-weighted canopy height aboveground live dry biomass for the conterminous United States. Development of this dataset used an empirical modeling approach that combined USDA Forest Service Forest Inventory and Analysis (FIA) data with high-resolution InSAR data acquired from the 2000 Shuttle Radar Topography Mission (SRTM) and optical remote sensing data acquired from Landsat-7 ETM+ data. There are 66 mapping zones in this dataset, and Florida is in zone 56. This dataset is open to the public at no cost at the Oak Ridge National Laboratory Distributed Active Archive Center (ORNL DAAC). A biomass map for the southern Everglades from this dataset is displayed in Figure 8.1. This biomass product only covers the woody plants; non-woody plants such as marshes have not been estimated. Biomass products of sawgrass in the Everglades have not existed. In this chapter, a methodology is presented for sawgrass biomass estimation using Landsat data.

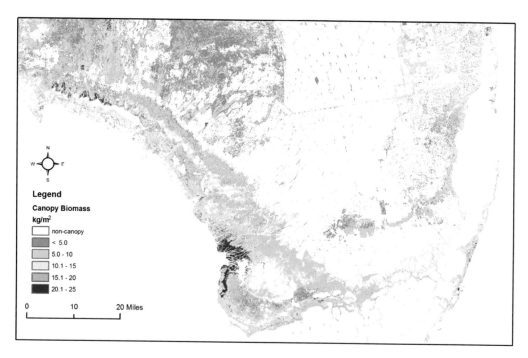

FIGURE 8.1
Estimated canopy aboveground live dry biomass in the southern Everglades from the NBCD 2000 project.

Research in biomass estimation of non-woody marshes in wetlands can be grouped into two categories (Zhang et al., 2018). The first is the application of point spectro-radiometer data (e.g., Kearney et al., 2009; Trilla et al., 2013), and the second is the application of airborne/spaceborne imaging sensors with the main objective of mapping biomass (e.g., Byrd et al., 2014; Kulawardhana et al., 2014; Ghosh et al., 2016; Kim et al., 2016; Lumbierres et al., 2017; Zhang et al., 2018). As mentioned in Chapter 2, each sensor/platform has its pros and cons. To monitor the spatial and temporal variation of coastal marsh biomass in the Everglades, spaceborne sensors are preferable to airborne sensors due to their regular cycle in data collection and broader spatial coverage. Landsat has suitable spatial resolution for coastal monitoring; it is used in this chapter.

8.2 Study Area and Data

To map sawgrass biomass, a Landsat-based remote sensing model should be developed first. This requires field biomass data to calibrate the model. Field data were collected in the coastal Everglades near the Turkey Point Nuclear Generating Station of Florida Power & Light (FPL) (Figure 8.2). The site is the region where fresh water from the Everglades (via the 50-foot wide and 20-mile long C-111 canal) meets the salt water in Florida Bay. It is an important transitional zone incubating a number of economically valuable crustaceans. The C-111 canal has modified the coastal wetlands by misdirecting

FIGURE 8.2
Study area and geographic locations of field plots for collecting sawgrass aboveground biomass data during 2011 to 2014.

freshwater flow since the 1960s. Restoration of the natural flow of water to Florida Bay is the key goal of CERP. The CERP C-111 restoration project seeks to fill the southern portion of the waterway and replace it with an east-west "spreader" canal. The new feature is expected to distribute floodwaters in a more natural fashion in northeastern Florida Bay. Monitoring the change of vegetation communities and biomass over this region is a critical indicator of the restoration success. This region is mainly dominated by sawgrass marshes, wet prairie, and mangroves.

During 2011 to 2014, FPL and the SFWMD conducted a monitoring plan to evaluate the impact of FPL uprate modifications on the ecosystem around the Turkey Point power plant. Field sawgrass biomass data were collected in the plan to identify baseline conditions and evaluate potential impacts of the uprate. In total, 13 plots (20m×20m) were established to measure sawgrass biomass changes each quarter from November 2011 to May 2014. The plots were distributed along six different transects to incorporate possible biomass variations resulting from environmental gradients perpendicular to the shoreline. The locations of these plots are shown in Figure 8.2. Nested in each plot are 4 subplots (1m×1m) located in each quadrant to measure the changes within the herbaceous community (Figure 8.3). Sawgrass biomass was measured using the procedures outlined in Childers et al. (2006) for each subplot. Each quarter, morphometric parameters representative of the height and size of plants in each subplot were measured, including the number of live leaves, dead leaves, longest leaf length, and culm diameter at the plant base. A bi-

FIGURE 8.3
A 1-meter field sawgrass subplot for measuring aboveground biomass (from SFWMD).

annual plant harvest was also conducted outside of the subplot for wet (May) and dry (November) seasons of the Everglades. Plants were harvested above the soil surface, placed individually in separate bags with a moist paper towel, and stored on ice in a dark cooler until they could be processed. The harvested plants were separated into live and dead components and wet weights were obtained. The live and dead leaves were placed into a drying oven for at least 2 weeks and then weighed. Parametric equations, to estimate the live and total (dead and live) aboveground biomass, were obtained using the field morphometric parameters and the sum of the live and dead biomass dry weights. The allometric equations were then applied to each subplot to quantify the biomass of each quarter. An average of biomass measures of 4 subplots nested in a plot was obtained and considered the biomass measures of this plot in the monitoring plan.

To spatially and temporally match the field biomass data with the Landsat data, cloud-free Landsat imagery needs to be acquired. Three dates of cloud-free Landsat data (November 10, 2011; October 30, 2013; and April 24, 2014; Path/Row: 15/42; Product: collection level-1) were obtained within a week of the field biomass data collection (November 2011, November 2013, and May 2014). They were downloaded and clipped to the study domain for model development. These images were collected by two Landsat sensors/missions: Landsat-5 TM and Landsat-8 OLI. Six spectral bands (three visible, one near-infrared (NIR), and two shortwave infrared (SWIR)) with a spatial resolution of 30 m were used, and the thermal channel was excluded. In addition, three dates of cloud-free Landsat-8 OLI scenes were processed to generate biomass map products in two seasons of 2014 and 2016 (November 2, 2014; May 15, 2016; and October 22, 2016). The natural color composites of these images are shown in Figure 8.4. The April 24, 2014, imagery was used for both model development and map generation. It was selected as the base imagery to normalize other images in imagery normalization, which is detailed in the next section.

8.3 Methodology

The methodology developed in Zhang et al. (2018) was used to generate sawgrass aboveground biomass products. In this methodology, an object-based sawgrass biomass modeling and mapping was conducted, and an ensemble analysis was used to estimate live biomass from the outputs of an Artificial Neural Network (ANN) algorithm and Support Vector Machine (SVM) algorithm. Total biomass was estimated from the ANN approach. The methodology framework is shown in Figure 8.5. To conduct multi-temporal image analysis, Landsat images collected from different dates needed to be normalized first. The normalized images were then segmented to generate image objects/segments and extract image features. An object-based classification procedure was used to extract sawgrass objects for each image, and non-sawgrass objects were not considered in further steps. The in-situ biomass measures were spatially and temporally matched to the image objects, leading to a matched dataset to be used for biomass model development. Four machine learning regression algorithms and one parametric regression approach were examined to identify a suitable biomass model for map production, including SVM, Random Forest (RF), k-Nearest Neighbor (k-NN), ANN, and Multiple Linear Regression (MLR). Machine learning has been proven valuable in image classification (Maxwell et al., 2018) but has been rarely used in biomass modeling. The performance of each model was assessed in

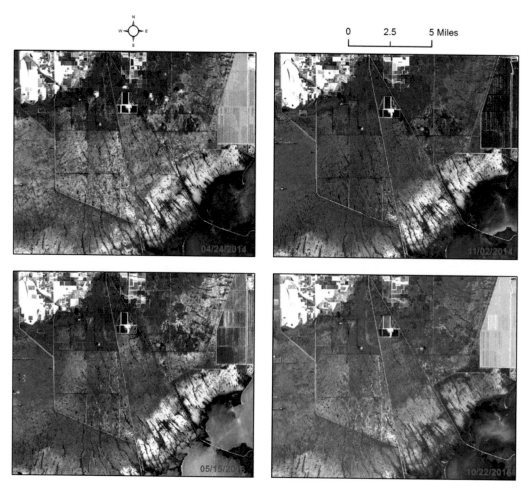

FIGURE 8.4
Landsat imagery used to generate sawgrass aboveground biomass products in the coastal Everglades.

terms of the correlation coefficient (r), mean absolute error (MAE), and root mean squared error (RMSE). Major steps in the framework include image normalization, image segmentation, data matching between in-situ measures and Landsat imagery, model development and biomass mapping, and model evaluation and accuracy assessment. These steps are detailed in the following subsections.

8.3.1 Image Normalization

Similar to chapter 6 for applying multi-date Landsat images for vegetation change analysis, image normalization is required to reduce the variation of pixel brightness values caused by factors in image acquisition. A base image needs to be selected. Here the April 2014 was used as the base image to which other images were normalized. Regression equations were established for each image to be normalized. To develop the

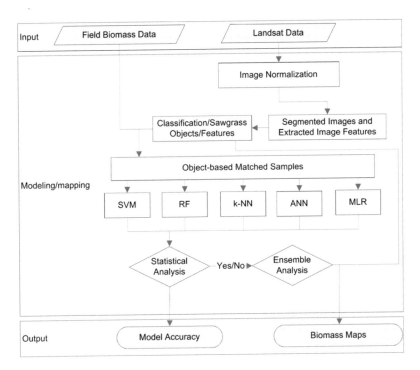

FIGURE 8.5
Methodology used to produce sawgrass biomass products using Landsat data (revised from Zhang et al., 2018).

regression equations, pseudo-invariant features (PIFs) were selected first. Here, 40 to 45 PIFs were selected for each scene, including unchanged water bodies, manmade homogeneous constructions, and road intersections. Based on the pixel BVs of the PIFs, regression equations were established for the corresponding bands of two scenes. Regression equations were developed for each band of five scenes, and a coefficient of determination (R^2) larger than 0.9 was achieved, indicating that robust PIFs were selected. These equations listed in Table 8.1 were then applied to normalize images.

8.3.2 Image Segmentation

The multiresolution segmentation algorithm in eCognition Developer 9.0 was used to generate image objects from each Landsat normalized scene to generate image objects. As described in previous chapters, for object-based image classification, this algorithm needs to set a scale parameter, as well as color/shape and smoothness/compactness weights. The scale parameter is an abstract term that determines the size of the objects with a small value generating more objects (small size) and a high value creating a smaller number of objects (big size). Using a smaller value of the scale parameter can produce more homogeneous objects than a higher value can; this should be used for matching in-situ measures with Landsat imagery. The biomass model can be applied to any objects generated from different scale parameters based on what level of detail is needed and is adaptable to different applications. For the study site, Zhang et al. (2018) have found that a scale parameter of 20 was optimal for developing the biomass models.

TABLE 8.1

Equations used to normalize Landsat images for biomass modeling and mapping (image collected on 24 April 2014 was used as the base image)

Date	Equations
November 10, 2011	$90.52 \times B1 + 4708.67$
	$174.64 \times B2 + 5213.02$
	$159.21 \times B3 + 5164.57$
	$222.48 \times B4 + 4400.25$
	$153.99 \times B5 + 4407.31$
	$211.01 \times B6 + 4420.41$
October 30, 2013	$1.35 \times B1 - 1559.09$
	$1.39 \times B2 - 1555.5$
	$1.44 \times B3 - 1938.78$
	$1.43 \times B4 - 2249.63$
	$1.49 \times B5 - 2463.41$
	$1.52 \times B6 - 2448.32$
November 02, 2014	$1.32 \times B1 - 1382.63$
	$1.32 \times B2 - 909.25$
	$1.36 \times B3 - 1195.51$
	$1.36 \times B4 - 1214.48$
	$1.40 \times B5 - 1632.27$
	$1.43 \times B6 - 2085.98$
May 15, 2016	$1.01 \times B1 - 418.15$
	$1.05 \times B2 - 469.29$
	$1.04 \times B3 - 459.29$
	$1.03 \times B4 - 844.67$
	$1.08 \times B5 - 1136.83$
	$1.17 \times B6 - 1376.58$
October 22, 2016	$1.23 \times B1 - 850.74$
	$1.22 \times B2 - 544.55$
	$1.31 \times B3 - 1138.25$
	$1.29 \times B4 - 1149.67$
	$1.30 \times B5 - 1417.19$
	$1.42 \times B6 - 2108.94$

B1-B6 refer to the blue, green, red, NIR, mid-IR 1, and mid-IR 2 Landsat sensors.

For biomass product generation, a scale of 150 was used. Color/shape weights were set to 0.9/0.1 so that spectral information would be considered most heavily for segmentation. Smoothness/compactness weights were set to 0.5/0.5 to not favor either compact or non-compact segments. Following the segmentations, the spectral features of each object were extracted. The Normalized Difference Vegetation Index (NDVI) was also calculated for each object. NDVI has proven valuable for marsh biomass estimation; thus, it was combined in the model. The six spectral bands, NDVI, and in-situ biomass measures were the inputs for each model with the in-situ data as the dependent variable, and spectral data as the independent variables.

8.3.3 Matching Field Data with Landsat Data

To conduct object-based modeling, field biomass data were spatially and temporally matched to an image object, rather than to a pixel. Four advantages were expected from the object-based matching scheme (Zhang et al., 2018). First, it can reduce the uncertainty of positional discrepancy between the image and field data. Second, a "pure" object is more representative for a plant community than any individual pixel within this community. Using features of an object is more robust than using a pixel. Third, the local noise and heterogeneity can be effectively reduced. Lastly, additional object-based spatial features (e.g., texture) can be extracted for each object; this may have the potential to improve biomass estimation by including the spatial attributes in the model. OBIA offers the capability to match the in-situ data to image objects, rather than pixels.

8.3.4 Object-based Biomass Model Development

Based on the object-based matched dataset, four machine learning regression algorithms, SVM, RF, k-NN, and ANN were evaluated and compared with the parametric regression approach MLR. SVM is a statistical learning approach (Vapnik 1995) that can be used for both classification and estimation. A review of SVM in remote sensing was provided by Mountrakis et al. (2011). An advantage of SVM is its capability to produce higher classification or more accurate estimation than other approaches using a small number of training samples. The SVM regression transforms the input data into a high-dimensional feature space using a nonlinear kernel function to minimize training error and the complexity of the model. Kernel based SVMs are commonly used and several parameters need to be tuned, including kernel to be used, precision, and penalty parameters. RF is a decision tree based ensemble approach that constructs numerous small regression trees contributing to the predictions. Detailed descriptions of RF can be found in Breiman (2001), and a recent review of RF in remote sensing was given in Belgiu and Drăguţ (2016). Two parameters need to be defined in RF: the number of decision trees to create and the number of randomly selected variables considered for splitting each node in a tree. k-NN is a relatively simple approach. The estimation is predicted as a weighted average value with k spectrally nearest neighbors using a weighting method. Again, parameters need to be tuned, including type of distance measures, weighted functions, and the choice of k value. A recent review of this technique for forest remote sensing was conducted by Chirici et al. (2016). ANN is also an important technique in image classification and biomass modeling. Various ANN algorithms have been developed and applied in remote sensing, as reviewed by Mas and Flores (2008). The multilayer perceptron algorithm of ANN was used. The algorithm needs to tune the learning rate and the numbers of hidden layers and training cycles. MLR has been commonly used in biomass estimation by assuming remotely sensed independent variables are linearly related to in-situ biomass measures. This approach requires suitable remote sensing variables to be used in the model. Previous research has focused on the comparative analysis of different models and selecting an optimal model for final mapping in terms of the statistical metrics.

8.3.5 Model Evaluation

The *k*-fold cross validation technique was used for model training and testing. This evaluation method has proven valuable in machine learning techniques for both classifications and estimation (Anguita et al., 2012). It splits the sampling data into *k*-subsets first, and then, iteratively, some of them are used to train the model and others are exploited to assess the model performance. Based upon the research from Anguita et al. (2012), *k* was specified as 4 because a relatively small number of matched samples were obtained (39), rather than the 10-fold cross validation, which was commonly used in the literature. In the iteration, 10 samples were used to calculate the errors, while 29 were used to train the model. Statistical metrics were used in this study, including r, MAE (g/ m^2), and RMSE (g/m^2). They were calculated by:

$$r = \frac{\sum_{i=1}^{N}(p_i - \overline{p_i})(p_{situ} - \overline{p_{situ}})}{\sqrt{\sum_{i=1}^{N}(p_i - \overline{p_i})^2}\sqrt{\sum_{i=1}^{N}(p_{situ} - \overline{p_{situ}})^2}}, \tag{8.1}$$

$$MAE = \frac{1}{N}\sum_{i=1}^{N}|p_i - p_{situ}|, \tag{8.2}$$

$$RMSE = \sqrt{\frac{1}{N}\sum_{i=1}^{N}(p_i - p_{situ})^2}, \tag{8.3}$$

where p_i is the model prediction; p_{situ} is the in-situ biomass measurement; $\overline{p_i}$ and $\overline{p_{situ}}$ are the mean of model predictions and mean in-situ biomass data, respectively. N is the total number of matched samples.

8.4 Results

The object-based model performance is shown in Table 8.2. For the live biomass modeling, both ANN and SVM produced a good result with an r more than 0.9. For the total biomass modeling, the ANN achieved the best result with an r of 0.94. In general, non-parametric machine learning algorithms had a better performance than the linear models, indicating biomass was nonlinearly related with remote sensing spectral data. To produce live biomass maps, the outputs of ANN and SVM were combined using an ensemble approach developed in Zhang et al. (2018), which assigns a weight to

TABLE 8.2

Model performance for live and total biomass estimation using different algorithms (Revised from Zhang et al., 2018)

Object-based Live Biomass Modeling					
Statistical Metrics	ANN	SVM	RF	k-NN	MLR
CC (r)	0.92	0.91	0.86	0.77	0.47
MAE (g/m^2)	15.74	16.32	18.54	24.38	36.10
RMSE (g/m^2)	20.35	21.31	27.35	38.87	49.56
Object-based Total Biomass Modeling					
CC (r)	0.94	0.87	0.75	0.53	0.44
MAE (g/m^2)	31.55	41.53	42.78	52.29	70.57
RMSE (g/m^2)	36.27	56.65	71.93	92.27	109.28

CC: Correlation Coefficient (r); MAE: Mean Absolute Error; RMSE: Root Mean Squared Error.
MLR: Multiple Linear Regression; SVM: Support Vector Machine; RF: Random Forest; k-NN: k-Nearest Neighbor; ANN: Artificial Neural Network.

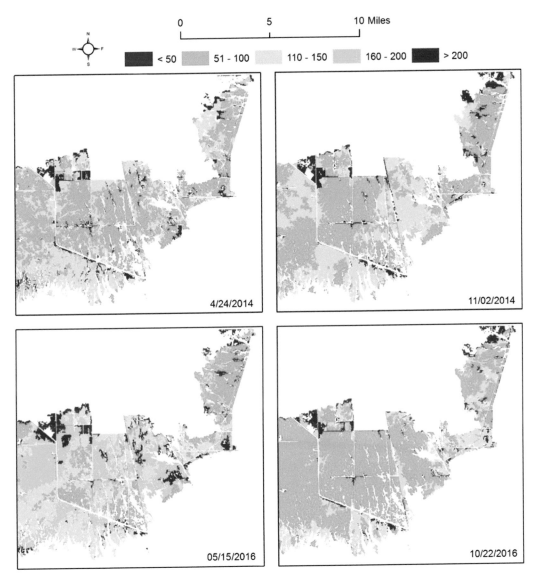

FIGURE 8.6
Estimated live sawgrass aboveground biomass (unit in g/m^2) from an ensemble estimation of ANN and SVM model outputs using Landsat data.

each output based on the value of r produced from each model. For the total biomass map production, the ANN model was used. The estimated sawgrass live biomass is displayed in Figure 8.6; the total biomass estimation is displayed in Figure 8.7.

In the maps, red represents a higher biomass and blue represents a lower biomass. Live biomass varied in the range of 50–250g/m^2, while total biomass was in the range of 50–500 g/m^2 based on the estimations. This result was consistent with the field measures collected during 2011 to 2014. In the wet season of 2014 (April 24, 2014), the northern part of the C-111 canal basin had a relatively lower biomass compared with

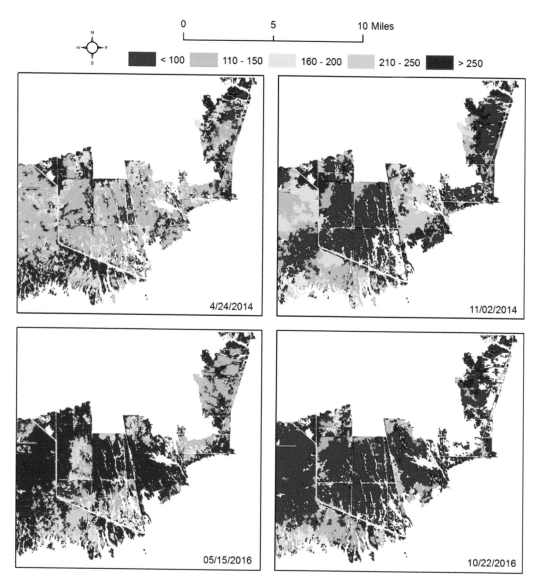

FIGURE 8.7
Estimated total sawgrass aboveground biomass (unit in g/m^2) from the ANN output using Landsat data.

its southern part, which is closer to the ocean. The highest biomass was found in the northernmost C-111 basin. In the dry season of 2014 (02/11/2014), in general, the live biomass was decreased, especially in the southern coastal regions (Figure 8.6). The live biomass map for the wet season of 2016 (May 15, 2016) showed that the biomass was much higher than in 2014, with most regions larger than 150 g/m^2, while in the dry season, a decreased pattern was clearly displayed (October 22, 2016). The seasonal variation of live biomass was stronger in 2016 than in 2014. Similarly, the total biomass was higher in the wet season than in the dry season, and again, a larger seasonal variation was observed in 2016 (Figure 8.7). These findings were consistent with the results in Childers et al. (2006) who investigated the temporal pattern of sawgrass biomass based on in-situ measures of 5 sites in the C-111 basin for 1998–2004. Both seasonal and interannual variations were found from the limited in-situ measures. The sawgrass biomass pattern over this region was controlled by salinity and hydrologic drivers such as hydroperiod (Childers et al., 2006). The spatial and temporal variation of sawgrass aboveground biomass was well demonstrated using the Landsat-based model estimations. The model performance and generated biomass map products confirmed the capability of Landsat for sawgrass marsh biomass modeling and monitoring. Further works should be directed to apply the models to estimate time series biomass products to investigate the variation of sawgrass biomass at different spatial and temporal scales to meet various needs in the restoration and carbon modeling.

References

Anguita, D., Ghelardoni, L., Ghio, A., et al., 2012. The 'K' in K-fold Cross Validation. *ESANN 2012 proceedings, European Symposium on Artificial Neural Networks*, Computational Intelligence and Machine Learning, Bruges, Belgium, April 25–27, 441–446.

Belgiu, M., and Drăguţ, L., 2016. Random forest in remote sensing: A review of applications and future directions. *ISPRS Journal of Photogrammetry and Remote Sensing*, 114, 24–31.

Breiman, L., 2001. Random forests. *Machine Learning*, 45, 5–32.

Byrd, K.B., O'Connell, J.L., Tommaso, S.D., and Kelly, M., 2014. Evaluation of sensor types and environmental controls on mapping biomass of coastal marsh emergent vegetation. *Remote Sensing of Environment*, 149, 166–180.

Childers, D.L., Iwaniec, D., Rondeau, D., et al., 2006. Responses of sawgrass and spikerush to variation in hydrologic drivers and salinity in southern Everglades marshes. *Hydrobiologia*, 569, 273–292.

Chirici, G., Mura, M., McInerney, D., et al., 2016. A meta-analysis and review of the literature on the k-Nearest Neighbors technique for forestry applications that use remotely sensed data. *Remote Sensing of Environment*, 176, 282–294.

Ghosh, S., Mishra, D.R., and Gitelson, A.A., 2016. Long-term monitoring of biophysical characteristics of tidal wetlands in the northern Gulf of Mexico – A methodological approach using MODIS. *Remote Sensing of Environment*, 173, 39–58.

Kearney, M.S., Stutzer, D., Turpie, K., and Stevenson, J.C., 2009. The effects of tidal inundation on the reflectance characteristics of coastal marsh vegetation. *Journal of Coastal Research*, 25, 1177–1186.

Kellndorfer, J., Walker, W., Kirsch, K., et al., 2013. NACP aboveground biomass and carbon baseline data, V. 2 (NBCD 2000), U.S.A., 2000. Data set. Available on-line from ORNL DAAC, Oak Ridge, Tennessee, USA.

Kim, J.Y., Im, R.-Y., Do, Y., et al., 2016. Above-ground biomass estimation of tuberous bulrush (Bolboschoenus planiculmis) in mudflats using remotely sensed multispectral image. *Ocean Science Journal*, 51, 151–158.

Klemas, V., 2013. Remote sensing of coastal wetland biomass: An overview. *Journal of Coastal Research*, 29, 1016–1028.

Kulawardhana, R.W., Popescu, S.C., and Feagin, R.A., 2014. Fusion of lidar and multispectral data to quantify salt marsh carbon stocks. *Remote Sensing of Environment*, 154, 345–357.

Lauck, M., and Benscoter, B., 2015. Non-destructive estimation of aboveground biomass in sawgrass communities of the Florida Everglades. *Wetlands*, 35, 207–210.

Loveless, C.M., 1959. A study of the vegetation in the Florida Everglades. *Ecology*, 40, 1–9.

Lu, D., Chen, Q., Wang, G., et al., 2014. A survey of remote sensing-based aboveground biomass estimation methods in forest ecosystems. *International Journal of Digital Earth*, DOI: 10.1080/17538947.2014.990526.

Lumbierres, M., Méndez, P.F., Bustamante, J., et al., 2017. Modeling biomass production in seasonal wetlands using MODIS NDVI land surface phenology. *Remote Sensing*, 9, 392.

Mas, J.F., and Flores, J.J., 2008. The application of artificial neural networks to the analysis of remotely sensed data. *International Journal of Remote Sensing*, 29, 617–663.

Maxwell, A.E., Warner, T.A., and Fang, F., 2018. Implementation of machine-learning classification in remote sensing: An applied review. *International Journal of Remote Sensing*, 39, 2784–2817.

Mountrakis, G., Im, J., and Ogole, C., 2011. Support vector machines in remote sensing: A review. *ISPRS Journal of Photogrammetry and Remote Sensing*, 66, 247–259.

Trilla, G.G., Pratolongo, P., Beget, M.E., et al., 2013. Relating biophysical parameters of coastal marshes to hyperspectral reflectance data in the Bahia Blanca estuary, Argentina. *Journal of Coastal Research*, 29, 231–238.

Vapnik, V.N., 1995, *The Nature of Statistical Learning Theory*, Springer-Verlag, New York.

Zhang, C., Denka, S., Cooper, H., and Mishra, D.R., 2018. Quantification of sawgrass marsh aboveground biomass in the coastal Everglades using object-based ensemble analysis and Landsat data. *Remote Sensing of Environment*, 204, 366–379.

9

Applying Landsat Products to Assess the Damage and Resilience of Mangroves from Hurricanes

David Brodylo and Caiyun Zhang

9.1 Introduction

Worldwide, mangrove forests are significant ecosystems that are found between the coastal lowlands and the ocean in both subtropical and tropical regions (Giri et al., 2011). They are important to both human society and the natural world as they provide multiple benefits such as stabilizing the shoreline and protecting wildlife and settlements from large storms; they are even capable of storing large quantities of carbon. Mangroves have proven to be reliable in mitigating the amount of coastal and inland storm surge flooding worldwide, along with limiting the damage caused by tropical cyclones as at many times they act as the first line of land-based defense against ocean-based flooding (Beever et al., 2016). Multiple studies have illustrated that the presence of mangroves can help reduce the water height and velocity of the storm surge caused by hurricanes and other natural disasters (Marois and Mitsch, 2015; Giri, 2016). Although mangroves provide plenty of beneficial goods and services, they are under threat by both human activities and natural disasters, especially with sea level rise and tropical cyclones. In the United States, the Florida Everglades contains the largest contiguous tracts of mangroves, most of them located within Everglades National Park. These mangroves are extremely vulnerable to hurricanes because Florida has been ranked as the number one location in which hurricanes made landfall in the USA since 1851 (NOAA/FAQ, 2019). Intensive storms can uproot trees, break branches, defoliate canopies, and alter forest structure and recovery (Smith et al., 1994, 2009; Doyle et al., 2003). In extreme cases, tropical cyclones can lead to a complete removal or large-scale mangrove loss and peat collapse (Cahoon et al., 2003), as well as long-term changes in mangroves (Whelan et al., 2009). A recent study of the effects of hurricanes Wilma and Irma, which made landfalls in South Florida, stated that these two hurricanes caused higher water levels, deposited sediment on the mangroves, and damaged the plants, which all hindered their growth (Lucas et al., 2019). According to Beever et al. (2016) and a recent review from Sippo et al. (2018), tropical cyclones are the primary natural factor that cause excessive mangrove damage worldwide. The increasing frequency, intensity, and destructiveness of cyclones, as well as other climate relevant events, have the potential to directly influence mangrove mortality and recovery.

South Florida is quite susceptible to hurricane-based impacts during the Atlantic hurricane season from June 1 to November 30. The National Oceanic and Atmospheric Administration (NOAA) estimates that between 1851 to 2017 there have been a combined total of 33 Category 3 or higher major hurricanes out of the total of 98 hurricanes that struck South Florida (NOAA/FAQ, 2019). In the past 30 years, there have been three such major hurricanes and one minor hurricane that have crossed

TABLE 9.1

Four hurricanes in the past 30 years that impacted mangroves in South Florida

Hurricane	Landfall date	Intensity
Andrew	24/08/1992	Category 5
Katrina, Wilma	25/082005, 24/10/2005	Category 1 and 3
Irma	10/09/2017	Category 4

through mangrove-heavy regions of South Florida (Table 9.1). The first major hurricane is Hurricane Andrew, which made landfall as a Category 5 hurricane on August 24, 1992, on the southeast coast of Florida near the city of Homestead before weakening to a Category 4 hurricane after passing through Everglades National Park. A minor hurricane, Hurricane Katrina, made landfall as a Category 1 hurricane on August 25, 2005, on the southeast coast of Florida along the border of Broward and Miami-Dade Counties. Two months later, Hurricane Wilma, the second major hurricane, made landfall as a Category 3 hurricane on October 24, 2005, on the southwest coast of Florida near Cape Romano located by Ten Thousand Islands. The third major hurricane is Hurricane Irma, which was a Category 4 hurricane when it crossed north of the Florida Keys before weakening to a Category 3 hurricane when it made landfall on September 10, 2017, on the southwest coast of Florida near Marco Island, which is close to where Hurricane Wilma made landfall. In this chapter, we applied Landsat products to look at the immediate effects of these hurricanes on mangroves in South Florida and investigated the recovery track of mangroves from each hurricane at a large scale.

9.2 Study Area and Data

The study area is located in South Florida, specifically between the Golden Gate Parkway in the city of Naples on the west coast of Florida, through Everglades National Park, to under Maule Lake in North Miami Beach on the east coast of Florida (Figure 9.1). It encompasses all of the mangroves within Everglades National Park, which contains the largest mangrove ecosystem in the Western Hemisphere. Mangrove forests for the study area lie on a carbonate platform between the freshwater marshes of the Everglades and the marine waters of the Gulf of Mexico and Florida Bay. The mangrove forests over this region are immediately exposed to sea level rise and hurricanes and have little threat from human activities (Doyle et al., 2003). Three mangrove species are common over this area, including black mangrove (*Avicennia germinans*), white mangrove (*Laguncularia racemosa*), and red mangrove (*Rhizophora mangle*).

Data used in this chapter include a mangrove mask and Landsat products, which are available in Google Earth Engine (GEE). The mangrove distribution data was from a global mangrove database of the year 2000, which was provided by the Center for International Earth Science Information Network (CIESIN) to GEE. This global mangrove forest distribution 2000 data set was originally produced through a compilation of the extent of mangrove forests from the global land survey and the Landsat archive with

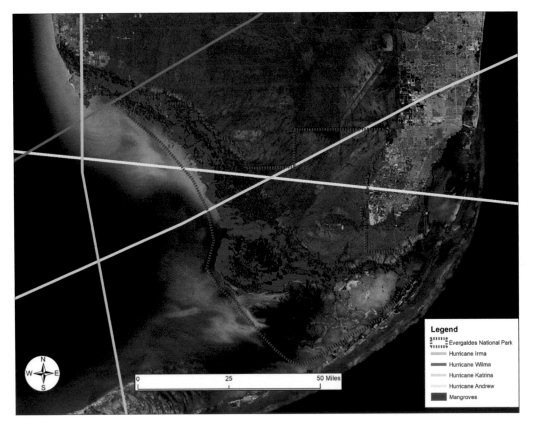

FIGURE 9.1
The study area located in the southern coastal Everglades where the largest mangrove forest in the Western Hemisphere is present alongside the hurricane tracks of the four hurricanes.

hybrid supervised and unsupervised digital image classification techniques (Giri et al., 2005). The data are available at 30-meter spatial resolution. The total area of mangroves was estimated at 137,760 km^2 in 118 countries and territories in the tropical and subtropical regions of the world.

Landsat level 2 products at a 30-meter resolution were selected in this chapter because they have already been atmospherically corrected compared to the level 1 and raw products. Landsat products were generated by the US Geological Survey (USGS). We used Landsat 5 Thematic Mapper (TM) products for damage and recovery analysis of Hurricanes Andrew, Katrina, and Wilma and applied Landsat 8 Operational Land Imager (OLI) products for Hurricane Irma. Application of one Landsat image scene collected in a specific date for damage and recovery analyses is a challenge using Landsat due to the heavy cloud cover in South Florida. Thus, using a composite dataset compiled from image scenes collected within a time window is an ideal solution for this issue. For a specific time window, only non-cloud pixels are analyzed; cloud contaminated pixels are dropped. The time windows selected for the mangrove damage analysis of each hurricane are listed in Table 9.2. An anniversary time window is ideal to avoid the phenological effects on the damage analysis; thus, it was considered in the time

TABLE 9.2

Time windows of Landsat products for mangrove damage analysis

Hurricane	Pre-hurricane	Post-hurricane
Andrew	25/08/1990 to 25/11/1990	25/08/1992 to 25/12/1992
Katrina/Wilma	26/08/2004 to 26/11/2004	25/10/2005 to 25/01/2006
Irma	11/09/2016 to 11/12/2016	11/09/2017 to 11/12/2017

TABLE 9.3

Time windows of Landsat products for mangrove recovery analysis

Post-hurricane time	6 months	12 months	18 months	24 months
Andrew	25/08/1992 to 24/02/1993	25/02/1993 to 24/08/1993	25/08/1993 to 24/02/1994	25/02/1994 to 24/08/1994
Katrina/Wilma	25/10/2005 to 24/04/2006	25/04/2006 to 24/10/2006	25/10/2006 to 24/04/2007	25/04/2007 to 24/10/2007
Irma	11/09/2017 to 10/03/2018	11/03/2018 to 10/09/2018	11/09/2018 to 10/03/2019	11/03/2019 to 12/07/2019

window selection. We specified the image collections to a three-month period one year prior to each of the hurricanes making landfall. The selection of a larger time window has a higher chance that each mangrove pixel can be analyzed in the study domain compared with a smaller time window, bearing in mind that Landsat has a temporal resolution of 16 days. For Hurricane Andrew, pre-hurricane images were selected from 1990 due to significant cloud cover during this period in 1991. Post-hurricane images were from the same time window immediately after each of the hurricanes made landfall for damage analysis. Note that Hurricanes Katrina and Wilma hit in the same year with a short time gap, these two hurricanes were thus combined for analysis. For the recovery analysis, the post-hurricane time windows were extended to four 6-month periods over 2 years immediately after each major hurricane landfall to analyze the mangrove recovery track. The selected time windows for mangrove recovery analysis are displayed in Table 9.3. The final 6-month time frame for Hurricane Irma was not completed at the time of this chapter; thus, it contains only the most recent 4 months of that period. Again, as both Hurricane Katrina and Wilma occurred in the same year, we decided to base the prior images 1 year before Hurricane Katrina for analysis, while the post images were after Wilma made landfall.

9.3 Methodology

The study was mainly conducted using the GEE, which is a platform for scientific analysis and visualization of geospatial datasets (https://earthengine.google.com/faq/). GEE hosts satellite imagery and stores it in a public data archive that includes historical earth images going back more than 40 years. The images, ingested on a daily basis, are available for global-scale data mining and visualization. GEE has massive computational and mapping capability (Gorelick et al., 2017). The mangrove dataset and Landsat products used in this chapter are both stored in GEE. This opens an opportunity to systemically analyze and map the damage and recovery of mangroves from major hurricanes.

9.3.1 Identifying Mangroves for the Selected Study Domain using GEE

To look at the damage and recovery of mangroves from each hurricane, the coverage of mangroves was delineated first so that the analysis would be constrained to mangrove pixels only. We determined the coverage of mangroves in GEE. We first manually defined a geometry polygon layer to cover the desired locations where the mangroves appeared, ranging from under Golden Gate Parkway in the city of Naples on the west coast of Florida, through southern and northern Everglades National Park, to under Maule Lake in North Miami Beach on the east coast of Florida. We then clipped the global mangrove forest distribution 2000 dataset to the defined polygon layer, leading to a new mangrove layer for the selected study domain. The final derived mangrove forests cover approximately 1634 km^2, as shown in Figure 9.1. This mangrove mask was used to constrain the study to the mangrove forest only, and other land cover types such as water bodies and marshes were not considered.

9.3.2 Damage and Recovery Analysis in GEE

Both damage analysis and recovery analysis were accomplished by comparing the Normalized Difference Vegetation Index (NDVI) values between the pre- and post-hurricane time periods, which would identify the healthiness of the mangroves and any changes that occurred. More specifically, we applied a Mangrove Hurricane Damage Index (MHDI) developed by Zhang et al. (2019) to assess mangrove damages from each hurricane as

$$MHDI = NDVI_{pre-hurricane} - NDVI_{post-hurricane} \tag{9.1}$$

where $NDVI_{pre\text{-}hurricane}$ and $NDVI_{post\text{-}hurricane}$ are NDVI values before and after a hurricane, respectively. Damage analysis can be considered as a remote sensing change detection procedure. Landsat-based vegetation indices have proven effective to characterize the mangrove change in South Florida due to episodic disturbances such as hurricanes. Here, we used the median NDVI in the selected time window to calculate the NDVI change before and after a hurricane. A high value of MHDI represents a high level of damage, while a low value infers a low level of damage from the hurricane. We identified mangrove pixels with a severe damage level by setting a threshold of 0.25 to the MHDI. For the recovery analysis, similarly, the median NDVI was calculated in four 6-month time windows after 6 months, 12 months, 18 months, and 24 months after landfall of Hurricanes Andrew, Wilma, and Irma. To look at the recovery track, the difference between each pre-hurricane and post time window NDVI was calculated.

To effectively calculate the median NDVI in the selected time window, a custom cloud composite algorithm was applied to the images for each of the image collections. An image collection in GEE refers to a set of images. For example, here, all Landsat images for a selected time window comprise an image collection. This algorithm can remove the vast majority of cloud-covered images and pixels and calculate the median values of the pixels in an image collection. Extreme values along with any leftover cloud pixels that would distort the NDVI readings of the mangroves are dropped. As with the mangrove layer, each of the separate image collections was clipped to the same geometry polygon layer. This was accomplished to keep all the layers uniform in the same study area and to assist in limiting the number of pixels displayed from including areas that did not contain mangrove values so as to keep the exported data to a manageable size.

We calculated the NDVI values for each of the 6 hurricane-based image layers (before and after) via $NDVI=(R_{NIR}- R_{red})/(R_{NIR+}R_{red})$, where R_{NIR} and R_{red} refer to the reflectance of near infrared (NIR) and red band, respectively. For Landsat 5 imagery, the NIR is Band 4 and red wavelength was Band 3, while for Landsat 8 imagery it is Band 5 for the NIR and Band 4 for the red wavelength. Six specific NDVI layers were determined to represent the healthy status of mangroves before and immediately after each hurricane. The same procedure was repeated for the twelve 6-, 12-, 18-, and 24-month recovery images for each of the hurricanes after landfall.

In total, 18 NDVI-based layers along with the mangrove layer were created in GEE and then exported to Google Drive, after which they were imported into ArcMap software for further analysis and mapping. MHDI was calculated in ArcGIS. The NDVI and MHDI layers were also re-projected from the WGS 1984 Geographic Coordinate System into NAD 1983 (2011) Florida GDL Albers (Meters) Projected Coordinate System so as to change the units of measurement from degrees to meters and to keep the accuracy of the shape and distance of the study area.

9.4 Results and Discussion

9.4.1 Mangrove Damage Analysis

The maps of pre-, immediate post-Hurricane Andrew NDVI values, along with the MHDI for Hurricane Andrew are displayed in Figure 9.2. Figure 9.2(a) and (b) and show a clear difference in the healthiness of the mangrove vegetation. The pre-Andrew mangrove forests were healthier, and most regions had a NDVI value more than 0.4. After Hurricane Andrew, a clear drop in mangrove greenness was observed and the NDVI values were significantly reduced. Figure 9.2(c) shows the damage pattern with red representing severe damages, orange and yellow showing moderate damages, and green and blue indicating minimal damages. Further statistical analysis showed that after Hurricane Andrew passed through, almost 260 km^2 of mangroves were severely damaged with a MHDI equal to and over 0.25.

The effects of Hurricane Katrina and Wilma on mangroves are shown in Figure 9.3. Mangroves had a higher NDVI before Hurricane Katrina and Wilma. Mangroves were even healthier before Katrina than before Hurricane Andrew with a higher NDVI observed (Figures 9.2(a) and 9.3 (a)). Figure 9.3(b) shows a noticeable drop in NDVI after Hurricane Wilma. Figure 9.3(c) further demonstrates where exactly the mangroves were hit the hardest after Hurricanes Katrina and Wilma. Roughly 216 km^2 of mangroves were severely damaged with a MHDI of over 0.25. The severe damage areas were mainly observed along the western and southern regions of the study area. In the case of Hurricane Irma, the pre-hurricane image Figure 9.4(a) shows that the vast

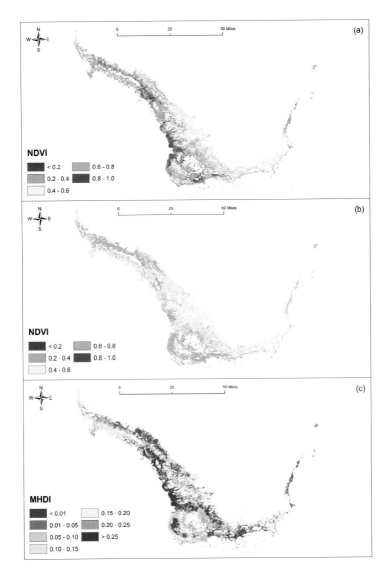

FIGURE 9.2
Maps of NDVI of (a) pre-Andrew, (b) immediate post-Andrew, and (c) MHDI to reveal the effects of
Hurricane Andrew on mangroves.

majority of the mangroves were in good to excellent condition with NDVI values over
0.6. However, as Figure 9.4(b) shows, these values have been changed dramatically to
mostly between 0.2 to 1.0 in NDVI along the western and southern coast. Figure 9.4(c)
reveals the damage pattern from Hurricane Irma. In total, 297 km^2 were severely
damaged with the MHDI over 0.25.

Similarity and difference were both observed among these mangrove damage maps
from hurricanes. There was a high degree of spatial variation in mangrove damage
across the study domain from each hurricane. Larger damages appeared at the fringe,
and the degree of damage decreased as the distance from the shoreline increased. This

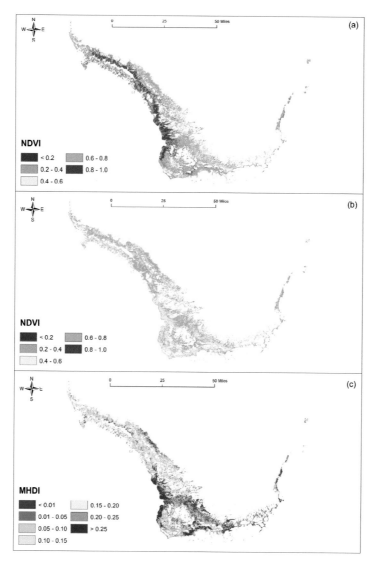

FIGURE 9.3
Maps of NDVI of (a) pre-Katrina/Wilma, (b) immediate post-Wilma, and (c) MHDI reveal the effects of hurricanes Katrina and Wilma on mangroves in 2005.

general damage pattern was expected. It has been reported that amplitudes of storm surge at the front of a mangrove zone increase by 10-30% because of the "blockage" of mangroves to surge water, which can cause greater impacts on structures at the front of mangroves than wetlands behind (Zhang et al., 2012). As a result, the general spatial pattern of damages was expected due to the protection function of the front mangroves and attenuated storm surge over the back mangroves (Zhang et al., 2019). The severe damage areas were also closely related to each hurricane track. It has been recognized that most tropical storms cause the greatest damage severity on the right and front side of the eye track, where wind and wave stress can be up to 25% greater than on the left

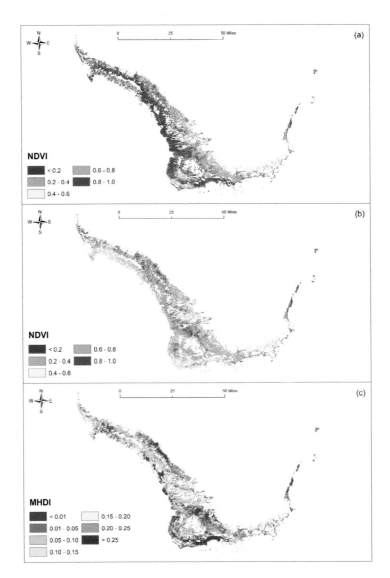

FIGURE 9.4
Maps of NDVI of (a) pre-Irma, (b) immediate post-Irma, and (c) MHDI reveal the effects of Hurricane Irma on mangroves in 2017.

side (Chen et al., 2013; Long et al., 2016). The severe damage areas observed from each hurricane were consistent with findings in the literature relating damage to aspects of the hurricane track. The degree of damage generally deceased with increasing distance from the eye path. However, a nonlinear spatial pattern of damages was also observed. For example, for Hurricane Irma, the Flamingo region also had severe damage while this area is far away from the track. This suggests that the degree of damage is related to factors besides the hurricane track. Table 9.4 compares the severe damage caused by each of the hurricanes, with Hurricane Irma causing the most immediate destruction to the mangroves and Hurricanes Katrina/Wilma the least damage.

TABLE 9.4

Severe damage mangroves from each hurricane

Hurricane	Severe damage to mangroves (km²)
Andrew	259
Katrina/Wilma	216
Irma	297

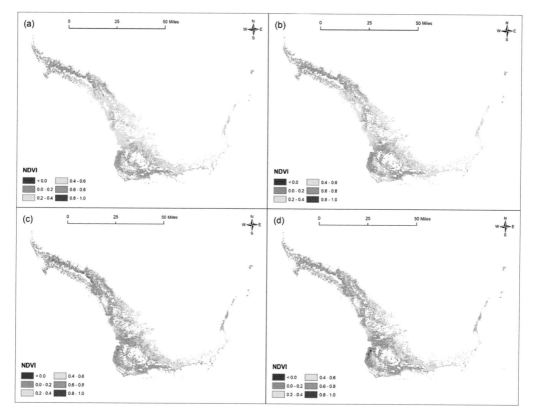

FIGURE 9.5
Maps of NDVI (a) 1 to 6 months, (b) 7 to 12 months, (c) 13 to 18 months, and (d) 19 to 24 months after Hurricane Andrew.

9.4.2 Mangrove Recovery Analysis

For the 6-month recovery time windows post Hurricane Andrew, Figure 9.5(a), (b), (c), and (d) show the 6-, 12-, 18-, and 24-month NDVI periods, respectively, after the hurricane made landfall. After 24 months, the NDVI seemed to be much recovered but not fully in comparison to after the initial 6-month period. Figures labeled (a) and (c) are relatively in the same time period as the pre-hurricane imagery, and as such are better indicators of mangrove damage and recovery due to the climate being mostly in the wet season where the temperature, humidity, and rainfall are all relatively high. Figures 9.5

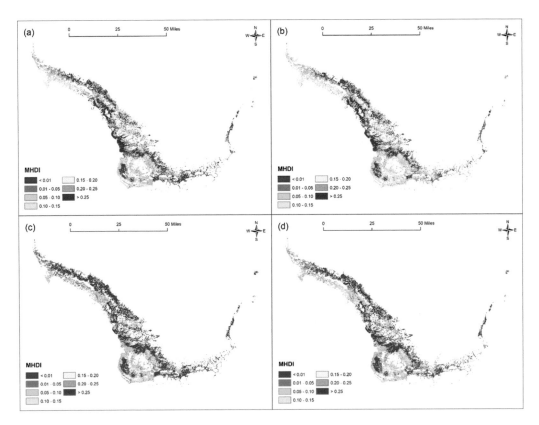

FIGURE 9.6
Maps of MHDI (a) 1 to 6 months, (b) 7 to 12 months, (c) 13 to 18 months, and (d) 19 to 24 months after Hurricane Andrew.

(b) and (d) generally show the climate opposite that of the pre-hurricane imagery, which is largely during the dry season, which experiences lower temperatures and less humidity and rainfall. All of these factors may impact the NDVI to reveal the recovery of mangroves from a hurricane, but changes of NDVI can indicate the recovery to some degree. These same factors also impact the NDVI and MHDI for recovery analyses from Hurricanes Katrina/Wilma, and Irma. Figure 9.6(a), (b), (c), and (d) show the 6-, 12-, 18-, and 24-month MHDI periods post Hurricane Andrew and show a clear reduction over time in severely damaged mangroves toward less-damaged MHDI values (under 0.25); also, more areas have seemed to have regained NDVI values over pre-hurricane levels as indicated in the MHDI values below 0.01.

For Hurricane Katrina/Wilma, Figure 9.7(a), (b), (c), and (d) show the same four 6-month NDVI periods, with clear improvement in the mangrove NDVI between the first 6 months and the final 6 months of the 2-year period. Figure 9.8(a), (b), (c), and (d) go more in-depth, showing the MHDI values between the post-hurricane images and the pre-hurricane Katrina/Wilma image. The severely damaged mangroves had largely dispersed into less-damaged values after only 1 and a half years post landfall, which indicates that while Hurricane Wilma did have a strong initial impact, its long-term impact is not as severe as Hurricane Andrew's. Lastly, with Hurricane Irma, Figure 9.9

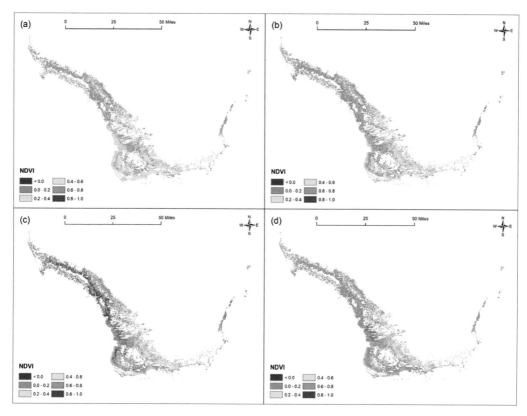

FIGURE 9.7
Maps of NDVI (a) 1 to 6 months, (b) 7 to 12 months, (c) 13 to 18 months, and (d) 19 to 24 months after
Hurricanes Katrina/Wilma.

(a), (b), (c), and (d) show the NDVI change over each of the separate 6-month periods,
with the final image showing the median NDVI values of 4 months instead. Like
Katrina/Wilma before it, there are definitive improvements in the mangrove NDVI
values over the course of a year; however, many more areas contain poor to average
NDVI values near the coast. Figure 9.10(a), (b), (c), and (d) help showcase where exactly
the MHDI calculated the greatest changes to the mangrove NDVI values, with a clear
pattern of severe mangrove damage seen along the western and southern portions of the
Everglades where Hurricane Irma would have had the greatest impact. Much of the
interior has seemed to recover to pre-hurricane levels, with MHDI being below 0.01, or
has begun to recover; the coastal mangrove forests have had difficulty recovering.

9.4.3 Discussion

Both internal and external factors attribute the varying degree of mangrove damage
patterns from tropical cyclones. Internal factors are related to physical properties of the
ecosystem, while external factors are related to damage drivers, in this case, hurricanes.
Zhang et al. (2019) identified three internal metrics and two external metrics to characterize mangrove damage patterns from hurricanes in South Florida. Internal metrics

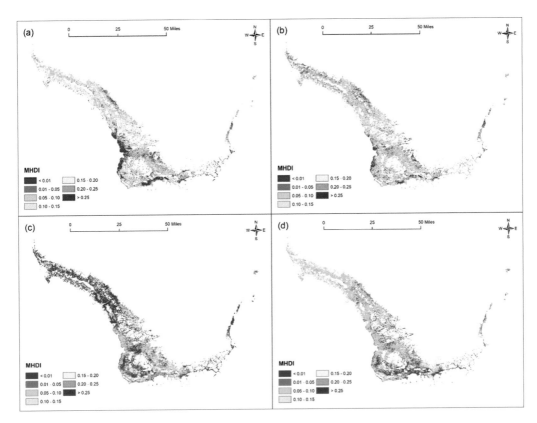

FIGURE 9.8
Maps of MHDI (a) 1 to 6 months, (b) 7 to 12 months, (c) 13 to 18 months, and (d) 19 to 24 months after Hurricane Katrina/Wilma.

include vegetation indices, canopy height, and geographical setting/locations of mangroves. Two external contributors include hurricane track and storm surge inundation. Vegetation indices derived from optical sensors are valuable variables for estimating the state of plant health (Jensen, 2015). Unhealthy vegetation leads to the breakdown of important stabilizing feedbacks necessary for wetland response to stressors, such as hurricanes. Thus, the physical health status of mangroves indicated from vegetation indices is an important factor for assessing their damage levels. Mangrove canopy height is also related to the vulnerability and resilience of mangroves (Smith et al., 1994; Zhang et al., 2016). Tall mangroves are more vulnerable to hurricanes. Fringe mangroves are also more vulnerable, which can be indicated from their distance to the open ocean. External factors such as hurricane track and storm surge inundation have proven important in characterizing mangrove damage from hurricanes. It has been demonstrated that the extent of mangrove damage exponentially decreases with increasing distance from the storm path (Doyle et al., 2003), indicating that hurricane path is important. Mangrove hurricane damage is believed to be mainly from strong wind and inundation caused by storm surge (Zhang et al., 2012), which are closely related to the intensity of the hurricanes. We have developed a machine learning-based mangrove risk model to project the mangrove damage pattern by connecting the satellite-observed

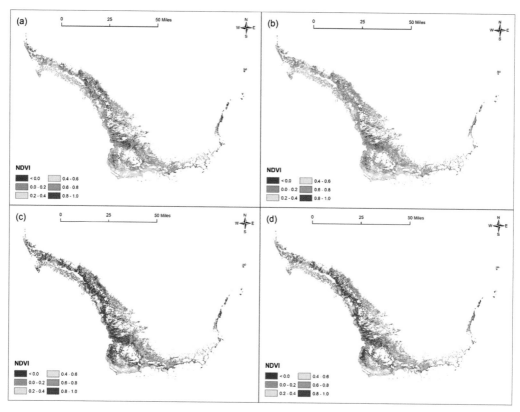

FIGURE 9.9
Maps of NDVI (a) 1 to 6 months, (b) 7 to 12 months, (c) 13 to 18 months, and (d) 19 to 22 months after Hurricane Irma.

damage data with the identified external and internal metrics of mangrove vulnerability to hurricanes (Zhang et al., 2019). The model can be further calibrated and validated by the damage analysis results presented in this chapter.

Wang (2012) found that mangrove recovery rate was fast in the first year after the hurricanes had passed and decreased considerably in the years after. This was confirmed in our study because the severely damaged area was largely reduced in the first year. It was also reported that recovery was related to mangrove species types. Based on Ward and Smith (2007), of the three mangrove species located in South Florida during Hurricane Wilma, the white mangroves appeared to increase the fastest in the most damaged areas, and black mangroves improved the most in lightly damaged areas. While red mangrove damage was linked with the hurricane damage, its recovery appeared to be independent of it. Radabaugh et al. (2019) noted in their study of Hurricane Irma that the mangrove canopy cover in Ten Thousand Islands and in the Lower Keys recovered from 40% to 60% within 4 months of the hurricane; after that, it seemed to increase very little. This suggests that the mangrove canopies need additional years for the new branches and leaves to fully develop and properly cover the holes in the mangrove canopies. Additionally, they found that while red mangroves had minimal leaf growth, the black and white mangroves were more successful in growing new

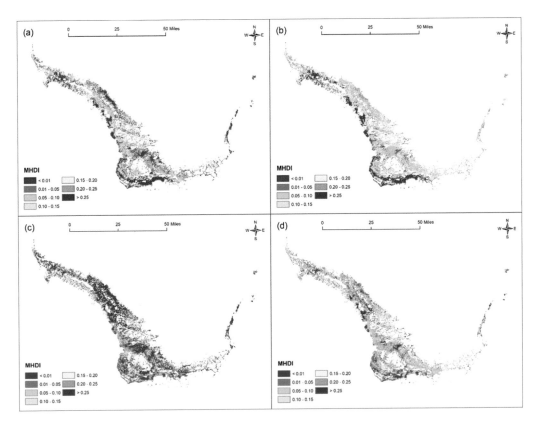

FIGURE 9.10
Maps of MHDI (a) 1 to 6 months, (b) 7 to 12 months, (c) 13 to 18 months, and (d) 19 to 22 months (d) after Hurricane Irma.

leaves, which would indicate that the red mangrove areas did not recover as quickly as those of the other two mangrove species. Parenti (2015) stated that the recovery of the mangrove communities can also be influenced by the quantity of mangrove seedlings that survived the hurricane and were able to successfully sprout, which could be related to how strong the storm surge was and where it might have been able to wash the seeds out into the ocean or perhaps even further inland. It is worth looking at the species-specific response of recovery by connecting the mangrove species map with our recovery maps. There is also a potential to develop a mangrove recovery model by linking the physical settings of mangroves and damage levels with environmental variables to assist with mangrove resilience analysis.

The Landsat data are valuable for large-scale damage and recovery analyses compared with the aerial data and field surveys. The same satellite specifications used between Landsat 5 and 8 are beneficial and lead to trouble-free comparisons, with virtually the same code in GEE. Additionally, the 16-day intervals and the satellite imagery being available as soon as possible meant that data could always be added and further analyzed. There are several limitations to using Landsat data for damage and recovery analysis, such as cloud cover issues and relatively low temporal resolution. A fusion of Landsat with other satellites such as Sentinel-2 and MODIS may mitigate this issue. The damage and recovery analysis results were not validated by any field data or other data sources for Hurricanes Andrew and Wilma. But validations from fine spatial resolution aerial photography, light detection and ranging (lidar) data, and field photos have confirmed that Landsat is effective for large-scale mangrove damage analysis after Hurricane Irma (Zhang et al., 2019).

9.5 Summary and Conclusions

In all instances, it was evident that mangrove communities largely decreased because of the passing of major hurricanes, and fringe mangroves were often observed to sustain more severe damage than mangroves offshore. The best recovery rates appear to occur further inland where the wind and wave influence of the hurricanes was less impactful. Google Earth Engine (GEE) proved to be a powerful tool for this study, as it easily processes the large volume of stored satellite data and has the flexibility of using the JavaScript language in the code editor to create parameters and functions that better suited the data collection and analysis. It is good for the purpose of change detection and large-scale analysis, as can be seen in the National Aeronautics and Space Administration (NASA) long-term vegetation change analysis of the Everglades from 1995 to 2015 (Kirk et al., 2016). The overall NDVI of the mangroves for each of the 3 years in this chapter (1992, 2005, and 2016) seems to have also increased, which could be due to a rise in sea level and salinization of former freshwater areas, which improved and expanded the coverage of mangroves by 172 km^2 between 1985 and 2017 within Everglades National Park (Han et al., 2018). As Fuller (2009) states, there was a 24 cm-rise in relative sea level in southwest Florida over a period of 75 years, which would encourage salinization of more inland areas of South Florida. This could greatly benefit the mangroves that previously lacked access to higher concentrations of salt deeper inland. However, improvements can be made to improve the efficiency of the study and help improve the clarity of the data. For better understanding about the NDVI influence the hurricanes had on the mangroves in the region, the data could include mangroves that are found on all of the islands in the southern Everglades, the upper Florida Keys, and the islands within Biscayne National Park. The study area could also differ between the paths of the hurricanes (e.g., Hurricane Irma would have had limited influence on the mangroves located on the east coast, while Hurricane Andrew certainly had much stronger influence as it passed right through them). Factors outside of the hurricane impacts, such as wildfire, controlled burn, disease, and other natural and human-based disturbances, were not taken into account. In sum, the study shows that Landsat is useful for large-scale mangrove damage and recovery analyses from hurricanes and that GEE is a powerful tool for enhancing the capability of Landsat products for such applications regionally and globally.

References

Beever, L.B., Beever, III, J.W., Lewis, III, R.R., et al., 2016. Identifying and diagnosing locations of ongoing and future Saltwater Wetland loss: Mangrove heart attack. Charlotte Harbor National Estuary Program, Punta Gorda, FL.

Cahoon, D.R., Hensel, P.R., Rybczyk, J., et al., 2003. Mass tree mortality leads to mangrove peat collapse at Bay Islands, Honduras after Hurricane Mitch. *Journal of Ecology*, 91, 1093–1105.

Chen, S.S., Zhao, W., Donelan, M.A., and Tolman, H.L., 2013. Directional wind–Wave coupling in fully coupled atmosphere–Wave–Ocean models: Results from CBLAST-Hurricane. *Journal of the Atmospheric Sciences*, 70, 3198–3215.

Doyle, T.W., Girod, G.F., and Books, M.A., 2003. Modeling mangrove forest migration along the southwest coast of Florida under climate change. In: Ning, Z.H., Turner, R.E., Doyle, T.W., and Abdollahi, K. (eds.) *Integrated Assessment of the Climate Change Impacts on the Gulf Coast Region*, GCRCC, Baton Rouge, LA, 211–221.

Fuller, D.O., 2009. Mapping cyclone damage to Mangrove habitats: An example from South Florida. Geospatial World, https://www.geospatialworld.net/.

Giri, C., 2016. Observation and monitoring of mangrove forests using remote sensing: Opportunities and challenges. *Remote Sensing*, 8, 1–8.

Giri, C., Ochieng, E., Tieszen, L.L., et al., 2005. *Global Mangrove Forests Distribution, 2000*, NASA Socioeconomic Data and Applications Center (SEDAC), Palisades, NY. DOI:10.7927/H4J67DW8. Accessed on 16 July 2019.

Giri, C., Ochieng, E., Tieszen, L.L., et al., 2011. Status and distribution of mangrove forests of the world using earth observation satellite data. *Global Ecology and Biogeography*, 20, 154–159.

Gorelick, N., Hancher, M., Dixon, M. et al., 2017. Google earth engine: Planetary-scale geospatial analysis for everyone. *Remote Sensing of Environment*, 202, 18–27.

Han, X., Feng, L., Hu, C., et al., 2018. Hurricane-induced changes in the everglades national park mangrove forest: Landsat observations between 1985 and 2017. *Journal of Geophysical Research: Biogeosciences*, 123, 3470–3488.

Jensen, J.R., 2015. *Introductory Digital Image Processing: A Remote Sensing Perspective*, 4th Edition, Prentice Hall, Upper Saddle River, NJ.

Kirk, D., Wolfe, A., Ba, A., et al., 2016. Everglades ecological forecasting II: Utilizing NASA earth observations to enhance the capabilities of Everglades National Park to monitor & predict Mangrove extent to aid current restoration efforts. Develop Technical Report, NASA Langley Research Center, Hampton, Virginia, 1–34.

Long, J., Giri, C., Primavera, J., and Trivedi, M., 2016. Damage and recovery assessment of the Philippines' mangroves following Super Typhoon Haiyan. *Marine Pollution Bulletin*, 109, 734–743.

Lucas, K.J., Watkins, A., Phillips, N., et al., 2019. The impact of hurricane Irma on population density of the black salt-marsh mosquito, aedes taeniorhynchus, in Collier County, Florida. *Journal of the American Mosquito Control Association*, 35, 71–74.

Marois, D.E., and Mitsch, W.J., 2015. Coastal protection from tsunamis and cyclones provided by mangrove wetlands – A review. *International Journal of Biodiversity Science, Ecosystem Services & Management*, 11, 71–83.

NOAA/FAQ, 2019. https://aoml.noaa.gov/hrd/tcfaq/E19.html, Accessed on 17 July 2019.

Parenti, M.S., 2015. Hurricane effects on mangrove canopies observed from MODIS and SPOT imagery. Strategy & Analytics, Saatchi & Saatchi Wellness.

Radabaugh, K.R., Moyer, R.P., Chappel, A.R., et al., 2019. Mangrove damage, delayed mortality, and early recovery following Hurricane Irma at two landfall sites in southwest Florida, USA. *Estuaries and Coasts*. DOI:10.1007/s12237-019-00564-8.

Sippo, J.Z., Lovelock, C.E., Santos, I.R., et al., 2018. Mangrove mortality in a changing climate: A review. *Estuarine, Coastal and Shelf Science*, 215, 241–249.

Smith, III, T.J., Anderson, G.H., Balentine, K., et al., 2009. Cumulative impacts of hurricanes on Florida mangrove ecosystems: Sediment deposition, storm surges and vegetation. *Wetlands*, 29, 24–34.

Smith, III, T.J., Robblee, M.B., Wanless, H.R., and Doyle, T.W., 1994. Mangroves, hurricanes, and lightning strikes. *BioScience*, 44, 256–262.

Wang, Y., 2012. *Detecting Vegetation Recovery Patterns after Hurricanes in South Florida using NDVI Time Series*. Open Access Thesis, 355.

Ward, G.A., and Smith, III, T.J., 2007. Predicting mangrove forest recovery on the southwest coast of Florida following the impact of Hurricane Wilma, October 2005. *Circular*, 1306, 175–182.

Whelan, K.R.T., Smith, T.J., Anderson, G.H., and Ouellette, M.L., 2009. Hurricane Wilma's impact on overall soil elevation and zones within the soil profile in a mangrove forest. *Wetlands*, 29, 16–23.

Zhang, C., Durgan, S., and Lagomasino, D., 2019. Modeling risk of mangroves to tropical cyclones: A case study of hurricane Irma. *Estuarine, Coastal, and Shelf Science*, 224, 108–116.

Zhang, K., Liu, H., Li, Y., et al., 2012. The role of mangroves in attenuating storm surges. *Estuarine, Coastal and Shelf Science*, 11-23, 102–103.

Zhang, K., Thapa, B., Ross, M., and Gann, D., 2016. Remote sensing of seasonal changes and disturbances in mangrove forest: A case study from South Florida. *Ecosphere*, 7, e01366.

Part III

Hyperspectral Remote Sensing Applications

10

Applying Point Spectroscopy Data to Assess the Effects of Salinity and Sea Level Rise on Canopy Water Content of Juncus roemerianus

Donna Selch, Cara J. Abbott and Caiyun Zhang

10.1 Introduction

Florida coasts have been threatened by the ongoing sea level rise (SLR). In 2015, Southeast Florida Regional Climate Change Compact estimated SLR of 6 to 10 inches by 2030. Predictions for the mid-term are between 11 and 22 inches of additional sea level rise by 2060 and longer term between 28 and 57 inches by 2100 (Compact, 2015). Sea level rise will modify the coastal hydrology such as tidal flushing and saltwater intrusion into coastal qualifiers, which in turn negatively affects productivity and community composition of previously freshwater systems (Cormier et al., 2013). An increase of 60 cm (~24 inches) will greatly impair the freshwater sources of South Florida, restructuring plant communities organized along the slight gradient in elevation within Florida's low-lying coasts (Saha et al., 2011; Guha and Panday, 2012). As sea levels continue to rise, many coastal ecosystems may experience inundation. This phenomenon will only be exacerbated in Florida as its low elevation habitats and extensive wetland networks are likely to experience an increase in intense storms due to global climate change (Martin et al., 2011). The projected damage of sea level rise presents a great challenge for conservation and management in Florida, especially in coastal regions.

One critical effect of sea level rise is the increase of salinity, which will impact plant health. Increased salinity can decrease the growth and net photosynthesis in plants by decreasing the canopy water content (CWC) (Long and Baker, 1986). A decrease in CWC restricts transpiration, induces closing of the stomata, and increases leaf temperature (Jackson, 1986; Ceccato et al., 2001). The amount of water a plant acquires also may cause stress: too much water, and the plant can actually drown; too little water, and the plant enters a drought cycle and may also die. Salinity and sea level rise are two major factors affecting the CWC of coastal plants. Rapid changes of salinity can reduce the spread and total numbers of plant in the water and on land; most species can only survive within a specific window of salinity.

Assessing plant health via CWC is important in the South Florida ecosystem because little can be done to reverse the results once the process causing water stress injury has transpired (Tucker, 1980). The best remedy is the prevention of the stress itself. For prevention purposes, the first step is to know where the stress is occurring. Field observation or measurement of plant stress is time consuming and labor intensive. In addition, field visits are restricted only to accessible areas. In contrast, remote sensing is a method that allows for rapid assessment of a large area, including remote sites, often at a free or reduced cost to other traditional methods of gauging CWC to evaluate vegetation stress.

In this study, we applied the spectroscopic data collected from an ASD spectroradiometer to quantify CWC of *J. roemerianus* under different treatments of water levels and salinity at a controlled environment. Acquiring field/laboratory hyperspectral data is more convenient and economical compared with the application of airborne hyperspectral sensors. Spaceborne hyperspectral sensors (e.g., EO-1/Hyperion) meet with heavy cloud coverage in Florida, which largely reduces their applications for vegetation stress studies. Point spectroscopic techniques rapidly collect hyperspectral data in the field. This differs from hyperspectral imagery in which typically numerous pixels are collected over a larger area, generating an image cube. Point spectroscopic data can also be collected in laboratory conditions where the type and amount of stress placed upon plant species can be controlled. The ASD field spectroradiometer is a portable device able to measure plants individually and supply observations and data that may be compiled on a singular basis so that a variety of stressors may be studied frequently and regularly on a particular plant in the lab.

To quantify CWC for vegetation stress analysis using remote sensing, multiple spectral indices have been developed, among which the Normalized Difference Vegetation Index (NDVI) developed by Rouse et al. (1973) is the most popular index. NDVI is calculated using the red and near-infrared (NIR) bands carried on most earth observation sensors; it has been frequently used to estimate leaf water content (Tucker, 1979). Another popular index is the Normalized Difference Water Index (NDWI), which determines water content in vegetation using the NIR and short-wave infrared (SWIR) bands located near two water absorption features (Gao, 1996). Spaceborne sensors such as Landsat and MODIS include SWIR bands to calculate such spectral indices (Jackson et al., 2004). A Water Band Index (WBI) was also developed using two features at bands 970 nm and 900 nm (Peñuelas et al., 1993). WBI has proven sensitive to changes in canopy water stress. Gamon et al. (1999) used the inverse of WBI and confirmed the findings of WBI for CWC characterization. Their research also indicates a strong relationship between WBI and NDVI. Note that the NDWI manipulates the similar (but slightly different) liquid water absorption properties of reflectance at 857 nm and 1241 nm. The scattering of light by vegetation canopies enhances the weak liquid water absorption at 1241 nm (Gao, 1996). Cole et al. (2014) used wavelengths 860 nm and 1240 nm for their analysis with NDWI. Wavelength peaks may differ due to other influences including atmospheric conditions. Other water indices were also developed. The Normalized Difference Infrared Index (NDII) is sensitive to changes in water content of plant canopies, and the index value results increase with increasing water content. The Moisture Stress Index (MSI) is sensitive to leaf water content; the strength of the absorption around 1599 nm increases with increasing water content, absorption at 819 nm is nearly unaffected. Uto and Kosugi (2012) used the MSI to estimate water content in vegetation for water stress detection. The Normalized Multi-Band Drought Index (NMBDI) uses three bands to monitor potential drought conditions. The difference between two water absorption features (bands 1640 nm and 2130 nm) in the SWIR region is used to measure water sensitivity in vegetation. This is combined with a NIR band to enhance sensitivity to drought severity. Wang and Qu (2007) developed this index to estimate both soil and vegetation moisture. The existing spectral indices for plant water content characterization in the literature are summarized in Table 10.1. Most vegetation indices were calculated using broadband multispectral sensors, rather than narrowband spectral sensors. The narrowband based spectral indices have proven more effective in CWC characterization. A comparison of applying broadband multispectral Landsat TM sensor and the ASD point spectroscopic data to display a healthy

TABLE 10.1

Spectral indices in the literature for plant water content quantification

Spectral Indices		Sensitive to	Meaning of Higher Values	References
Water Band Index (WBI a)	$WBI = R900/R970$	changes in canopy water status	decreased water content of vegetation canopies	Peñuelas et al., 1993
Water Band Index (WBI b)	$WBI = R970/R900$	changes in canopy water status	increased water content of vegetation canopies	Gamon et al., 1999
Normalized Difference Water Index (NDWI a)	$NDWI = (R860\text{-}R1240)/(R860\text{+}R1240)$	changes in vegetation canopy water content	healthier vegetation	Cole et al., 2014
Normalized Difference Water Index (NDWI b)	$NDWI = (R857\text{-}R1241)/(R857\text{+}R1241)$	changes in vegetation canopy water content	healthier vegetation	Gao, 1996
Normalized Difference Infrared Index (NDII)	$NDII = (R819\text{-}R1649)/(R819\text{+}R1649)$	changes in water content of plant canopies	increasing water content	Hunt et al., 1987; González-Fernández et al., 2014
Moisture Stress Index (MSI)	$MSI = R1599/R819$	increasing leaf water content	greater water stress and less water content	Uto and Kosugi, 2012
Normalized Multi-band Drought Index (NMBDI)	$NMDI = [R860\text{-}(R1640\text{-}R2130]/[R860\text{+}(R1640\text{-}R2130]$	dryer soil	decreased moisture	Wang and Qu, 2007

Juncus (J.) roemerianus plant is shown in Figure 10.1. The broadband Landsat TM has 7 bands (six of which are shown in Figure 10.1) with a spectral resolution bandwidth ranging from 60 nm (red) to 270 nm (SWIR 2). The spectroscopic data in Figure 10.1 has a 1 nm resolution and ranges from 350 nm to 2500 nm.

Derivative analysis (DA) is another promising method in CWC analysis. DA is an established technique in remote sensing for elimination of background signals and resolving overlapping spectral features. It was proposed for tackling analogous problems such as interference from soil background reflectance in the remote sensing of vegetation or for resolving complex spectra of several target species within individual pixels in remote sensing (Demetriades-Shah et al., 1990). Spectral derivatives at the slope of the red edge region (680–800 nm) can be used for chlorophyll estimation (Clevers et al., 2008). Zarco-Tejada et al. (2002) suggested a double peak in the derivative reflectance indicates stress conditions caused by low chlorophyll content and chlorophyll fluorescence. Derivative analysis of the red edge region has been used to describe changes due to stress. Rock et al. (1988) used this methodology to detect stress in spruce trees due to air pollution before visible symptoms were identified. Derivatives applied to the 970 nm and 1200 nm water absorption features were found to correlate higher to CWC than spectral indices (Clevers et al., 2008). The amplitude of the derivative is highly dependent upon the shape and height of the reflectance peak. For example, healthy plants show accentuated areas in the green and NIR regions due to elevated chlorophyll levels. In this study, we tested the first and second derivatives in CWC estimation.

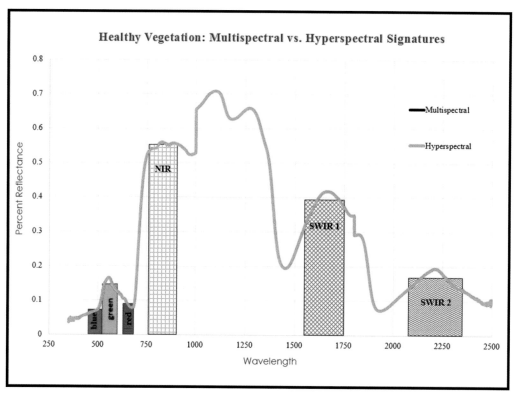

FIGURE 10.1
Spectral response of a healthy Juncus roemerianus plant measured by an ASD spectroradiometer with 1 nm in spectral resolution and the broadband Landsat TM.

10.2 Study Site and Data Collection

This study was conducted at the Florida Atlantic University (FAU) Boca Raton Greenhouse. We established various treatments on *J. roemerianus* in an enclosed greenhouse rather than in the natural environment because the greenhouse offers a controlled environment conducive to testing specific variables in isolation while maintaining a natural photoperiod (Lytle and Lytle, 1998). *J. roemerianus* is a dominant coastal salt marsh plant; it expands along the Atlantic Ocean and Gulf of Mexico coasts of the United States and thrives with completely emerged shoots. The special growing characteristic of *J. roemerianus* is its ability to grow in both freshwater and saltwater habitats (Touchette et al., 2012). In fresh water, *J. roemerianus* grows to roughly 2 meters in height and has a deep green color in the shoot. In saltwater levels of 5–20 parts per thousand (ppt), the plant will not exceed roughly 1 meter in height at maturity and often exhibits a grayish coloration throughout the shoot (Christian et al., 1990). This gray coloration notably gave rise to the plant's common name, the black needlerush. These differing physical traits in varying levels of salinity have made this plant a potential saltwater intrusion indicator species (Abbott, 2015). The species is ecologically important because it provides substantial primary production, helps to stabilize the shoreline as it buffers impacts of coastal storms, and serves as a wildlife habitat for many fauna species, including nursery grounds for fisheries (Wang et al., 2005). *J. roemerianus* was therefore selected in this study due to (1) its unique growing characteristics in freshwater and saltwater habitats, (2) its being native to Florida, (3) the fact that it is an economical and robust plant used consistently in restoration projects throughout Florida, (4) the fact that the exact tolerance ranges of *J. roemerianus* are currently unknown and highly disputed (Touchette et al., 2012; Sparks et al., 2013), and (5) its spectral response to stress remaining unclear.

Assessing *J. roemerianus* responses to stressors such as salinity and water levels will aid in the monitoring and mapping of a changing coastal environment. In order to examine the tolerance ranges of *J. roemerianus* to salinity and determine how shoot height changes with increasing salinity, 4 main treatments were established: 0 ppt, 20 ppt, 30 ppt, and 40 ppt levels. These salinity levels were the whole plot factor within our split plot design. The 0 ppt treatment level represents freshwater habitats where it is well known *J. roemerianus* thrives. The 20 ppt treatment represents the upper end of brackish-water habitats found within the historic home range of *J. roemerianus* (Brinson and Christian, 1999). The 30 ppt treatment represents the salinity level of sea water under sea level rise scenarios, and the 40 ppt treatment represents the salinity level of hypersaline habitats that are possible with accelerated evaporation with climate change. It is currently not well known whether *J. roemerianus* can tolerate and survive in either of these salinity levels for a prolonged period of time (Touchette et al., 2012). To simulate the effects of water level on plant health, half of the plants were raised approximately 10 cm (~4 inches) more than the others. This could simulate current levels of the seas, with the lower pots representing potential water-inundated plants. The design of these experiments in bins and pots is displayed in Figure 10.2. Figure 10.3 shows 1 example of 5 pots in a bin with a filtration system. In total, 1600 mature, living plants were planted into 160 pots that were then topped off with more potting soil so that each pot was just over ¾ full. A thin layer of small rock was added on top of the soil to prevent soil loss into the bins but still allow recruitment of new shoots if necessary. Once planting was completed, each pot was randomly assigned to one of the 32 whole plots spread throughout the greenhouse on 8 plywood sheets. Due to the higher temperatures and humidity levels in greenhouses in June through August 2014 in Florida, this study began in late September 2014. It was carried out through February 2015, encompassing 23 weeks for set up, equilibration, and treatment. All plants in all treatment groups were grown along the west side of the greenhouse to ensure all plants received the same degree of light throughout the day.

Data was collected from both the plants themselves and the surrounding water. Every week each shoot height was measured; shoot density from each pot was then calculated as shown in Figure 10.4. Plants were measured from the point of soil emersion to the top tip of the plants. In addition, water quality data of pH, salinity, redox, and total dissolved solids were also collected each week using probes. Every two weeks, hyperspectral data was collected using an ASD spectrometer for every pot. On the last day of the experiment, all plants were removed from their environments and aboveground biomass, CWC, and longest root length were measured. For the spectroscopic data collection, the ASD spectroradiometer was set up using a fiber optic cable with a 25° field of view (Figure 10.5). Measurement height above the pots was approximately 0.5 m; the resulting field of view at the plot level was circular with a radius of about 3.5 mm. Five measurements per pot were performed and averaged for further analysis.

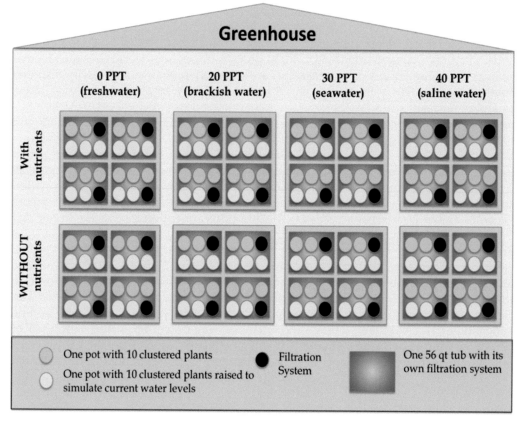

FIGURE 10.2

Example of the breakdown of pots and bins and the way the plants were treated in the study site. Note that each filtration system keeps a steady flow of water moving at a small flow rate similar to that of the Everglades in South Florida. Plants were either elevated (in ~3 cm of water) or inundated (~13 cm of water) at varying salinity levels (ppt).

Calibration was done using a Spectralon white reference panel. At the conclusion of the study, plants were sacrificed for final measurements and analysis. After cutting, every viable shoot was measured along a yardstick for the final height measurements. For CWC measurements, we first separated above- and belowground mass. All of the aboveground biomass (both dead and alive) was then cut, measured, and weighed to determine the fresh weight (FW), sometimes identified as wet weight. Next, the labeled bags containing the aboveground biomass were dried for 48 hours at 150°F and then weighed to get the dry weight (DW). To calculate the estimated CWC, a simple subtraction was performed using the formula: $CWC=FW-DW$ (Clevers et al., 2010). This followed the traditional lab procedure to determine the CWC of plants. The complex belowground matter typical of *J. roemerianus* meant that some roots would be left behind in the soil and that soil would cling to the roots; both factors would prevent accurate root biomass measurements. Thus, we did not determine belowground biomass in this study.

FIGURE 10.3
Five pots were placed in each bin with half the plants elevated from the water to represent current sea levels and the rest inundated to represent future sea levels. Each bin was replicated 4 times. Water cycled through a closed system. The environment for the bin shown was fresh water (0 ppt). Image was taken on February 26, 2015, at the conclusion of the experiment.

10.3 Methodology

To determine whether salinity, inundation, or a combination of the two had a significant physical effect on the plants, averaged data within each pot were analyzed by split plot design using a linear model. We first analyzed the data for weeks 0, 5, 10, 15, and 20 to determine differences among treatments throughout time. We then examined the relationship between plant stress and plant spectral response with an aim to developing a spectral model to predict CWC. We applied spectral indices and derivative analysis for CWC estimation.

Spectral indices are a more commonly used method for CWC analysis. They utilize combinations of surface reflectance measurements at two or more wavelengths that indicate relative abundance for the features of interest. Previous research has used them to estimate water in vegetation as a rapid, non-destructive method (Gutierrez et al., 2010). Indices for vegetation are the most popular type, but other indices are available for targets such as manmade materials, water chemistry, and geologic features. The algorithms used for the established indices to derive spectral properties of water content in the plants are displayed in Table 10.1. Note that several indices have the same name but use different equations.

FIGURE 10.4
Weekly measurements of plant height for each shoot.

For the derivative analysis, we calculated the first and second derivatives based on the algorithms developed by Tsai and Philpot (1998) as

$$\frac{dR}{d\lambda} = \frac{R(\lambda_i) - R(\lambda_j)}{\Delta\lambda}, \tag{10.1}$$

where $\Delta\lambda$ is the separation between adjacent bands $\Delta\lambda = \lambda_j - \lambda_i$, and $\lambda_j > \lambda_i$, and intervals between bands are assumed to be constant. $R(\lambda_i)$ and $R(\lambda_j)$ are the reflectance of bands i and j, respectively. The second derivative can be derived from the first derivative as

$$\frac{d^2R}{d^2\lambda} = \frac{R(\lambda_i) - 2R(\lambda_j) + R(\lambda_k)}{\Delta\lambda^2}, \tag{10.2}$$

where $\Delta\lambda = \lambda_j - \lambda_i = \lambda_k - \lambda_j$ and $\lambda_k > \lambda_j > \lambda_i$.

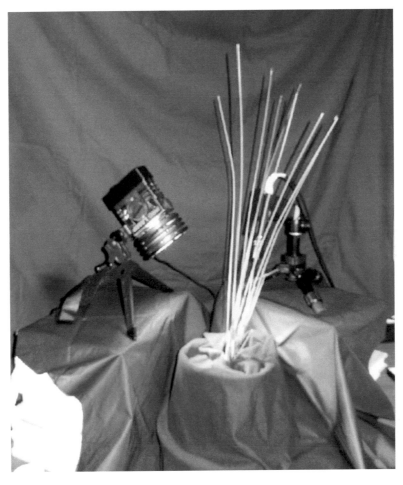

FIGURE 10.5
Spectroscopic measurements of *J. roemerianus* for each pot every two weeks. The halogen light was placed to reduce shadows.

10.4 Results and Discussion

10.4.1 Effects of Salinity and Water Level on the Shoot Height

Figure 10.6 shows shoot change after different treatments of *J. roemerianus*. Overall, *J. roemerianus* exhibited a rapid growth rate in the 0 ppt water, had reduced growth rate in the 20 ppt water, maintained viability with little growth in the 30 ppt water, and senesced with no growth in the 40 ppt water treatment. Despite our statistical analyses showing that salinity had the strongest effect on average mature shoot height, water level also played an important role by having a significant interaction effect with salinity. Previous research suggested that prolonged inundation would adversely affect the growth of *J. roemerianus* (Christian et al., 1990). This was proven accurate in the 20 ppt, 30 ppt, and 40 ppt treatments where the final average plant heights were greater in

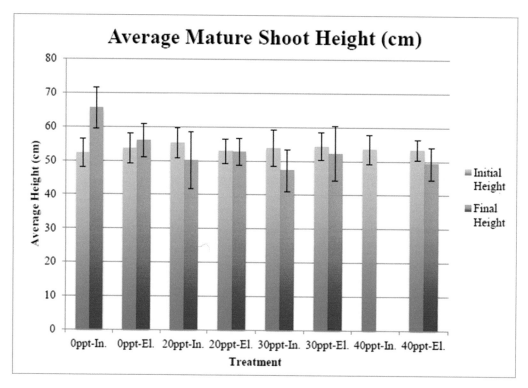

FIGURE 10.6

Initial and final average shoot height of mature *J. roemerianus* in salinity treatments (0 ppt, 20 ppt, 30 ppt, and 40 ppt) and water level treatments (elevated (El.) and inundated (In.)). The elevated condition represents current sea level, while inundated condition represents a sea level rise of 3 cm.

the elevated plants compared to the inundated plants (Figure 10.6). However, higher water levels seemed to have a positive effect on plants grown in the 0 ppt treatment as inundated plants had significantly greater average heights than the elevated plants. It is likely that the higher water levels allowed for greater nutrient exposure to the soil and roots, which in turn resulted in higher aboveground growth. Overall, the data collected on average height suggests that under sea level rise scenarios, *J. roemerianus* stands growing in eutrophic environments will be able to withstand salinity levels up to 30 ppt. Further, inundation will negatively affect the average height in brackish water and seawater conditions and positively affect the average height in fresh water.

10.4.2 Effects of Salinity and Water Level on Aboveground Biomass and CWC

Table 10.2 lists the average of aboveground biomass and CWC under different treatment of salinity and water levels. Since most of the matter in each replication in the 30 ppt and 40 ppt treatments was comprised of mostly senesced shoots, there was little difference between the wet and dry biomass in comparison to the 0 ppt treatment (Table 10.2). Although the plants with the 20 ppt treatment had high survivability and showed growth throughout the experiment, biomass values were closer to the 30 ppt and 40 ppt treatments than to the 0 ppt treatment. All three biomass measurements were

TABLE 10.2

Average aboveground biomass and CWC under different treatment of salinity and water levels

Treatment	Wet biomass (g)	Dry biomass (g)	CWC (g)
0 ppt inundated	26.48	14.71	11.77
0 ppt elevated	27.54	15.25	12.29
20 ppt inundated	17.89	12.30	5.59
20 ppt elevated	18.59	12.66	5.92
30 ppt inundated	17.63	12.73	4.89
30 ppt elevated	17.65	12.54	5.11
40 ppt inundated	15.97	11.66	4.31
40 ppt elevated	16.52	11.94	4.57

much higher for the 0 ppt inundated plants than the 0 ppt elevated plants. In combination with the average height, as well as other analyses in Abbott (2015), these biomass measurements signify the degree to which *J. roemerianus* thrived in the 0 ppt inundated treatment. Further, all three biomass measurements were in general higher in the 20 ppt, 30 ppt, and 40 ppt elevated plants compared to the inundated plants. The results also showed the strongest effect of salinity on CWC, though little effects of water level on CWC were revealed.

10.4.3 Spectral Response to Plant Stress Caused by Changes of Salinity and Water Levels

Figure 10.7 shows the spectral response of *J. roemerianus* under varying salinity and sea levels. The measured spectral signals less than 400 nm and more than 1200 nm were too weak to be considered in the analysis; thus, spectra over these regions

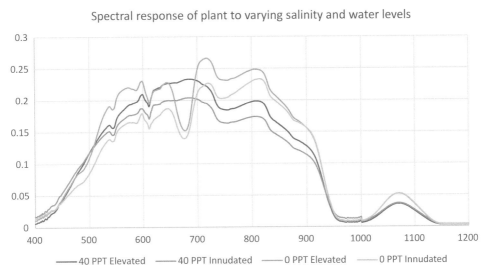

FIGURE 10.7
Spectral response of *J. roemerianus* with varying salinity and water levels.

were kept aside for further analysis. A reduced spectral reflectance over the NIR regions and an increased reflectance over the red regions were revealed when *J. roemerianus* was treated with a high salinity such as 40 ppt compared with the plants in fresh water with 0 ppt in salinity. Plants with an increased water level were observed to have a reduced reflectance compared with the elevated plants. This indicates that sea level does impact plants, which could be observed from their spectral response.

From the collected spectroscopic data, we calculated the WBI for each pot using the equation in Table 10.2. We then averaged it with the WBI for other pots having the same experimental variables ($N = 10$). Using biological data and known spectral ranges, "Good," "Caution," and "Dreadful" outlooks were determined based on the WBI numerical values and stress factors, as shown in Table 10.3. Again, the spectral signals of each pot above 1200 nm were too low. Other indices were not calculated because they use spectral reflectance more than 1200 nm. The plants did not thrive under higher salinity values, although some were still alive. A salinity of 40 ppt was too high for the plants, and most of the plants in this salinity environment appeared dead or almost dead. In the freshwater (0 ppt) environments, the plants revealed no signs of stress, and recruitment numbers over the study period indicated a very healthy community. The plants in the 20 ppt environments appeared to be able to tolerate higher salinity only if the plants were elevated above the water. The current sea level height is thus represented in the "elevated" plants in this experiment. The inundated plants anticipate water conditions with approximately 10 cm (~4 inches) of sea level rise, which falls within conservative estimates by the Intergovernmental Panel on Climate Change (IPCC) indicating ~8 to 18 cm (~3-7 inches) of SLR by 2030 (Warrick et al., 2012). With higher seas, the salinity tolerance levels of the *J. roemerianus* plants will decrease.

TABLE 10.3

The WBI calculated based on the spectroscopic data to determine the condition of plants

Salinity	Inundated/Elevated	WBI a	Overall condition
0	Inundated with SLR	1.059	Good
0	Elevated as current sea level	1.051	Good
0	Inundated with SLR	1.045	Good
0	Elevated as current sea level	1.054	Good
20	Inundated with SLR	0.98	Caution
20	Elevated as current sea level	1.011	Good
20	Inundated with SLR	0.978	Caution
20	Elevated as current sea level	1.01	Good
30	Inundated with SLR	0.967	Dreadful
30	Elevated as current sea level	0.995	Caution
30	Inundated with SLR	0.967	Dreadful
30	Elevated as current sea level	0.982	Caution
40	Inundated with SLR	0.954	Dreadful
40	Elevated as current sea level	0.979	Dreadful
40	Inundated with SLR	0.974	Caution
40	Elevated as current sea level	0.965	Caution

10.4.4 Identifying the Optimal Spectral Indices for CWC Estimation of *J. Roemerianus*

The WBI had a low correlation coefficient with the lab measured CWC; thus, we developed a new index to estimate CWC. We first conducted correlation analysis between bands and lab measured plant properties (wet biomass, dry biomass, and CWC), and the results are displayed in Figure 10.8. NIR had a positive correlation with these properties, while red and other visible regions had a negative correlation. This was expected because healthy green vegetation has a high reflectance over NIR. The highest correlation coefficient was 0.55 between CWC and bands around 1070 nm. All three plant properties had a consistent pattern in correlation with spectral reflectance values. To find the best band ratio between 400 nm and 1200 nm for CWC characterization, we calculated the correlation coefficients between each band ratio and CWC, and the results are shown in Figure 10.9. The best regions for CWC estimation are shown in bright blue in Figure 10.9. Ratios calculated using bands between 700 and 900 nm had a high correlation with CWC. A band ratio of 677 nm and 801 nm produced a correlation coefficient of -0.74 for CWC estimation. Meanwhile, spectral values around 1000 nm and 425 nm were also useful. Thus, for this study, a band ratio of $R677/R801$ was identified as the best ratio for CWC quantification with a correlation coefficient of 0.74 to CWC. We also examined other index calculation methods such as $(R_i + R_j)/(R_i - R_j)$, $j > i$, and the results are displayed in Figure 10.10. Again, regions shown in blue or red indicate the optimal spectral regions for CWC estimation using similar index algorithms. The best bands in this study were identified as 597 nm and 806 nm, which produced a correlation coefficient of 0.72 using $(R_{597} + R_{806})/(R_{597} - R_{806})$ equation. Both band ratio and other similar spectral index algorithms indicate that the red edge is valuable for plant stress identification, which confirmed the findings in the literature.

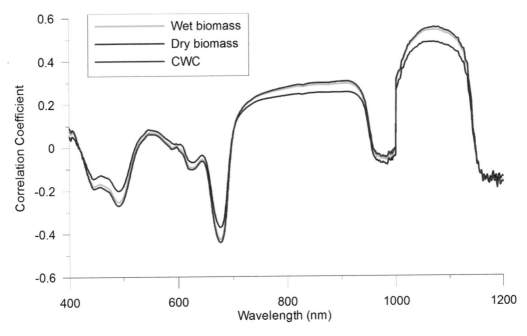

FIGURE 10.8
Correlation analysis between each spectral reflectance and wet biomass, dry biomass, and CWC.

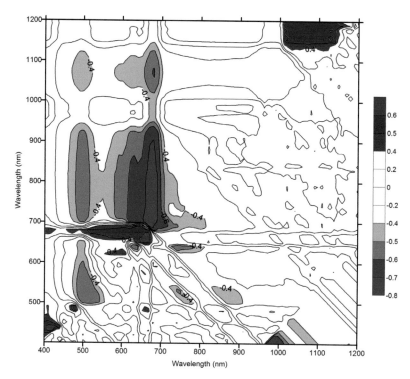

FIGURE 10.9
The identified spectral regions valuable for CWC characterization. The contour lines are calculated correlation coefficients between band ratios and CWC. The best regions are within the 700 to 900 nm shown in blue with a correlation coefficient of less than -0.7.

10.4.5 Derivative Analysis for CWC Estimation

Previous research indicates that the first derivative around the slopes of the water absorption features (at approximately 970 nm and 1200 nm) is more highly associated with CWC because of its ability to remove leaf structure effect on reflectance (Clevers et al., 2008). We conducted the correlation analysis between the first derivative and CWC, as well as the second derivative and CWC, and the results are shown in Figure 10.11 and 10.12. It was found that the first derivative had the highest correlation with CWC at the 680 nm wavelength with a correlation coefficient of 0.78, which was better than the application of spectral indices algorithms. The highest correlation coefficient was found at band 671 nm using the second derivative with a value of 0.74. The results of this analysis again demonstrated the capability of red edge in CWC characterization. Unfortunately, in our experiments, spectral values were very low and considered useless for regions above 1200 nm. This resulted in an exclusion of SWIR for plant stress analysis in this study. An inclusion of spectra over SWIR might increase the estimation of CWC and wet biomass. The performance of identified optimal spectral indices and derivative analysis of the spectroscopic data for CWC estimation is summarized in Table 10.4. For this study, the best algorithm was the first derivative, which produced a correlation coefficient of 0.78 at band 680 nm for CWC estimation.

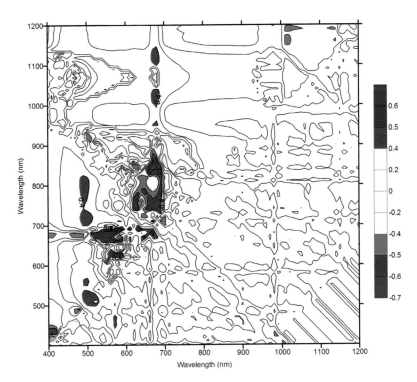

FIGURE 10.10
Same as Figure 10.9 but using band index $(R_i+R_j)/(R_i-R_j)$, $j>i$.

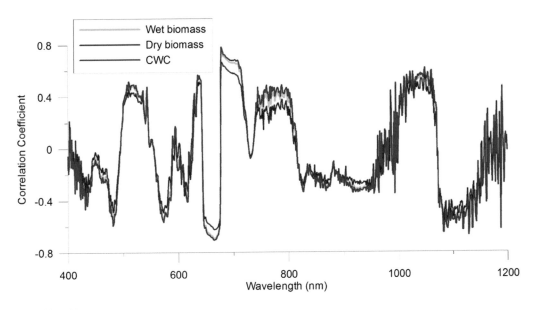

FIGURE 10.11
Correlation analysis between the first derivative of spectral reflectance and wet biomass, dry biomass, and CWC.

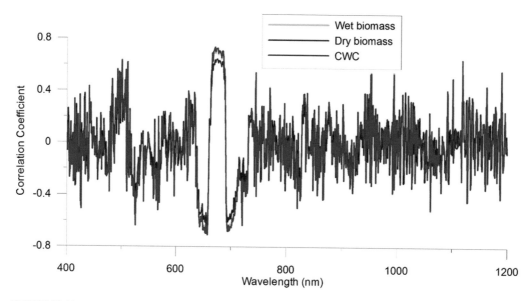

FIGURE 10.12

Correlation analysis between the second derivative of spectral reflectance and wet biomass, dry biomass, and CWC.

TABLE 10.4

Performance of spectral indices and derivative analysis of point spectroscopic data for quantification of CWC

Algorithms	Correlation coefficient
Band ratio: $R677/R801$	0.74
Spectral index: $(R_{597}+R_{806})/(R_{597}-R_{806})$	0.72
First derivative: 680 nm	0.78
Second derivative 671 nm	0.74

10.5 Summary and Conclusions

We implemented a 23-week greenhouse experiment to examine the effects of salinity and sea level rise on *J. roemerianus* and examined the spectral response of its stress caused by these two factors. The overall results of this study indicate that this species appears to be more tolerant to inundation than other common marsh species found in coastal wetlands of southeastern US such as *S. alterniflora*, and may prove to withstand and flourish under certain sea level rise scenarios. Salinity significantly affected the average mature shoot height, wet biomass, dry biomass, and CWC in the experiment. In all of these variables measured, increasing salinity decreased the measurements. Water level also significantly affected these plant variables. In all measured variables, inundation negatively affected *J. roemerianus* in the 20 ppt, 30 ppt, and 40 ppt treatments and positively affected *J. roemerianus* in the 0 ppt treatment. The results of this study have major implications for the future of coastal marsh ecosystems that are dominated by stands of *J. roemerianus*. These findings can be used in conjunction with studies on bordering marsh plants to predict shifts in the ecosystems of Florida that are experiencing SLR. For example, if an ecosystem is experiencing salt levels greater than 20 ppt, then it is likely that *J. roemerianus* will be replaced with more halophytic plants such as mangroves.

We also assessed plant stress by connecting plant CWC with point spectroscopic data. The spectral indices in the literature produced a low correlation with the lab-measured plant CWC. We thus developed a new ratio using bands 677 nm and 801 nm, which produced a high correlation coefficient of 0.74 for CWC estimation. The slope region of the water absorption feature at approximately 680 nm proved to be the most correlated with CWC using the first derivative, which achieved the best result in this study with a correlation coefficient of 0.78. The derivative approach can also minimize the impact of light variation. In this research, the field spectroscopic data was collected in a controlled laboratory setting using artificial light as the main source of illumination. However, the experiment was still influenced by external sunlight, helping to explain why the derivative approach proved to be more highly correlated than the spectral indices. Mid-infrared spectra have proven valuable for plant stress observation, but in this study, our measured spectral signals over mid-infrared regions were very low. An inclusion of this spectral region may increase the estimation result. Future research will explore methods to improve the correlation to CWC while expanding specific stressors: (1) assessing other stressors and parameters measured during the greenhouse experiment and (2) comparing the results and methodology using the spectroscopic imaging data collected by satellite and aerial (both multispectral and hyperspectral) for continuous health assessment of *J. roemerianus* plants in the coastal zones.

References

Abbott, C.J., 2015. The effects of sea level rise on *Juncus roemerianus* in a higher nutrient environment. M.S. Thesis, Florida Atlantic University.

Brinson, M.M., and Christian, R.R., 1999. Stability of *Juncus roemerianus* patches in a salt marsh. *Wetlands*, 19, 65–70.

Ceccato, P., Flasse, S., Tarantola, S., et al., 2001. Detecting vegetation leaf water content using reflectance in the optical domain. *Remote Sensing of Environment*, 77, 22–33.

Christian, R.R., Bryant, W.L., and Brinson, M.M., 1990. *Juncus roemerianus* production and decomposition along gradients of salinity and hydroperiod. *Marine Ecology Progressive Series*, 68, 137–145.

Clevers, J.G.P.W., Kooistra, L., and Schaepman, M.E., 2008. Using spectral information from the NIR water absorption features for the retrieval of canopy water content. *International Journal of Applied Earth Observation and Geoinformation*, 10, 388–397.

Clevers, J.G.P.W., Kooistra, L., and Schaepman, M.E., 2010. Estimating canopy water content using hyperspectral remote sensing data. *International Journal of Applied Earth Observation and Geoinformation*, 12, 119–125.

Cole, B., McMorrow, J., and Evans, M., 2014. Spectral monitoring of moorland plant phenology to identify a temporal window for hyperspectral remote sensing of peatland. *ISPRS Journal of Photogrammetry and Remote Sensing*, 90, 49–58.

Compact, 2015. Southeast Florida Regional Climate Change Compact Sea Level Rise Work Group (Compact). Unified Sea Level Rise Projection for Southeast Florida. A document prepared for the Southeast Florida Regional Climate Change Compact Steering Committee, 35.

Cormier, N., Krauss, K.W., and Conner, W.H., 2013. Periodicity in stem growth and litterfall in tidal freshwater forested wetlands: Influence of salinity and drought on nitrogen recycling. *Estuaries and Coasts*, 36, 533–546.

Demetriades-Shah, T.H., Steven, M.D., and Clark, J.A., 1990. High resolution derivative spectra in remote sensing. *Remote Sensing of Environment*, 33, 55–64.

Gamon, J.A., Qiu, H., Roberts, D. A., et al., 1999. Water expressions from hyperspectral reflectance implications for ecosystem flux modeling. Summaries of the 8th Annual JPL, Earth Science Workshop, Pasadena, CA, USA.

Gao, B., 1996. Normalized difference water index for remote sensing of vegetation liquid water from space. *Proceedings of SPIE*, 2480, 225–236.

Guha, H., and Panday, S., 2012. Impact of sea level rise on groundwater salinity in a coastal community of South Florida. *Journal of the American Water Resources Association*, 8, 510–529.

Gutierrez, M., Reynolds, M.P., and Klatt, A.R., 2010. Association of water spectral indices with plant and soil water relations in contrasting wheat genotypes. *Journal of Experimental Botany*, 61, 3291–3303.

González-Fernández, A.B., Rodríguez-Pérez, J.R., Marcelo, V., and Valenciano, J.B., 2014. Application of field spectrometry to the estimation of leaf water content: A case study with vitis vinifera L. cv Tempranillo in El Bierzo D.O (North-Western Spain). *Proceedings of the International Conference of Agricultural Engineering*, Zurich, Switzerland.

Hunt, Jr., E.R., Rock, B.N. and Nobel, P.S., 1987. Measurement of leaf relative water content by infrared reflectance. *Remote Sensing of Environment*, 22, 429–435.

Jackson, R.D., 1986. Remote sensing of biotic and abiotic plant stress. *Annual Review of Phytopathology*, 24, 265–287.

Jackson, T.J., Chen, D., Cosh, M., et al., 2004. Vegetation water content mapping using Landsat data derived normalized difference water index for corn and soybeans. *Remote Sensing of Environment*, 92, 475–482.

Long, S., and Baker, N., 1986. *Saline Terrestrial Environments. Photosynthesis in Contrasting Environments*, Elsevier, New York, 63–102.

Lytle, J.S., and Lytle, T.F., 1998. Atrazine effects on estuarine marophytes *Spartina alterniflora* and *Juncus roemerianus*. *Environmental Toxicology and Chemistry*, 17, 1972–1978.

Martin, J., Fackler, P., Nichols, J., et al., 2011. Structured decision making as a proactive approach to dealing with sea level rise in Florida. *Climatic Change*, 107, 185–202.

Peñuelas, J., Filella, I., Biel, C., et al., 1993. The reflectance at the 950–970 region as an indicator of plant water status. *International Journal of Remote Sensing*, 14, 1887–1905.

Rock, B.N., Hoshizaki, T., and Miller, J.R., 1988. Comparison of *in situ* and airborne spectral measurements of the blue shift associated with forest decline. *Remote Sensing of Environment*, 24, 109–127.

Rouse, Jr., J.W., Haas, R.H., Schell, J.A., and Deering, D.W., 1973. Monitoring vegetation systems in the great plains with ERTS. In: Freden, S. C., Mercanti, E. P., and Becker, M.A. (eds), NASA SP-351, NASA, Washington, DC, 309–317.

Saha, A., Saha, S., Sadle, J., et al., 2011. Sea level rise and South Florida coastal forests. *Climatic Change*, 107, 81–108.

Sparks, E., Cebriana, J., Biber, P., et al., 2013. Cost-effectiveness of two small-scale salt marsh restoration designs. *Ecological Engineering*, 53, 250–256.

Touchette, B.W., Adams, E.C., and Laimbeer, P., 2012. Age-specific responses to elevated salinity in the coastal marsh plant Black Needlerush (*Juncus roemerianus* Scheele) as determined through polyphasic Chlorophyll a Fluorescence Transients (OJIP). *Marine Biology*, 159, 2137–2147.

Tsai, F., and Philpot, W., 1998. Derivative analysis of hyperspectral data. *Remote Sensing of Environment*, 66, 41–51.

Tucker, C.J., 1979. Red and photographic infrared linear combinations for monitoring vegetation. *Remote Sensing of Environment*, 8, 127–150.

Tucker, C.J., 1980. Remote sensing of leaf water content in the near infrared. *Remote Sensing of Environment*, 10, 23–32.

Uto, K., and Kosugi, Y., 2012. Hyperspectral manipulation for the water stress evaluation of plants. *Contemporary Materials*, III–1, 18–25.

Wang, L., and Qu, J., 2007. NMDI: A normalized multi-band drought index for monitoring soil and vegetation moisture with satellite remote sensing. *Geophysical Research Letters*, 34, L20405.

Wang, J., Seliskar, D., and Gallagher, J., 2005. Tissue culture and plant regeneration of the salt marsh monocots *Juncus roemerianus* and *Juncus gerardi*. *In-vitro Cellular & Developmental Biology*, 4, 274–280.

Warrick, R., Oerlemans, J., Beaumont, P., et al., 2012. Chapter 9: Sea Level Rise. IPCC: Fifth Assessment Report (AR5).

Zarco-Tejada, P.J., Miller, J.R., Mohammed, G.H., et al., 2002. Vegetation stress detection through chlorophyll a+b estimation and fluorescence effects on hyperspectral imagery. *Journal of Environmental Quality*, 31, 1433–1441.

11

Applying Point Spectroscopy Data to Characterize Sand Properties

Molly E. Smith, Donna Selch and Caiyun Zhang

11.1 Introduction

Sand is a grouping of unconsolidated rock and mineral particles, typically sourced from the weathering of a host rock. Although classification schemes vary, sand is most widely accepted as having a size finer than gravel and coarser than silt and ranging in size from 0.0625 mm to 2 mm (Udden, 1914; Wentworth, 1922). It can vary in composition and be transported by wind or water and be deposited in various forms such as beaches, dunes, sand spits, and sand bars. In geological context, studying sand characteristics can provide important information about source materials, depositional environment, and other physical and chemical factors that impact properties of sands. For example, in inland continental settings and non-tropical coastal settings, the most common constituent of sand is silica (silicon dioxide), usually in the form of quartz; the bright sands found in tropical and subtropical coastal settings are made up of a combination of weathered limestone, shell, and coral fragments; dark sands are commonly from volcanic basalts containing magnetite, olivine, glauconite, and chlorite. In beach sands, the most common constituent is quartz, but other common materials will also include carbonate shell and skeletal fragments, heavy minerals, and other rock fragments (Friedman and Sanders, 1978). Some generalizations about sand color can be made first on the basis of mineralogy and second from organic coatings, iron-staining, or other environmental influences (Berkowitz et al., 2018). Quartz (and to a lesser extent, feldspar) is most resistant to the forces of erosion and abrasion in coastal environments, meaning that this mineral is more commonly found in sands, and depending on chemical composition, typically will give sand lighter coloration where it appears in sufficient quantities (Cardenes and Rubio, 2017). Additionally, because shell fragments will become quickly bleached when sub-aerially exposed, sands rich in calcium carbonate will show lighter coloration. Heavy and ferromagnetic minerals, such as magnetite and garnet, contribute to dark grains found in some sands (Hobbs, 2012).

Geologists normally collect sand information using the microscopic and sieving techniques in which a set of sand attributes (e.g., grain shape, grain size, texture, and composition) are derived (Microbus, 2019). Analyzing sand using a geological/microscopic approach is a tedious and labor-intensive procedure, and the analysis result is often subjective. Remote sensing spectroscopy techniques can identify and characterize materials by measuring a continuum of spectral data covering visible, near infrared, and short-wave infrared wavelengths (Clark, 1999). These techniques have a broad application such as soil and vegetation analysis, mining exploration, and calibration and validation of remote sensing imagery. The US Geological Survey (USGS) and other research institutes have built a range of spectral libraries of minerals, rocks, liquids, organic compounds, vegetation, and many other natural and manmade materials using spectroscopic techniques (Kokaly et al., 2017). However, to date, a spectral library of sands has not existed. In addition, few efforts have been made to explore the capability of spectroscopy techniques for characterizing sands. Sands are unconsolidated materials and may have a completely different spectral behavior from minerals and rocks, which have been extensively studied using spectroscopic techniques. To this end, the main objective of this chapter is to evaluate whether spectroscopy techniques can be used to characterize sand properties (i.e., color, grain size, and composition) using qualitative and quantitative methods by combining the geological/microscopic and remote sensing spectroscopy techniques. For the first time, we built a comprehensive sand spectral library using sand samples collected worldwide. This large volume of new-to-science data is expected to benefit both the geology and remote sensing research communities.

11.2 Data Collection, Processing, and Analysis

An anonymous benefactor, who collected over 1000 sand samples worldwide, donated these specimens to the John D. Macarthur Beach State Park. These samples were loaned to the Department of Geosciences at Florida Atlantic University for this study. We also collected sand samples in local beaches, such as Fort Lauderdale Beach, and included these samples in the analysis. Among all these samples, a total of 113 representative samples with varying size, color, composition, and grain shape were visually identified and selected for both geological/microscopic and spectroscopic analysis. We did not

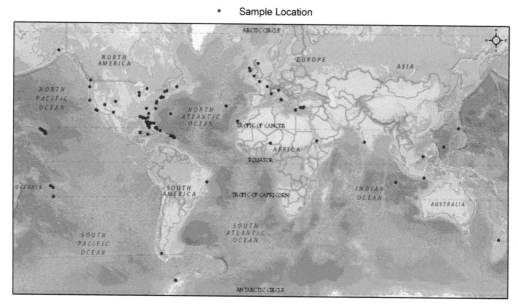

FIGURE 11.1
The geographic locations of sand samples used in this study.

analyze all of the John D. Macarthur Beach State Park collection due to the time-consuming and labor-intensive geological/microscopic procedure. The approximate geographic locations of the selected samples are displayed in Figure 11.1.

We first conducted a visual examination of these samples under an Olympus SZX12 binocular microscope, which is equipped with a digital camera system and ring illumination. We put each sand sample in an opaque black dish with 10 cm squares marked throughout and took a digital image of each sample under the microscope with varying magnifications. The samples were then grouped into three group types (classes 1–3) based on colors. Class 1 is generally white to beige and has a low iron content. Most samples in this class were found to be composed of biogenic carbonates. Class 3 is generally black with a high iron and ferromagnetic content. Class 2 is an intermediate color group consisting mostly of mixtures of classes 1 and 3. Samples in class 2 had colors ranging from brown to dark gray with multiple components such as iron-stained quartz. We determined the grain size under the microscope using the chart developed by Emrich and Wobber (1963), as shown in Figure 11.2. The selected samples were grouped into 4 categories of grain size: fine (~0.11 mm), medium (~0.25 mm), coarse (~0.5 mm), and very coarse (>0.5 mm). We estimated sand composition using 2 options.

Grain Size Description	Examples (Under 15x magnification)
Fine (~0.11 mm)	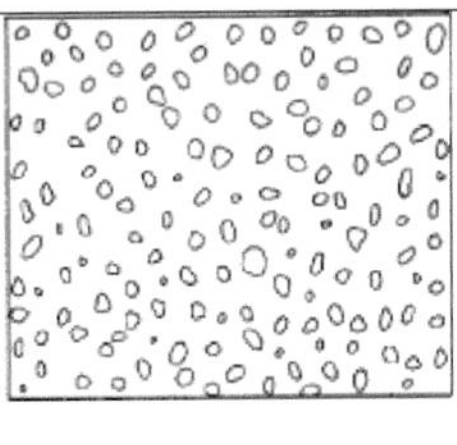
Medium (~0.25 mm)	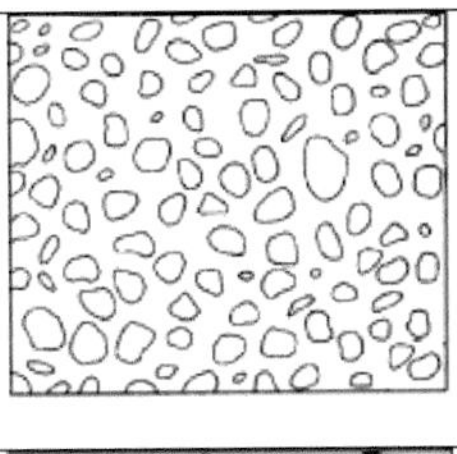
Coarse (~0.5 mm)	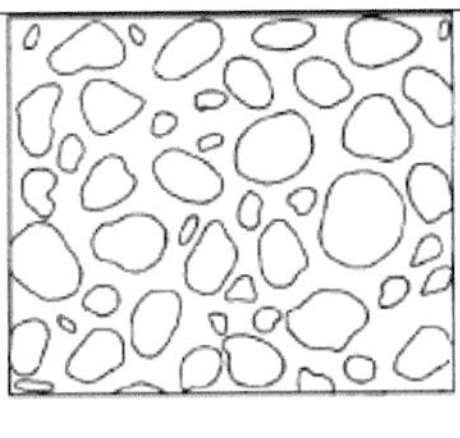

FIGURE 11.2
Grain size estimation chart adapted from Emrich and Wobber (1963). Sediments greater than 0.5 mm were defined as "very coarse" in this study.

The first was to use a percentage estimation chart adapted by Folk (1951) and Terry and Chilingar (1955), and the second was to manually separate each grain from a sample. The first is more time saving, and most of our samples were estimated using this approach. Additional tools were used in this procedure. For example, a small magnet was used to test ferromagnetic properties of the sand and an eye-drop bottle with HCL to test for carbonate content when necessary. Four major components (i.e., total carbonates, silicates, magnetite, and total volcaniclastics) were estimated for each selected sample. In initial microscope analysis, carbonate minerals were broken into subgroups: carbonate rock fragments, coral fragments, large shell fragments, shell hash, and skeletal fragments. These subgroups were combined to form an estimation of the more generalized carbonate content.

We collected spectroscopic data of sand samples using an ASD Spectroradiometer FieldSpec Pro (350–2500 nm with a spectral resolution of 1 nm) in a laboratory environment with a controlled light source. Each sand sample was placed in a black, opaque petri dish, with enough sand to provide a sample thickness at approximately 3.5 mm. We measured 5 spectra for each sample and used the average of these 5 measurements as the final spectra of the sample to reduce noise during the data collection. The instrument was calibrated using a white board that is included with the ASD bundle. Spectral reflectance can be directly determined using the ASD bundle. After spectral data collection, we imported the data into ENVI's Spectral Library Builder, leading to a sand spectral library to be used in the sand and remote sensing study community.

To assess the efficacy of spectroscopic data for sand property estimation, the sand attributes derived from the geological/microscopic procedure of each sample were matched with their corresponding spectroscopic data, resulting in a matched dataset with spectra and sand properties. We applied remote sensing classification algorithms to predict sand colors (3 classes) and grain size (4 classes) and explored modeling algorithms to estimate sand compositions. Here, only components appearing in high quantities from the microscopic analysis were estimated (i.e., carbonate and silicate materials). We examined machine-learning algorithms for sand attribute prediction. The k-fold cross-validation, error matrix, and statistical metrics including correlation coefficient (r), Root Mean Squared Error (RMSE), and Mean Absolute Error (MAE) were used to evaluate the performance of spectroscopic data for predicting sand properties. The description of these methods can be found in previous chapters. We also identified the optimal bands for carbonate and silicate estimation using a single band approach and a band ratio approach.

11.3 Results and Discussion

11.3.1 Geological/Microscopic Analysis Results

Among the 113 selected samples, 65 were classified into class 1 with a light color, 16 were grouped as class 3 with a dark color, and 32 were put into class 2 with a color intermediate between light and dark. The histogram of the distribution of samples in color is displayed in Figure 11.3. Similarly, the histogram of the distribution of samples in grain size is displayed in Figure 11.4. A majority of the samples were fine-grained. The diversity in color and grain size of the selected sand samples provided a good

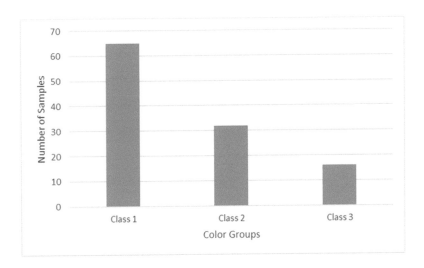

FIGURE 11.3
The histogram of samples in color.

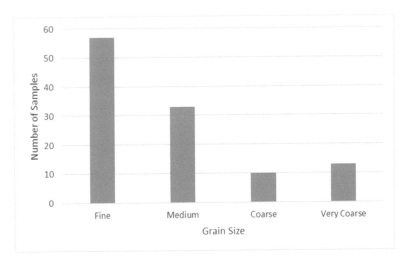

FIGURE 11.4
The histogram of samples in grain size.

opportunity to test the capability of spectroscopic techniques for assessing sand properties. The sand composition analysis results showed that 42 samples were comprised of 50% or more carbonate, 50 samples contained 50% or more quartz, and only 11 samples included of 50% or more volcanoclastic material. A majority of carbonate samples were sourced from the Caribbean, Florida, and the Pacific/Oceania regions. Quartz content was more or less distributed through North America and Europe. Figure 11.5 shows the majority of sand samples with their composition and geographic distribution.

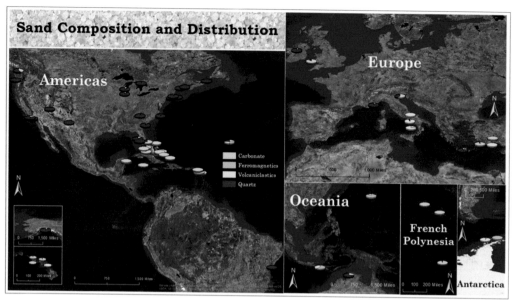

FIGURE 11.5
Major sand compositions and their geographic distributions.

11.3.2 Qualitative Analysis of Spectroscopy Data

To help the reader appreciate the combined microscopic and spectroscopic study of sands, a sample of the digital images collected under the microscope and spectral reflectance collected from the spectroradiometer is displayed in Figure 11.6. A comparison of spectral reflectance of sand samples with a similar color but different grain sizes is shown in Figure 11.7, while sands with a similar grain size but different colors is shown in Figure 11.8. The spectra acquired from the ASD FieldSpec Pro used in this study often exhibit steps at the splice of the three built-in sensors, positioned at 1000 nm (end of VNIR detector) and 1830 nm (end of SWIR 1 detector) (Figures 11.7 and 11.8). This is a noted instrument sensitivity drift that can be corrected, if necessary, using an interpolation algorithm (Beal and Eamon, 2010), but this splice problem would not impact the analysis results; thus, corrections were not conducted here. In general, darker sands were observed to have a lower reflectance, while brighter sands had a higher reflectance value (Figure 11.6). Fine-grain sands showed a higher reflectance than sands with a larger grain size (Figure 11.7). In the light color samples with varying grain sizes, reflectance differences were clearly demonstrated between 1850–2030 nm, where the fine-grained samples were observed being ~0.1 higher in reflectance than the other samples. Coarse and very coarse sands had a similar spectrum in class 1. This is unusual given that Beer's Law specifies that larger grain sizes should have lower reflectance (as larger grain sizes increase the length of travel for photons); thus, this is likely an outlier. The very coarse samples often showed average grain size just 0.1 mm larger than the largest average grain size in the coarse samples, leading to confusion in spectra of two groups. More samples in this size range need to be analyzed to test this finding. The difference in reflectance between coarse and very coarse sands in class 2 was also small,

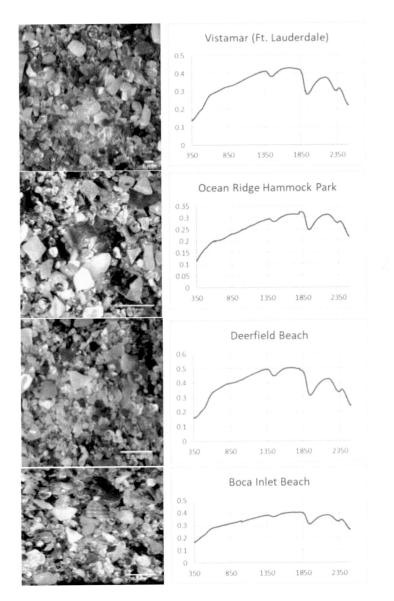

FIGURE 11.6
Collected sand imagery using a microscope and its corresponding spectral reflectance measured from an ASD spectroradiometer.

while the inconsistencies in reflectance among fine, medium, and coarse sands was distinct across the entire spectrum for both class 2 and class 3. Figure 11.8 illustrates that sands with similar size but different color have a different reflectance; light/bright sand had a higher reflectance than dark sands. Darker color sands showed a stronger absorptive effect compared to lighter color sands that are more reflective.

Figure 11.9 displays the reflectance of sands with similar carbonate percentage. This illustrates several absorption features of carbonate in sands such as the one around 1250 nm shown by the arrow in Figure 11.9. Sand composition is not independent from its

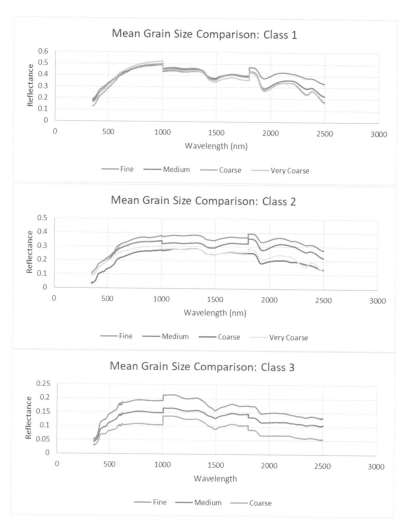

FIGURE 11.7
Comparison of spectral reflectance of sand samples with different grain sizes but a similar color.

color. Sand color generally is determined by its composition and particularly mineral content. For example, biotite mica, which can be found in immature and un-reworked sands, has brownish-black coloration due to its iron content (Klein et al., 2008). Sands with similar color or composition can show different spectral reflectance caused by difference in size, shape, or texture. In summary, these qualitative analysis results suggest that the spectral reflectance is closely related to sand properties. This drove us to develop models to quantify sand properties based on their spectral signatures.

11.3.3 Classification of Sand Color and Grain Size

Three modern machine learning classifiers, k-Nearest Neighbor (k-NN), Support Vector Machine (SVM), and Random Forest (RF), integrated in an open-source data mining software

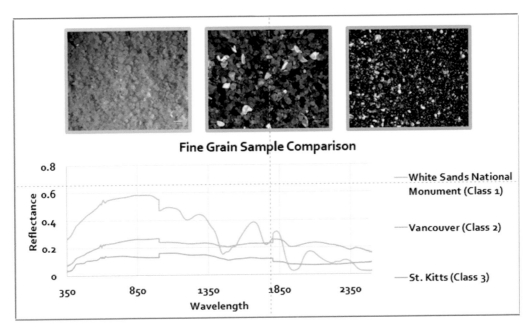

FIGURE 11.8
Comparison of spectral reflectance of three sand samples with similar grain size but different colors.

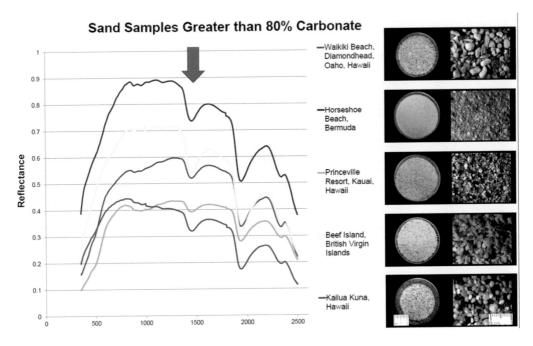

FIGURE 11.9
Spectral reflectance of sand samples with a high carbonate percentage.

package WEKA (Frank et al., 2016), were used to classify sand color and grain size using a 4-fold cross-validation procedure. The overall accuracies and Kappa values from the best-tuned classifier of each algorithm are displayed in Table 11.1. For sand color classification, three classifiers produced a comparable accuracy with an overall accuracy (OA) of 76 to 77% and Kappa value of 0.57 to 0.58. This is encouraging. The error matrix for color classification from the k-NN classifier is displayed in Table 11.2. As expected, a low classification accuracy was produced for identifying class 2 because this class was a mixture of classes 1 and 3 and had a high diversity in spectra within the samples. The results indicate that the color of sand is a major factor in determining the spectral profile of sands.

Color is commonly used to classify sand types, though to date, the number of sand types in the world remains unknown. There are no official sand types used in sand studies. Some soil scientists have preferred the Munsell color classification system for soil color, but there are other classification systems. However, the problem with these systems is that they are subject to interpretation, and classification can vary somewhat due to differing perceptions among operators. Whether the spectroscopic data can provide more details in color of sands needs more investigation because, in this study, all of the sand samples were generalized into three basic color groups only. In the real world, more colors are used for sand description—such as "pink" sand, "red-orange" sand, and "light-brown." In the future, the efficacy of spectroscopic data for classifying sand into the Munsell Soil System would be important to test.

TABLE 11.1

Classification accuracies of sand color and grain size using different classifiers

	Sand Color		
	k-NN	SVM	RF
Overall Accuracy (OA, %)	77.0	76.1	76.1
Kappa Value	0.58	0.57	0.58
	Sand Grain Size		
Overall Accuracy (OA, %)	51.3	51.3	49.6
Kappa Value	0.17	0.12	0.15

k-NN: k-Nearest Neighbor; SVM: Support Vector Machine; RF: Random Forest.

TABLE 11.2

Error matrix of sand color classification using the k-NN classifier and spectroscopic data

Classes	1	2	3	Column Total	UA (%)
1	58	7		65	89.2
2	15	15	2	32	46.9
3		3	13	16	81.3
Row Total	73	25	15	OA: 77.0%	
PA (%)	79.5	60.0	86.7	Kappa: 0.58	

PA: Producer's Accuracy; UA: User's Accuracy. Class 1 is bright sand, class 3 is dark sand, and class 2 is a mixture of classes 1 and 3.

TABLE 11.3

Error matrix of sand grain size classification using the k-NN classifier and spectroscopic data

Classes	Fine	Medium	Coarse	Very Coarse	Column Total	UA (%)
Fine	47	8		2	57	82.5
Medium	18	11	2	2	33	33.3
Coarse	3	5		2	10	0.0
Very Coarse	4	9			13	0.0
Row Total	72	33	2	6	OA: 51.3%	
PA (%)	65.3	33.3			Kappa: 0.17	

PA: Producer's Accuracy; UA: User's Accuracy.

For sand grain size classification, an OA of around 50% was produced with a low Kappa value, suggesting that grain size is more difficult to identify using hyperspectral reflectance. The error matrix of grain size classification from the k-NN classifier is displayed in Table 11.3. Spectroscopic data completely failed to identify coarse and very coarse sand, while fine sand was easily distinguished. Though the qualitative analysis results showed that fine-grain sands had a higher reflectance than sands with a coarse grain size, prediction of grain size using quantitative approaches is challenging. Similar grain size sand samples can have a very different spectral profile if their colors are different. This causes difficulty in classifying grain size using spectral data alone.

In addition, the surface texture and sphericity of sands also impacts the spectral reflectance of sand. This makes grain size identification more complicated. Surface texture determines light interaction with a sand sample, such as whether it will be scattered, absorbed, or entirely reflected. Sand mixtures are prone to showing non-Lambertian scattering behaviors, where increases in surface texture roughness will also influence light backscattering (Herold et al., 2004). For example, angular surface textures will increase the number of surfaces light will reflect on. Because angular textures increase reflective surfaces, sands with this texture may have higher reflectance values than those with smooth textures.

11.3.4 Prediction of Sand Composition using Spectroscopy Data

In most of the sand samples analyzed, a high percentage of carbonate and silicate minerals were found. We thus only quantitatively analyzed these two components using the spectroscopic data. We first conducted the correlation analysis between the reflectance of each band and carbonate/silicate with an aim to identifying the best narrowband for quantifying sand composition. The derived correlation coefficient (r) against the wavelength of each band is plotted in Figure 11.10. The highest r was observed at wavelength 880 nm between carbonate and spectral reflectance ($r = 0.53$), while the highest r was found at wavelength of 2500 nm between silicate and spectral reflectance ($r = 0.75$). Carbonate and silicate had opposite trends, which was expected because sands with more carbonate would have a relatively smaller percentage of silicate. Reflectance of wavelengths between 400 and 1000 nm was negatively correlated with silicate (i.e., high reflectance within this range indicates a low percentage of silicate in the sand samples).

We also conducted analysis of spectral indices to examine whether a combination of two bands would have a better performance in predicting carbonate and silicate content in sands. It was found that the application of band ratios did not significantly improve

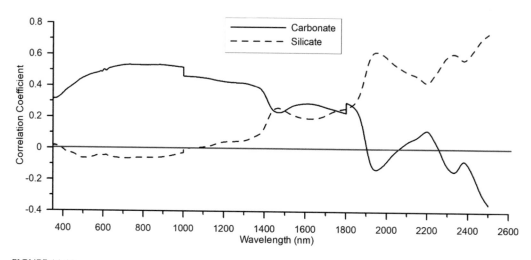

FIGURE 11.10

The correlation coefficient (r) of carbonate and silicate and the spectral reflectance of each band. Wavelengths 880 nm and 2500 nm were observed to have the highest correlation with carbonate (r=0.53) and silicate (r=0.74), respectively.

the prediction capability compared with the application of one band. Two optimal bands were identified at wavelengths 1044 nm and 1047 nm, which produced the highest r of 0.56 between the band ratio (R_{1044}/R_{1047}) and carbonate. The most correlated spectral regions using a band ratio to predict carbonate are highlighted in purple and blue, shown in Figure 11.11, which is a contour map of r produced using the ratio of two bands across the entire spectrum. Less correlation was found using a band ratio than using the best single narrowband for silicate prediction; thus, the results were not included here. We also found a combination of two bands using equations like Normalized Difference Vegetation Index (NDVI) did not improve the correlation coefficient compared over using the best-identified single band for both carbonate and silicate. Thus, methods of spectral indices were not recommended for sand composition prediction.

We explored regression algorithms to assess the suitability of spectroscopic data for carbonate and silicate estimation including the k-NN, SVM, RF, and Multiple Linear Regression (MLR) algorithms. As described in previous chapters, the three machine learning algorithms can be used for both classification and regression modeling. The performance of these algorithms in predicting the percentage of carbonate and silicate based on 4-fold cross-validation is shown in Table 11.4. As expected, since an encouraging result was produced in the color classification using spectroscopic data, 4 algorithms had a good performance in the prediction of carbonate with an r of 0.79 to 0.9, RMSE of 18.9 to 27.3%, and MAE in the range of 13.5 to 20.6%. SVM achieved the best result for estimating the percentage of carbonate in sand samples in terms of all the statistical metrics (r, RMSE, and MAE). For the silicate quantification, a high accuracy was also achieved using spectroscopic data with the r in the range of 0.80 to 0.85, RMSE varying between 22.7 and 25.8%, and MAE within 16.3 and 18.5%. The MLR produced a result comparable to the advanced machine learning algorithms, suggesting there was a strong linear relationship between the sand composition and multiple reflectance. Clark (1999) reported that the diagnostic absorption features of quartz was around 10,000 nm, which is far beyond the spectral range of the ASD spectroradiometer. This

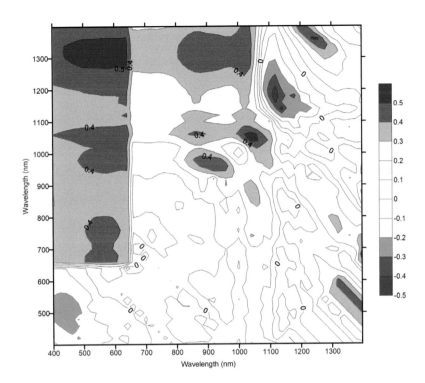

FIGURE 11.11

The identified spectral wavelengths valuable for sand carbonate characterization using band ratio techniques. The contour lines are correlation coefficients between each band ratio (R_i/R_j, $j > i$) and sand carbonate, where R_i, R_j are reflectance of bands i, and j. The best band ratio is R_{1044}/R_{1047} with a r of 0.56 to carbonate.

TABLE 11.4

Estimation accuracy of sand composition using different models

	k-NN	SVM	RF	MLR
Carbonate Estimation				
r	0.86	0.90	0.80	0.79
RMSE (%)	22.4	18.9	26.4	27.3
MAE (%)	13.5	13.5	16.9	20.6
Silicate Estimation				
r	0.82	0.85	0.80	0.85
RMSE (%)	24.5	23.0	25.8	22.7
MAE (%)	16.3	16.8	17.2	18.5

k-NN: k-Nearest Neighbor; SVM: Support Vector Machine; RF: Random Forest; MLR: Multiple Linear Regression; RMSE: Root Mean Squared Error; MAE: Mean Absolute Error.

is likely the reason the best narrowband with the highest correlation to silicate was at 2500 nm (Figure 11.10). An inclusion of spectral measurements around the diagnostic features may further improve the prediction of silicate in sands.

11.4 Summary and Conclusions

For the first time, we built a sand spectral library by measuring spectral reflectance of over 100 sand samples collected worldwide using an ASD spectroradiometer. We also examined the capability of narrowband spectroscopic data for characterizing fundamental sand properties including sand color, grain size, and sand composition. The results illustrate that spectroscopic techniques are powerful for sand color and composition analysis but difficult for grain size determination. Narrowband spectroscopic data achieved an overall accuracy of more than 80% for sand color classification, and predicted the percentage of carbonate and silicate in sands with a correlation coefficient of more than 0.8. Sand color and its composition are highly related and can be accurately predicted through spectroscopic techniques. This is encouraging given the fact that determination of sand composition from traditional geological approaches is time consuming and labor intensive. A spectroscopic technique is an attractive tool compared to or used in tangent with traditional sand analysis methods. Future studies will be oriented toward using the sand spectral library to quantify and map sand properties worldwide by connecting the library with remote sensing imagery. Linking the sand spectral library with digital imagery collected under the microscope to enhance the quantification of sand properties is another effort to be made in the future.

References

Beal, D., and Eamon, M., 2010. Preliminary results of testing and a proposal for radiometric error correction using dynamic, parabolic linear transformations of "stepped" data PCORRECT.EXE (available upon request to ASD Inc.).

Berkowitz, J.F., VanZomeren, C.M., and Priestas, A.M., 2018. Potential color change dynamics of beneficial use sediments. *Journal of Coastal Research*, 34, 5, 1149–1156.

Cardenes, V., and Rubio, Á., 2017. Measure of the color of beach nourishment sands: A case study from the Belgium coast. *Trabajos de Geología*, 35, 35, 7–18.

Clark, R.N., 1999. Spectroscopy of rocks and minerals, and Principles of spectroscopy. In: Rencz, A. N. (ed.) *Manual of Remote Sensing, Remote Sensing for the Earth Sciences*, John Wiley & Sons Inc., New York, 3–58.

Emrich, G.H., and Wobber, F.J., 1963. A rapid visual method for estimating sedimentary parameters. *Journal of Sedimentary Research*, 33, 831–843.

Folk, R.L., 1951. A comparison chart for visual percentage estimation. *Journal of Sedimentary Petrology*, 21, 32–33.

Frank, E., Hall, M.A., and Witten, I.H., 2016. *The WEKA Workbench*. In: Witten, I.H., Frank, E., Hall, M.A., and Pal, C. J., Data Mining: Practical Machine Learning Tools and Techniques, 4th Edition, Elsevier, Cambridge, UK.

Friedman, G., and Sanders, J., 1978. Complement A. In: Friedman, G., Principles of Sedimentology, Wiley, New York, USA.

Herold, M., Roberts, D.A., Gardner, M.E., and Dennison, P.E., 2004. Spectrometry for urban area remote sensing – Development and analysis of a spectral library from 350 to 2400 nm. *Remote Sensing of Environment*, 91, 304–319.

Hobbs, C.H., 2012. *The Beach Book: Science of the Shore*, Columbia University Press, New York.

Klein, M., Aalderink, B., Padoan, R., et al., 2008. Quantitative hyperspectral reflectance imaging. *Sensors*, 8, 5576–5618.

Kokaly, R.F., Clark, R.N., Swayze, G.A., et al., 2017. USGS spectral library version 7. *U.S. Geological Survey Data Series*, 1035, 61.

Microbus, 2019. www.microscope-microscope.org/applications/sand/microscopic-sand.htm. Accessed on 28 May 2019.

Terry, R.D., and Chilingar, G.V., 1955. Summary of "concerning some additional aids in studying sedimentary formations" by M.S. Shvetsov. *Journal of Sedimentary Petrology*, 25, 229–234.

Udden, J.A., 1914. Mechanical composition of clastic sediments. *Bulletin of the Geological Society of America*, 25, 655–744.

Wentworth, C.K., 1922. A scale of grade and class terms for clastic sediments. *Journal of Geology*, 5, 30, 377–392.

12

Land Cover-level Vegetation Mapping using AVIRIS

Caiyun Zhang

12.1 Introduction

Hyperspectral data have proven more effective for vegetation mapping than multi-spectral data due to their rich spectral content. They have been applied in wetlands for mapping purposes using moderate spatial resolution imagery (i.e., 20–30 m) collected by either spaceborne sensors such as EO-1/Hyperion or airborne sensors such as AVIRIS (e.g., Hirano et al., 2003; Rosso et al., 2005; Pengra et al., 2007; Zhang and Xie, 2012). Fine spatial resolution hyperspectral imagery (i.e., 4 m or smaller) is mainly collected by airborne sensors and has been applied in wetlands (e.g., Artigas and Yang, 2005; Belluco et al., 2006; Jollineau and Howarth, 2008). For the Greater Everglades, Hirano et al. (2003) applied 20-m AVIRIS imagery to map vegetation in a portion of Everglades National Park and obtained an overall accuracy of 66%. Our research group examined the capability of fine spatial resolution hyperspectral imagery for vegetation mapping in the Everglades at different detail levels (Zhang and Xie, 2012, 2013) and explored the potential of fusing moderate spatial resolution hyperspectral imagery with other remote sensing data sources to improve vegetation mapping (Zhang, 2014; Zhang and Xie, 2014; Zhang et al., 2016). In this chapter, 4-m AVIRIS imagery was applied to map land cover-level vegetation types in the central Greater Everglades using modern machine learning classifiers and object-based mapping techniques. To examine the value of hyperspectral imagery for wetland mapping, classification results were compared with the application of multispectral Ikonos imagery.

12.2 Study Area and Data

The study area is same as the study site in our previous study (Zhang and Xie, 2012) but covers a smaller region. It is a portion of Caloosahatchee River watershed in the central Everglades (Figure 12.1). In the Greater Everglades, Lake Okeechobee serves as the "water heart" for the Everglades, and the Caloosahatchee River functions as a primary canal that conveys basin runoff and regulatory releases from Lake Okeechobee. Caloosahatchee watershed is an important environmental and economic resource in the Everglades. The hydrology of this region has been severely changed because many canals were constructed along the banks of the river to support the agricultural communities associated with the river. Response of the plant community in this region is a crucial indicator of restoration success, and detailed vegetation maps can guide the path of restoration. The study area was selected because of the availability of hyperspectral imagery. It is a mosaic of common wetland vegetation community, agricultural

FIGURE 12.1
Study area in the Caloosahatchee River watershed shown as a natural color and a false color composite
generated from AVIRIS imagery.

plant community, exotic species, water bodies, and manmade concrete features. This is
a challenging site for vegetation characterization from remote sensing.

Data sources include the hyperspectral imagery collected by Airborne Visible/Infrared
Imaging Spectrometer (AVIRIS) and a reference map manually interpreted from National
Aerial Photography Program (NAPP) color infrared (CIR) images. AVIRIS is a premier
instrument in the realm of Earth Remote Sensing. It delivers calibrated hyperspectral
images in 224 contiguous spectral channels with wavelengths from 400 to 2500 nan-
ometers. The South Florida Water Management District (SFWMD) coordinated the collection of AVIRIS data over the Caloosahatchee watershed on November 16, 1998, by
working with the Jet Propulsion Laboratory (JPL) at the California Institute of Technology.
The instrument was on board the NASA's Twin Otter aircraft with low flying altitude.
This resulted in an acquisition of high-spatial resolution hyperspectral data with a pixel
size of 4.0 m. All AVIRIS data are available at AVIRIS's website (https://aviris.jpl.nasa.
gov/). A natural color and a false color composite for the study site produced from the
AVIRIS imagery is shown in Figure 12.1. JPL conducted the geometrical correction for the
deliverables. The reference map was from SFWMD and available at the District's Open
Data Portal (https://geo-sfwmd.opendata.arcgis.com/). The closest Land Cover Land Use
data to the collected AVIRIS imagery is the SFWMD Land Cover Land Use 1999. This

dataset was photo-interpreted from 1999 NAPP CIR 1:40,000 aerial photography and classified using the SFWMD modified Florida Land Use, Land Cover Classification System. Features were stereoscopically interpreted using a stereo plotter and calibrated from field surveys through a project known as "Land Cover/Land Use Mapping Project" conducted at the SFWMD. In the project, the data was compiled on screen over corresponding USGS Digital Orthophoto Quadrangles. The positional accuracy of the data meets the National Map Accuracy Standards adopted by USGS. The SFWMD reports the dataset has a minimum accuracy of 90%. The reference map for the study site is shown in Figure 12.2, which was used to assist with reference data collection to calibrate and validate the classifiers. In total, 331 reference samples were randomly selected over the study area. For the selected study area, 12 major vegetation covers were found: dry prairie, palmetto prairies, improved pastures, woodland pastures, pine flatwood, upland hardwood forest, Brazilian pepper (exotic species), Melaleuca (exotic species), hardwood/coniferous mix, mangrove swamp, mixed wetland hardwoods, and wetland forested mixes. These vegetation types were distinguished from the collected hyperspectral imagery. In addition, water and urban built-up (e.g., roads, driveways, buildings) were presented and classified. A summary of the classes to be mapped in this chapter is listed

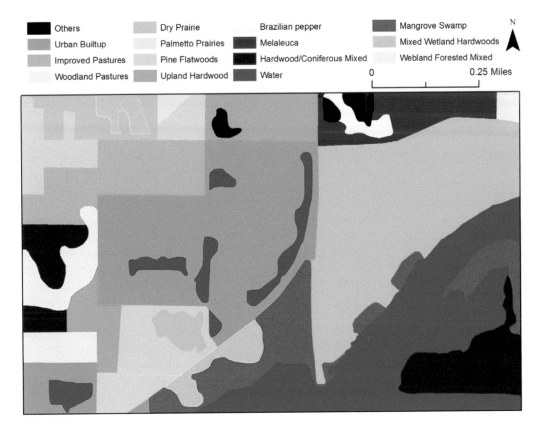

FIGURE 12.2
The land cover reference map used to select reference samples in the classification.

TABLE 12.1

Classes to be mapped using AVIRIS imagery in the central Everglades

Class #	Class name	Reference size
1	Urban built-up	60
2	Improved pastures	22
3	Woodland pastures	16
4	Herbaceous or dry prairie	42
5	Palmetto prairies	4
6	Pine flatwoods	25
7	Upland hardwood	22
8	Brazilian pepper	7
9	Melaleuca	12
10	Hardwood/Coniferous mixed	27
11	Water	33
12	Mangrove swamp	30
13	Mixed wetland hardwoods	13
14	Wetland forested mixed	18

in Table 12.1. Minor plants such as freshwater marshes were not included, although they also appeared in the study domain.

12.3 Methodology

The methodology flowchart is displayed in Figure 12.3. Hyperspectral imagery commonly has a tremendous amount of redundant spectral information, and thus, the Minimum Noise Fraction (MNF) transformation (Green et al., 1988) was used to reduce the high dimensionality of the data and computation requirements of further processing. MNF has proven valuable for vegetation mapping using hyperspectral imagery (Zhang and Xie, 2012). It applies two cascaded principal component analyses, the first transformation de-correlating and rescaling noise in the data, and the second transformation creating coherent eigenimages that contain useful information and generate noise-dominated eigenimages. The transformation generates the eigenvalues and corresponding eigenimages, both of which are used to determine the true dimensionality of the data. Useful MNF eigenimages typically have an eigenvalue with an order of magnitude greater than those that contain mostly noise. The MNF transformation in ENVI 4.7 was used. A plot of eigenvalues against the output MNF layers was generated, as shown in Figure 12.4. A due inspection of this plot and visual inspection of the eigenimages revealed that 15 MNF layers should be selected for the study site.

After data dimensionality reduction, the MNF imagery was segmented to produce image objects for object-based mapping. Our previous studies have proven object-based image analysis (OBIA) is more effective for vegetation mapping and biomass modeling in the Everglades than traditional pixel-based analysis. Thus, OBIA was applied. Image

FIGURE 12.3
Methodology flowchart to map land cover-level vegetation type using AVIRIS imagery.

segmentation was conducted using eCognition Developer 8.64.1, in which the multi-resolution segmentation algorithms was applied. This algorithm starts with 1-pixel image segments and merges neighboring segments together until a heterogeneity threshold is reached (Benz et al., 2004). The heterogeneity threshold is determined by a user-defined scale parameter, as well as color/shape and smoothness/compactness weights. The image segmentation is scale-dependent, and the quality of segmentation and overall object-based classification are largely dependent on the scale of the segmentation (Liu and Xia, 2010). Here, the scale parameter was set based on the identified scale used in Zhang and Xie (2012) because the same hyperspectral imagery was used. Similarly, the first 5 MNF layers were given equal weights with a value of 5 and the last 10 MNF layers were given equal weights with a value of 1. Color/shape weights were set to 0.9/0.1 so that spectral information would be considered most heavily for segmentation. Smoothness/compactness weights were set to 0.5/0.5 so as to not favor either compact or non-compact segments. The segmentation result of the MNF imagery is displayed in Figure 12.5. After segmentation, the mean spectrum, and standard deviation of each image object were determined and used for classification.

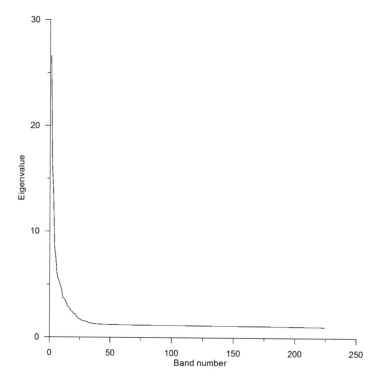

FIGURE 12.4
Eigenvalue of each band in the MNF transformation.

To identify the best classifier for vegetation mapping, the reference image objects (i.e., their class labels were already known based on the reference map) were used to calibrate and validate 4 machine learning classifiers including Random Forest (RF), Support Vector Machine, k-Nearest Neighbor, and Artificial Neural Network (ANN). These machine learning algorithms have been popular in remote sensing classifications (Maxwell et al., 2018) and have had a good performance for vegetation mapping in the Everglades using hyperspectral imagery (Zhang and Xie, 2013; Zhang, 2014). Thus, here these 4 classifiers were tested. RF is a decision-tree based ensemble classifier. Detailed descriptions of RF can be found in Breiman (2001), and a recent review of RF in remote sensing was given in Belgiu and Drăguţ (2016). Two parameters need to be defined in RF: the number of decision trees to create and the number of randomly selected variables considered for splitting each node in a tree. SVM is a supervised classifier with the aim of finding a hyperplane that can separate the input dataset into a discrete predefined number of classes in a fashion consistent with the training samples (Vapnik, 1995). A detailed review of SVM in remote sensing was provided by Mountrakis et al. (2011). Researchers commonly use the kernel-based SVM algorithms in the classification, among which the radial basis function and polynomial kernels are the most applied kernel functions. Both kernels were tested in this chapter. k-NN identifies objects based on the closest training samples in the feature space. It searches away in all directions until it encounters k user-specified training objects and then assigns the object to the

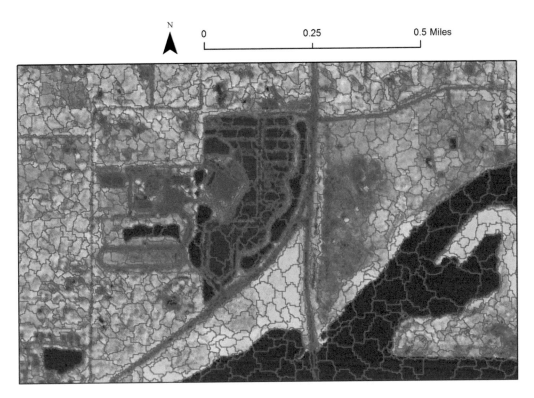

FIGURE 12.5
Segmentation result overlaid on a color composite produced from the first 3 MNF layers.

class with the majority vote of the encountered objects. k-NN needs to set the type of distance measures and the choice of k value. A recent review of this technique in remote sensing was published by Chirici et al. (2016). ANN is also an important technique in image classification. Various ANN algorithms have been developed and applied in remote sensing, as reviewed by Mas and Flores (2008). Here the multilayer perceptron algorithm of ANN was used. This algorithm needs to tune the number of hidden layers, learning rate, and the number of training cycles.

The performance of each classifier was assessed using the traditional error matrix and Kappa statistics techniques. The error matrix is summarized as an overall accuracy and Kappa value. The overall accuracy is defined as the ratio of the number of correctly classified validation samples to the total number of validation samples irrespective of class. The Kappa value describes the proportion of correctly classified validation samples after random agreement is removed. If classifiers have a comparative accuracy in terms of overall accuracy and Kappa value, by analyzing their classification outputs, a spatial uncertainty map can be generated. In this chapter, the k-fold cross-validation technique was used for classifier calibration and validation. This evaluation method has proven valuable in machine learning techniques for both classifications and estimation (Anguita et al., 2012). It splits the sampling data into k-subsets first, and then, iteratively, some of them are used to train a classifier and the others are exploited to assess the performance of this classifier. Based upon research from Anguita et al. (2012), k was specified as 4 because a relatively small number of reference samples were selected (331), rather than the 10-fold cross-validation, which was commonly used in the literature. Once the best classifier is identified, the final vegetation map can be produced by classifying all the image objects using the best classifier.

To assess the value of hyperspectral data for vegetation mapping in the Everglades, the results from AVIRIS were compared with the application of Ikonos imagery. Since there was no real Ikonos imagery for the study domain, the AVIRIS imagery was spectrally resampled to Ikonos spectral configuration in ENVI 4.7, resulting in a simulated Ikonos imagery with the same spatial resolution of AVIRIS. Similar processing procedure was conducted to the simulated Ikonos imagery including segmentation, spectral mean and standard deviation extraction, reference data collection, and classification. To make the results comparable, the same reference samples should be used for each class. The reference objects of AVIRIS were first converted from polygon features to point features in ArcGIS, and then the point features were spatially joined to the segmentation polygons of Ikonos. In this way, polygons of Ikonos with known labels were then exported out as a text file to be used as training and testing file for classification.

12.4 Results and Discussion

The performance of each algorithm to classify the MNF transformed AVIRIS imagery and the simulated Ikonos imagery at the object-level is displayed in Table 12.2. RF and SVM produced a comparable accuracy with an overall accuracy of 74% and Kappa value of 0.71, while k-NN generated the lowest accuracy with an overall accuracy of 68.9% and Kappa value of 0.66 in classifying the MNF transformed imagery. When the AVIRIS was resampled to Ikonos with 4 spectral bands, the accuracy was largely reduced, with the highest overall accuracy of 61.7% obtained from RF. This illustrates that fine spectral resolution hyperspectral data does produce a better result for wetland mapping than multispectral imagery if spatial resolution is consistent. For the study site, AVIRIS hyperspectral imagery produced acceptable accuracy from the RF and SVM classifiers. Thus, land cover-level vegetation maps were produced from these two classifiers, as shown in Figure 12.6. In general, two classifications illustrate a consistent spatial pattern

TABLE 12.2

Performance of each classifier and each dataset

	MNF transformed AVIRIS data			
	RF	**SVM**	**k-NN**	**ANN**
Overall accuracy (%)	74.0	74.0	68.9	71.3
Kappa value	0.71	0.71	0.66	0.68
	Ikonos data resampled from AVIRIS			
Overall accuracy (%)	61.7	50.3	51.3	49.4
Kappa value	0.58	0.44	0.46	0.43

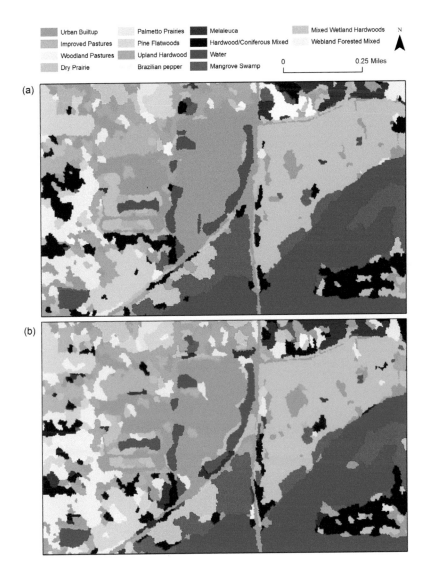

FIGURE 12.6
Classification map from (a) RF classifier and (b) SVM classifier.

of vegetation distribution in identifying major vegetation types such as dry prairie, pine flatwood, and mangrove swamps. Difference in classifications has also been demonstrated such as the delineation of Melaleuca, an exotic species in Florida. The reference map in Figure 12.2 shows that Melaleuca only appears on the northeastern region of the study domain, but SVM classified other land cover types into Melaleuca shown in red in Figure 12.6 (b). The classification maps present more details of vegetation patches than the reference map.

The error matrix of two classifications from RF and SVM is displayed in Tables 12.3 and 12.4, respectively. Although the overall accuracy and Kappa values of two classifiers were consistent (74% and 0.71), their performance in classifying each individual class

TABLE 12.3

Error matrix of RF classification using MNF transformed AVIRIS imagery

Class	1	2	3	4	5	6	7	8	9	10	11	12	13	14	Row Total	UA (%)
1	59	1													60	98.3
2	1	18	1	2											22	81.8
3	1		9	2		1		1		2					16	56.3
4				39			1			1		1			42	92.9
5	1			1		2									4	0.0
6			1			19				3		1		1	25	76.0
7	1	1	4			4	6			4		1	1		22	27.3
8							2			1			4		7	0.0
9				2			2	1	5	1		1			12	41.7
10			2			3	2		1	15		1	1	2	27	55.6
11	2										31				33	93.9
12										1	1	28			30	93.3
13							1	1		2		1	8		13	61.5
14						5	3		1	1				8	18	44.4
Col. Total	65	20	17	46	0	34	17	3	7	31	32	34	14	11	OA: 74%	
PA (%)	90.8	90.0	52.9	84.8	0.0	55.9	35.3	0.0	71.4	48.4	96.9	82.4	57.1	72.7	Kappa: 0.71	

TABLE 12.4

Error matrix of SVM classification using MNF transformed AVIRIS imagery

Class	1	2	3	4	5	6	7	8	9	10	11	12	13	14	Row Total	UA (%)
1	58	1	1												60	96.7
2		19	2				1								22	86.4
3			8			1	2		1	4					16	50.0
4	2			38			1			1					42	90.5
5	1				2				1						4	50.0
6			1			18	1			1			2	2	25	72.0
7	1		2	1		2	7		2	5			2		22	31.8
8						1		3	2				1		7	0.0
9		1					2		6	1			2		12	50.0
10			2			1	4		3	12			2	3	27	44.4
11	1									1	31				33	93.9
12	1			1			1			1		26			30	86.7
13										1		1	11		13	84.6
14						7			1	1				9	18	50.0
Col. Total	64	21	16	40	2	29	20	0	17	30	31	27	20	14	OA: 74%	
PA (%)	90.6	90.5	50.0	95.0	100.	62.1	35.0	0.0	35.3	40.0	100.	96.3	55.0	64.3	Kappa: 0.71	

was different. The producer's accuracy (PA) of RF varied from 0.0% for discriminating palmetto prairies and Brazilian pepper to 96.9% for classifying water. From the user's perspective, the accuracies of RF changed from 0.0% for mapping Brazilian pepper to 98.3% for mapping urban built-up (Table 12.3). The highest PA among vegetation identification was to classify improved pastures (90.0%), while the highest user's accuracy (UA) was to map the mangrove swamp (93.3%). For SVM classification, the PA was in the range of 0% to 100%, while UA varied between 0.0% and 96.7%. Both classifiers completely failed to identify Brazilian pepper (exotic species) in terms of the reference samples. However, the final classification map in Figure 12.6(a) does show the presence of Brazilian pepper in the location consistent with the position of Brazilian pepper in the reference map (Figure 12.2). Thus, though the RF classifier failed to identify the reference objects of Brazilian pepper, it still could map this exotic species. The major factor impacting the failure of classifying Brazilian pepper of the reference samples was the low number of reference samples (i.e., 7). A misleading accuracy assessment result can be derived due to a small number of reference samples caused by the minor presence of a vegetation type in a project domain. The statistical result may not be reliable due to limited samples. The same logic occurred to the classification of palmetto prairies, which had only 4 reference samples. In the calibration and validation procedure, the k-fold cross-calibration and validation approach randomly groups the reference samples into training and testing data and determines the error matrix based on the testing data within the reference samples. However, to generate a more robust classification map, all reference samples were used to calibrate the classifier. Thus, the error matrix may not have the true accuracy of final classification maps.

Lower accuracy was obtained in this chapter than in the study of Zhang and Xie (2012) who used the same hyperspectral imagery but mapped a larger area for the study site. They achieved an overall accuracy of 94% in discriminating 15 land cover-level vegetation types. The lower accuracy was attributed to several factors. First, the aim of our previous study in Zhang and Xie (2012) was to develop a new neural network classifier capable of modeling the characteristics of multiple spectral and spatial signatures within a class and characterizing spectral and spatial differences between classes. This neural network performed well in classifying hyperspectral imagery using spatial and spectral features derived from the imagery for vegetation mapping in a complex wetland environment. However, a newly developed classifier has not always been integrated into commonly used software packages, such as ENVI and the open source WEKA used in this chapter. This limits their applications. Selection of classifiers impacts the mapping result. Second, the previous study combined spatial features (i.e., texture measures) in the classification, while this chapter only integrated the spectral standard deviation. A synergy of texture measures in the classification has a potential to improve 8% in terms of overall accuracy. Lastly, the number of reference samples influences the classification result. Our previous study selected 2580 reference objects; in this chapter only 331 reference objects were selected. As discussed above, a small number of reference samples will impact the performance of the classifiers. Although research has shown that a key advantage of SVMs is that they can achieve better results than other classifiers when a limited amount of reference data is available. However, these reference samples need to be representative, which might be difficult in the reference sample selection. More reference samples can produce a more robust classification map. For minor plants present in a project domain, reference samples should be selected as often as possible.

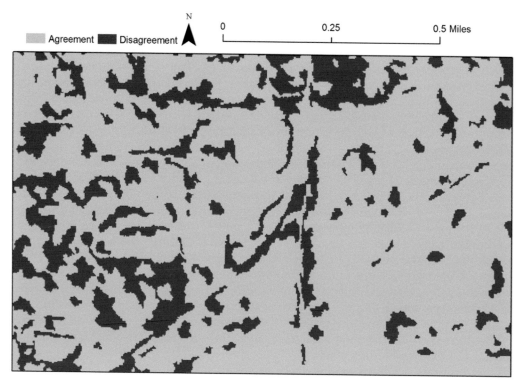

FIGURE 12.7
Classification consistence and difference between RF and SVM classifiers. Green refers to agreement between two classifications, while red refers to disagreement between them.

As demonstrated above, SVM and RF produced comparative accuracy but showed difference in classifying. The consistence and difference of the classification outputs from two classifiers was mapped in Figure 12.7 by simply comparing the two outputs. Agreement was achieved for a majority of the study site (shown in green), while inconsistence was observed for many regions (shown in red). These regions in disagreement had a higher probability of misclassification. To improve classification for these regions, an ensemble analysis of the outputs from each classifier using the scheme developed in Zhang (2014) can be used, but the ensemble classification was not explored. The uncertainty map compares the two outputs and provides some complementary information to the error matrix of the classified map, which can guide the post-classification fieldwork.

12.5 Summary and Conclusions

This chapter presented an application of hyperspectral imagery for vegetation mapping in the central Everglades. Four machine-learning classifiers were examined, and object-based vegetation maps were produced. An overall accuracy of 74% and a Kappa value of 0.71 were obtained using the RF and SVM classifiers. An improved classification could be produced if spatial features such as texture measures were combined in the classification and more reference samples were selected. The experimental results confirmed findings in the literature that hyperspectral data are more valuable in vegetation classification than multispectral data. Hyperspectral imagery is powerful for classifying a large number of land cover-type vegetation types. A broad application of hyperspectral data is still constrained due to the higher cost in data collection using airborne platforms. It is anticipated that a spaceborne hyperspectral sensor with fine spatial resolution can be launched and become operational for complicated wetland mapping.

References

Anguita, D., Ghelardoni, L., Ghio, A., et al. 2012. The 'K' in *k*-fold cross validation. In: *ESANN 2012 Proceedings, European Symposium on Artificial Neural Networks, Computational Intelligence and Machine Learning*, Bruges, Belgium, April 25–27, 441–446.

Artigas, F.J., and Yang, J.S., 2005. Hyperspectral remote sensing of marsh species and plant vigour gradient in the New Jersey Meadowlands. *International Journal of Remote Sensing*, 26, 5209–5220.

Belgiu, M., and Drăguţ, L., 2016. Random forest in remote sensing: A review of applications and future directions. *ISPRS Journal of Photogrammetry and Remote Sensing*, 114, 24–31.

Belluco, E., Camuffo, M., Ferrari, S., et al. 2006. Mapping salt-marsh vegetation by multispectral and hyperspectral remote sensing. *Remote Sensing of Environment*, 105, 54–67.

Benz, U., Hofmann, P., Willhauck, G., et al. 2004. Multiresolution, object-oriented fuzzy analysis of remote sensing data for GIS-ready information. *ISPRS Journal of Photogrammetry and Remote Sensing*, 58, 239–258.

Breiman, L., 2001. Random forests. *Machine Learning*, 45, 5–32.

Chirici, G., Mura, M., McInerney, D., et al. 2016. A meta-analysis and review of the literature on the k-Nearest Neighbors technique for forestry applications that use remotely sensed data. *Remote Sensing of Environment*, 176, 282–294.

Green, A.A., Berman, M., Switzer, P., and Craig, M.D., 1988. A transformation for ordering multispectral data in terms of image quality with implications for noise removal. *IEEE Transactions on Geoscience and Remote Sensing*, 26, 65–74.

Hirano, A., Madden, M., and Welch, R., 2003. Hyperspectral image data for mapping wetland vegetation. *Wetlands*, 23, 436–448.

Jollineau, M.Y., and Howarth, P.J., 2008. Mapping an inland wetland complex using hyperspectral imagery. *International Journal of Remote Sensing*, 29, 3609–3631.

Liu, D., and Xia, F., 2010. Assessing object-based classification: Advantages and limitations. *Remote Sensing Letters*, 1, 187–194.

Mas, J.F., and Flores, J.J., 2008. The application of artificial neural networks to the analysis of remotely sensed data. *International Journal of Remote Sensing*, 29, 617–663.

Maxwell, A.E., Warner, T.A., and Fang, F., 2018. Implementation of machine-learning classification in remote sensing: An applied review. *International Journal of Remote Sensing*, 39, 2784–2817.

Mountrakis, G., Im, J., and Ogole, C., 2011. Support vector machines in remote sensing: A review. *ISPRS Journal of Photogrammetry and Remote Sensing*, 66, 247–259.

Pengra, B.W., Johnston, C.A., and Loveland, T.R., 2007. Mapping an invasive plant, Phragmites australis, in coastal wetlands using the EO-1 Hyperion hyperspectral sensor. *Remote Sensing of Environment*, 108, 74–81.

Rosso, P.H., Ustin, S.L., and Hastings, A., 2005. Mapping marshland vegetation of San Francisco Bay, California, using hyperspectral data. *International Journal of Remote Sensing*, 26, 5169–5191.

Vapnik, V.N., 1995. *The Nature of Statistical Learning Theory*, Springer-Verlag, New York.

Zhang, C., 2014. Combining hyperspectral and lidar data for vegetation mapping in the Florida Everglades. *Photogrammetric Engineering & Remote Sensing*, 80, 733–743.

Zhang, C., Selch, D., and Cooper, H., 2016. A framework to combine three remotely sensed data sources for vegetation mapping in the central Florida Everglades. *Wetlands*, 36, 201–213.

Zhang, C., and Xie, Z., 2012. Combining object-based texture measures with a neural network for vegetation mapping in the Everglades from hyperspectral imagery. *Remote Sensing of Environment*, 124, 310–320.

Zhang, C., and Xie, Z., 2013. Object-based vegetation mapping in the Kissimmee River watershed using HyMap data and machine learning techniques. *Wetlands*, 33, 233–244.

Zhang, C., and Xie, Z., 2014. Data fusion and classifier ensemble techniques for vegetation mapping in the coastal Everglades. *Geocarto International*, 29, 228–243.

13

Species-level Vegetation Mapping in the Kissimmee River Floodplain using HyMap Data

Caiyun Zhang

13.1 Introduction

Both multispectral and hyperspectral imagery have been widely used to map wetlands, as reviewed by Adam et al. (2010). However, wetland mapping down to the plant species-level is rare. Efforts have been made to map salt and freshwater marsh species using hyperspectral remote sensing (e.g., Artigas and Yang, 2005; Li et al., 2005; Belluco et al., 2006; Judd et al., 2007; Wang et al., 2007) with varying degrees of accuracy. A comparison study of 5 data sources by Belluco et al. (2006) demonstrated that the spatial resolution is more important than the spectral resolution in mapping marsh species; multispectral sensors such as Ikonos and QuickBird can achieve accuracy similar to hyperspectral sensors for mapping saltwater marshes. Zhang and Xie (2013) made a first effort to map a large number of vegetation species in the Greater Everglades using hyperspectral imagery and achieved high accuracy. In this chapter, hyperspectral data was compared with multispectral data to evaluate the benefits of hyperspectral imagery for wetland vegetation species identification and mapping in the Kissimmee River Floodplain. A species-level map was produced. As in the last chapter, modern machine learning algorithms and object-based image analysis (OBIA) were integrated to map wetlands down to the species level.

13.2 Study Area and Data

The study area is located in the lower basin of the Kissimmee River watershed (Figure 13.1). Historically, Kissimmee River flowed south from Lake Kissimmee to Lake Okeechobee over a 103-mile path, which created a vast floodplain and supported a diverse mosaic of wetland communities. In the late 1960s, the Kissimmee River was transformed into a 56-mile-long canal, called Canal-38, for flood control. The channelization disrupted the entire riverine ecosystem. Dominant wetland vegetation types such as broadleaf marsh, wet prairie, and wetland shrub were replaced by dry pasture and other upland vegetation types (Bousquin et al., 2005). In 1992, the US Congress approved the Kissimmee River Restoration Project (KRRP) to restore the original river ecosystem and its associated plant communities. The upper basin of the river contains a series of connected lakes, and the lower basin contains Canal-38, remnants of the Kissimmee River, and 6 water control structures. The river and floodplain slope to the south from an elevation of 15.5 m at Lake Kissimmee to an elevation

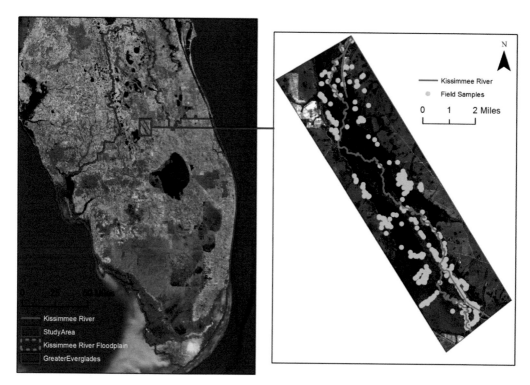

FIGURE 13.1
Geographic location of Kissimmee River Floodplain in the Greater Everglades, and study area shown as a nature color composite generated from HyMap hyperspectral imagery.

of 4.6 m at Lake Okeechobee. As the headwaters of Lake Okeechobee and Everglades National Park, the river plays a critical role in the Greater Everglades by controlling the water quality and quantity, as well as the integrity of the river and floodplain within the landscape. The selected study area is the region known as Pool C in CERP and consists of a portion of Canal-38 that was backfilled in 2001 with a new river channel carved to allow the river to flow naturally to Lake Okeechobee (Bousquin et al., 2005). This particular region is currently in its post-construction monitoring phase. The degree of change in the river and floodplain vegetation community is one of the factors used to judge the success of the restoration.

Data used in this chapter includes hyperspectral imagery and field collected species data. The South Florida Water Management District (SFWMD) provided the hyperspectral data and field surveyed data for the study in Zhang and Xie (2013), which was also used in this chapter. SFWMD conducted a Kissimmee River Restoration Remote Sensing Pilot Study Project during 2001 to 2003, in which hyperspectral and field data were collected. This pilot project collected the hyperspectral data (3 flightlines, 107 km²) over the Pool C area on September 21, 2002, using the HyMap sensor. HyMap is an airborne imaging spectrometer that provides 128 spectral bands across the wavelength region of 0.45 to 2.5 μm with bandwidths between 15 and 20 nm. The spatial and spectral

TABLE 13.1

Spatial and spectral configuration of HyMap in imagery collection

Spatial Configuration	
IFOV	2.5 mr along track, 2.0 mr across track
FOV	61.3 degrees (512 pixels)
GIFOV	3–10 m (typical operational range)
Swath	2.3 km at 5m IFOV (along track)
	4.6 km at 10m IFOV (along track)

IFOV: instantaneous field of view; FOV: field of view; GIFOV: ground instantaneous field of view.

Spectral configuration			
Module	Spectral range (µm)	Bandwidth (nm)	Sampling interval (nm)
VIS	0.45–0.89	15–16	15
NIR	0.89–1.35	15–16	15
SWIR 1	1.40–1.80	15–16	13
SWIR 2	1.95–2.48	18–20	17

configuration of HyMap for data collection is displayed in Table 13.1. In the pilot project, imagery was collected with a spatial resolution of 3.5 m. Images were atmospherically and geometrically corrected by the provider (Lowe Engineers and SAIC, 2003) and provided in the projection of Universal Transverse Mercator (UTM) Zone 17N. The imagery was aligned to the US Geological Survey (USGS) orthophotos. A nature color composite generated from the HyMap data for the study domain is displayed in Figure 13.1. To compare the capability of HyMap hyperspectral imagery with the multispectral data for species mapping, the collected HyMap data was spectrally resampled into the Ikonos spectral configuration using the spectral resample function in ENVI, leading to a simulated Ikonos image with the sample spatial resolution of HyMap data. A false color composite of the simulated Ikonos imagery for the study area is displayed in Figure 13.2.

Field data were collected from September 23 to October 6, 2002. A stratified random sampling scheme augmented with several rules was adopted in the field surveys. Sampling points were required to (1) not be within 25 meters of each other, (2) exhibit major variation within each class, and (3) capture larger and more homogenous areas. It was also important that the personnel understand the variation within each class. In total, 361 samples were collected and screened using the field photos and georeferenced image data. Suspect samples were removed, leading to a final sample of 340 points. The field-surveyed locations are shown in Figure 13.1. Each sample represented a 5-m radius circle ($78.5m^2$) within which the location, plant community, and other information (e.g., date and time) were recorded. These samples were labeled based on B-Codes, the finest level of the KRREP vegetation classification system described in Chapter 3 (Table 13.2). The B-Codes in the KRREP represent the percent cover of dominant species and co-occurring species or other land covers (Bousquin et al., 2005) and were used for the species-level classification in this chapter. In total, 57 B-Codes were found, 55 of them represented by 4 or more samples. The number of field samples for each of the 55 B-Codes is listed in Table 13.4.

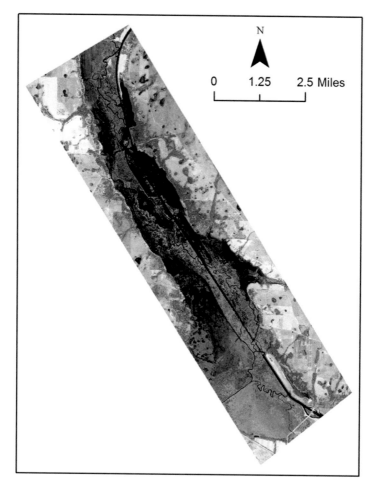

FIGURE 13.2
A false color composite of Ikonos imagery for the study area produced by spectrally resampling HyMap data.

TABLE 13.2

B-Codes defined in the KRREP vegetation classification system. B-Codes were considered the species-level mapping in this chapter. Only B-Codes observed in the study area are listed

B-Codes	Description
1. H.AF	*Axonopus fissifolius* dominant, usually with mixtures of *Paspalum notatum* and other species
	Axonopus fissifolius herbaceous vegetation
2. H.AG	*Andropogon glomeratus* dominant
	Andropogon glomeratus herbaceous vegetation
3. H.CD	*Cynodon dactylon* dominant
	Cynodon dactylon herbaceous vegetation
4. H.CJ	*Cladium jamaicense (sawgrass)* dominant
	Cladium jamaicense herbaceous vegetation

(Continued)

TABLE 13.2 (Cont.)

B-Codes	Description
5. H.CS	*Cyperus* spp. dominant *Cyperus* spp. herbaceous vegetation
6. H.EC	*ichhornia crassipes* dominant *Eichhornia crassipes E* herbaceous aquatic vegetation
7. H.EC-PST	*Eichhornia crassipes* and *Pistia stratiotes* codominant *Eichhornia crassipes-Pistia stratiotes* herbaceous aquatic vegetation
8. H.ES	*Eleocharis* spp. dominant *Eleocharis* spp. herbaceous vegetation
9. H.HA	*Hemarthria altissima* dominant *Hemarthria altissima* herbaceous vegetation
10. H.HG	*Hibiscus grandiflorus* cover equal to or greater than 50% *Hibiscus grandiflorus* herbaceous vegetation
11. H.HU	*Hydrocotyle umbellata* dominant *Hydrocotyle umbellata* herbaceous aquatic vegetation
12. H.JEd	*Juncus effusus* dominant in ponds or depressions that are inclusions within otherwise upland habitats *Juncus effusus* herbaceous vegetation (upland depressions)
13. H.JEp	*Juncus effusus* cover equal to or greater than 30%, not within isolated ponds or depressions *Juncus effusus* herbaceous vegetation
14. H.LH	*Leersia hexandra* dominant *Leersia hexandra* herbaceous vegetation
15. H.MFM	*Scirpus* mats with other herbaceous species dominant Miscellaneous herbaceous floating mat vegetation
16. H.MxE	Invasive exotics dominant (levees, abandoned pastures) Miscellaneous exotic herbaceous vegetation
17. H.MxFA	Aquatic communities dominated by combinations of floating species Miscellaneous aquatic vegetation dominated by floating species
18. H.MxFN	Communities dominated by fern species Miscellaneous fern-dominated herbaceous vegetation
19. H.MxM	Littoral vegetation dominated by unclassified combinations of species including *Sagittaria lancifolia, Pontederia cordata,* and others Miscellaneous littoral marsh vegetation
20. H.MxN	Native terrestrial grasses dominant, usually with scattered shrubs and upland forbs Miscellaneous native herbaceous vegetation
21.H.MxW	Invasive, weedy native species dominant Miscellaneous invasive herbaceous vegetation
22. H.MxWP	Native terrestrial grasses dominant, usually with scattered shrubs and upland forbs Miscellaneous native herbaceous vegetation
23. H.MxWT	Other mixtures of native wetland grasses (e.g., *Phragmites australis, Paspalidium* spp.) and/or graminoids (*Cyperus* spp., *Scirpus californicus, Juncus* spp.) or dominance not clear Miscellaneous native wetland graminoid vegetation
24. H.NL	*Nuphar lutea* dominant *Nuphar lutea* herbaceous aquatic vegetation
25. H.PD	*Polygonum densiflorum* dominant *Polygonum densiflorum* herbaceous aquatic vegetation
26. H.PH	*Panicum hemitomon* cover equal to or greater than 50% *Panicum hemitomon* herbaceous vegetation
27. H.PN	*Paspalum notatum* cover equal to or greater than 50%, usually with mixtures of upland species *Paspalum notatum* herbaceous vegetation

(Continued)

TABLE 13.2 (Cont.)

B-Codes	Description
28. H.PP	*Polygonum punctatum* dominant *Polygonum punctatum* herbaceous vegetation
29. H.PR	*Panicum repens* dominant *Panicum repens* herbaceous vegetation
30. H.PS	*Sagittaria lancifolia* and/or *Pontederia cordata* combined or individual cover equal to or greater than 50%. If present, *Cephalanthus occidentalis* cover less than 5% *Pontederia cordata-Sagittaria lancifolia* herbaceous vegetation
31. H.PS-CO	*Cephalanthus occidentalis* cover 5%-25% in otherwise H.PS herbaceous vegetation *Pontederia cordata-Sagittaria lancifolia-Cephalanthus occidentalis* herbaceous vegetation
32. H.PS-HG	*Hibiscus grandiflorus* cover 30-45% in otherwise H.PS vegetation *Hibiscus grandiflorus-Pontederia cordata-Sagittaria lancifolia* herbaceous vegetation
33. H.PS-PH	*Sagittaria lancifolia* and/or *Pontederia cordata,* and/or cover 10-45%, *Panicum hemitomon* cover equal to or greater than 10%; these 3 species combined making up equal to or greater than 40% cover *Pontederia cordata-Sagittaria lancifolia-Panicum hemitomon* herbaceous vegetation
34. H.PS-PH-CO	*Cephalanthus occidentalis* cover 5%-25% in otherwise H.PS-PH herbaceous vegetation *P. cordata-S. lancifolia-P. hemitomon-C. occidentalis* herbaceous vegetation
35. H.PST	*Pistia stratiotes* dominant *Pistia stratiotes* herbaceous aquatic vegetation
36. H.RN	*Rhynchospora* spp. dominant (usually *R. inundata*) *Rhynchospora* spp. herbaceous vegetation
37. H.SB	*Spartina bakeri* (sand cordgrass) cover equal to or greater than 30% *Spartina bakeri* herbaceous vegetation
38. H.SCF	*Scirpus cubensis* dominant *Scirpus cubensis* herbaceous floating mat vegetation
39. H.SS	*Sacciolepis striata* dominant *Sacciolepis striata* herbaceous aquatic vegetation
40. H.TY	*Typha domingensis* (southern cattail) cover equal to or greater than 50% *Typha dom*ingensis herbaceous vegetation
41. S.CO	*Cephalanthus occidentalis* cover 50% or greater, understory like H.PS herbaceous vegetation *Cephalanthus occidentalis* shrubland
42. S.CO.PS	*Cephalanthus occidentalis* cover 30%-45% cover in otherwise H.PS herbaceous vegetation *Cephalanthus occidentalis-Pontederia cordata-Sagittaria lancifolia* shrubland
43. S.CO-PS-PH	*Cephalanthus occidentalis* cover 30%-45% in otherwise H.PS-PH herbaceous vegetation, understory sometimes composed primarily of wet prairie species (e.g., *Panicum hemitomon*) *C. occidentalis-P. cordata-S. lancifolia-P. hemitomon* shrubland
44. S.HF	*Hypericum fasciculatum* the dominant shrub species *Hypericum fasciculatum* shrubland
45. S.LS	*Ludwigia* spp. (*L. peruviana* and/or *L. leptocarpa*) the dominant shrub, often with *Salix caroliniana, Baccharis halimifolia,* or other shrub species *Ludwigia* spp. shrubland
46. S.LSF	*Ludwigia* spp. (*L. peruviana* and/or *L. leptocarpa)* dominant *Ludwigia* spp. floating mat shrubland
47. S.MC	*Myrica cerifera* (waxmyrtle) usually the dominant shrub species, occasionally approximately codominant with *Ludwigia peruviana, Baccharis halimifolia,* or other woody mesophytes or hydrophytes; not on floating mat vegetation *Myrica cerifera* shrubland
48. S.MCF	*Myrica cerifera* the dominant shrub species *Myrica cerifera* floating mat shrubland

(Continued)

TABLE 13.2 (Cont.)

B-Codes	Description
49. S.MxFS	Other shrub-dominated communities on floating mats Miscellaneous floating mat shrubland
50. S.SC	*Salix caroliniana* the dominant shrub species, sometimes associated with *Ludwigia peruviana* *Salix caroliniana* shrubland
51. S.SR	*Serenoa repens* (saw palmetto) the dominant shrub species *Serenoa repens* shrubland
52. S.ST	*Schinus terebinthifolius* (Brazilian pepper) the dominant shrub species *Schinus terebinthifolius* shrubland
53. V.LM	*Lygodium microphyllum* cover equal to or greater than 30%, typically on living trees or shrubs *Lygodium microphyllium-dominated* communities
54. V.MxV	Other vine species with cover exceeding 30%, typically on living trees or shrubs Miscellaneous vine-dominated communities
55. XUNCL	Unclassified combinations of species Unclassified
56. H.LF	*Luziola fluitans* dominant *Luziola fluitans* herbaceous vegetation
57. H.IV	*Iris virginica* dominant *Iris virginica* herbaceous vegetation

13.3 Methodology

Similar to the land cover-level mapping in the last chapter, OBIA and machine learning classifiers were combined to map vegetation species using the HyMap data. Before segmentation, the low signal-to-noise bands of HyMap imagery were dropped by a visual examination of each band in ENVI. In total, 119 bands remained for further analysis. Non-vegetated areas were masked out using a threshold of NDVI derived from the HyMap data so that the mapping would constrain to the vegetated areas. An open water mask was derived from the Land Cover Land Use 2004 dataset to limit the analysis to the vegetated areas only. The Minimum Noise Fraction (MNF) method was used to reduce the high dimensionality and inherent noise of hyperspectral data. The first 20 MNF eigenimages were spatially coherent and selected. The MNF imagery was then segmented using a relatively smaller scale parameter in the multi-resolution segmentation algorithm (Trimble, 2014) due to the high spatial heterogeneity of species in the study domain. A scale of 6 was found to produce the optimal segmentation (Zhang and Xie, 2013). The weights of the MNF layers were set based on their eigenvalues. Color and shape weights were set to 0.9 and 0.1 so that spectral information would be considered more heavily for segmentation. Smoothness and compactness weights were set to 0.5 and 0.5 so as not to favor either compact or non-compact segments. Part of the segmentation results is displayed in Figure 13.3.

Following segmentation, object boundaries, spatial and spectral attributes of each object were exported from the eCognition software package into ArcGIS. Field surveyed plots were spatially matched with the image objects in ArcGIS, rather than an individual

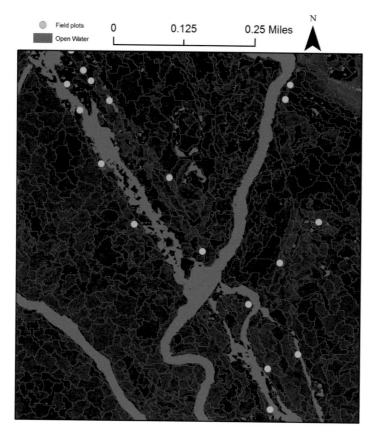

FIGURE 13.3
A small coverage of segmentation in the study domain for object-based species classification and mapping.

pixel where the samples were located. This strategy can reduce the positional discrepancy between the image data and the field data (Zhang et al., 2018). Additionally, it can increase the robustness of object-based image classification since the reference data collected from an image object is more representative than any pixel within this object. The matched dataset was then used to calibrate and validate a classifier for species identification. Final classification for the entire study domain was conducted by calibrating a classifier using all the reference samples. Here, the Support Vector Machine (SVM) and Random Forest were used. Error matrix and Kappa statistics were used for accuracy assessment. Training, testing, and classification includes a data export procedure from ArcGIS to ASCII format, an import of ASCII format into WEKA software package, and an import of outputs from WEKA into ArcGIS for mapping.

The number of species-level field samples is limited for each B-Code, which may cause statistical problems for accuracy assessment. It is difficult to expand the reference data through the photo interpretation technique. To solve this problem, an artificial sample interpolation approach (Demir and Ertürk, 2009) was used to increase the number of reference samples. The first new sample of a class was generated using the mean spectrum calculated from the first sample and all other samples belonging to this class. The second new sample was generated using the mean spectrum calculated from the second sample and the remaining samples with the first sample excluded. The procedure was repeated until the last sample of the corresponding class was processed. Previous samples were not considered in the derivation of new samples. In this way, N(N-1)/2 new samples could be produced for a class, where N is the initial number of samples for this class. Demir and Ertürk (2009) have shown that the interpolation approach is effective for hyperspectral data classification if a limited number of training samples are available. Accuracy assessment was performed on the first 55 B-Codes, which had at least 4 original samples. After interpolation, each B-Code class had at least 10 samples.

To compare the classification results from HyMap data and Ikonos imagery, the simulated Ikonos imagery was segmented and spatially matched to the field plots. Matched reference samples were interpolated to expand the number of reference samples to be used in the Ikonos imagery classification. The same classification and accuracy assessment procedure was followed as the classification of MNF transformed HyMap data to make the results comparable.

13.4 Results and Discussion

The performance of SVM and RF for species classification using HyMap data and Ikonos data is displayed in Table 13.3. Both classifiers achieved a high accuracy in classifying 55 species with SVM produced an overall accuracy of 85% and RF generated an overall accuracy of 79% using the MNF transformed HyMap data. In contrast, low accuracy was produced using the simulated Ikonos data with an overall accuracy of 37% and Kappa value of 0.36 from RF, and an overall accuracy of 30% and Kappa value of 0.28 from SVM, suggesting that spectral resolution is critical for a large number of wetland species classification. This is different from the findings in Belluco et al. (2006) who demonstrated that the spatial resolution is more important than the spectral resolution in mapping marsh species. In their study, multispectral sensors such as Ikonos and Quick-Bird could achieve accuracy similar to hyperspectral sensors for mapping saltwater marshes. Overall accuracy and Kappa value were reported as 36% and 0.36,

TABLE 13.3

Classification accuracies using HyMap data, Ikonos data, and different classifiers

Classifier	MNF transformed HyMap data	
	Overall accuracy (%)	Kappa
SVM	85	0.85
RF	79	0.79
	Ikonos data	
SVM	30	0.28
RF	37	0.36

SVM: Support Vector Machine; RF: Random Forest.

respectively, using the same hyperspectral and field datasets in the pilot project report of SFWMD (Lowe Engineers and SAIC, 2003). The project applied the Spectral Mixture Analysis (SMA) by identifying the endmember of each species first. SMA assumes each species has only one representative spectral signature. This is commonly not true in the Everglades because multiple spectral signatures appear for the same species due to the diversity of local conditions, as demonstrated in Figures 13.4. Three samples were collected for B-Code class H.ES under different flood conditions (dry, partially

FIGURE 13.4

B-Code class H.ES appeared in the study domain under different flooding conditions (from SFWMD).

flooded, and nearly submerged). At 3.5-meter spatial resolution, the spectral signatures of this class are distinctly different but have to be combined to produce one spectral signature in SMA. A similar situation occurs when vegetation exists in different states across the area, as shown in Figure 13.5. The same B-Code H.AG can have markedly different physical appearance and thus appear distinct from each other in the spectral image data. When background and plant conditions are consistent, however, the spectral signature from one class to another can be similar. The application of endmember-based classifiers such as SMA could result in low accuracy. For such a complex situation, machine learning-based classifiers such as SVM and RF used here are more appropriate.

SVM produced a higher accuracy than RF for species-level classification. Thus, the final species classification was conducted from the MNF transformed data and the SVM algorithm, which was calibrated by all the field surveyed data. The object-based species map is displayed in Figure 13.6. Due to the high diversity of B-Codes and spatial heterogeneity of the study domain, the species map was divided into 4 maps with each map showing a limited number of B-Codes. The object-based species map is more informative and useful than a pixel-based map, which may be noisy due to the high degree of spatial and spectral heterogeneity of the Kissimmee River floodplain. A land cover map with a nested species map was also produced, as shown in Figure 13.7. The land cover map was generated using the same hyperspectral imagery with overall accuracy of 90% and Kappa value of 0.89 from the RF method (Zhang and Xie, 2013). The B-Codes were aggregated into B-Code Group level, as shown in Figure 13.8. Land cover types and B-Code group were well delineated through classification of HyMap data.

Table 13.4 shows the producer's and user's accuracies for the classified species map. For comparison, the producer's and user's accuracies from RF method are also displayed. The producer's accuracy using SVM was in the range of 36% to 100%, and user's accuracy was in the range of 41% to 100%. For the RF classifier, both the producer's and user's accuracies ranged from 0% to 100%. The RF completely failed to identify H.MxM (class 19, miscellaneous marsh vegetation) that was misclassified as S.MCF (class 48, floating mat shrubland) or as S.MxFS (class 49, miscellaneous floating mat shrubland). For most of the B-Codes, both

FIGURE 13.5
B-Code class H.AG for two different locations in the study domain (from SFWMD).

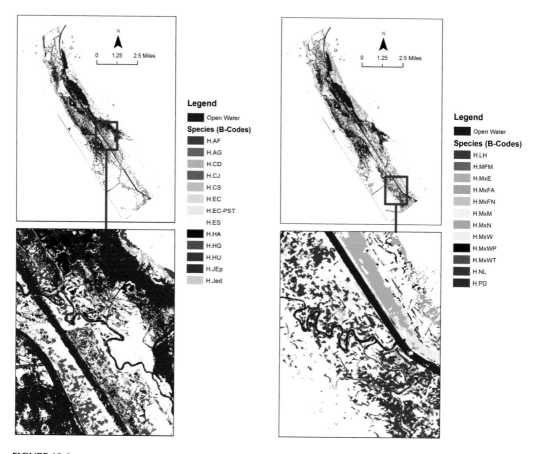

FIGURE 13.6
Species-level map produced using HyMap data and SVM classifier.

SVM and RF performed well. Comparison of the SVM and RF results shows that some species are more easily identified by RF, indicating an ensemble of these 2 classifiers may improve the mapping results. But here the ensemble analysis was used for species mapping.

It is difficult to map heterogeneous wetlands, especially at the species level. Here an encouraging result was achieved for species-level mapping using fine spatial resolution hyperspectral imagery by combining MNF transformation, OBIA, and machine learning algorithms. The study demonstrates the potential of fine spatial resolution hyperspectral data for species mapping in heterogeneous wetland environments. Belluco et al. (2006) have indicated that the use of high spatial resolution dataset for vegetation mapping is particularly advantageous in heterogeneous wetland environments where such datasets can reduce the mixed pixel problem in classification. Although high accuracies were obtained using HyMap data, it is still difficult to detect and map species that are in small or narrow patches with a width of less than

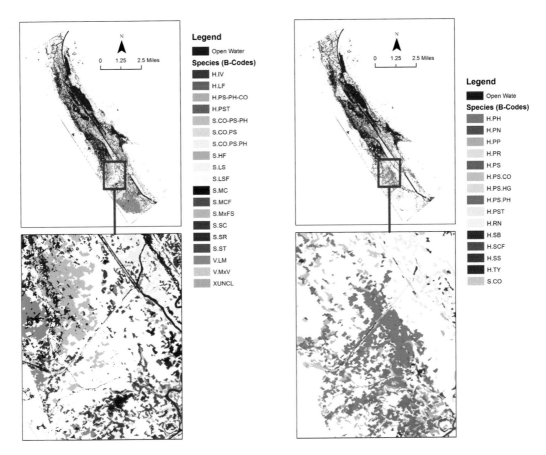

FIGURE 13.6 (Cont.)

3 meters using the 3.5-meter resolution of HyMap data. This could be mitigated by fusing high spatial resolution aerial photography and hyperspectral imagery. The performance may also be improved by increasing the number of field samples to account for the heterogeneity of the study area. Another issue in applying airborne hyperspectral imagery for wetland mapping is the higher cost in data collection. To mitigate this, the data collected can be dissected and reassembled in many different ways, allowing the same data collection effort to be used to address different questions. For example, data collected to assess the impacts of restoration on the vegetation community may also be used for monitoring the spread of exotic vegetation. This would allow cost reduction through economy of scale and cost sharing. Fine spatial resolution hyperspectral imagery has a large data volume, which is a challenge in data processing for a normal computation and processing facility. To solve this, the imagery can be segmented into several small tiles to reduce the data volume, and data processing will be then conducted tile by tile.

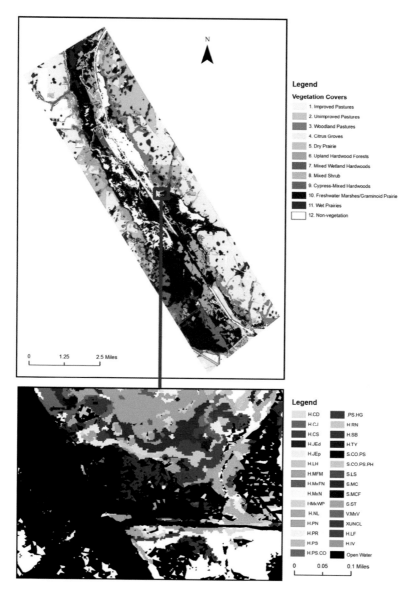

FIGURE 13.7
Land cover-level mapping and species-level mapping using HyMap data.

13.5 Summary and Conclusions

In this chapter, fine spatial resolution hyperspectral imagery was examined to map a large number of plant species in the complex Kissimmee River floodplain by combining OBIA and machine learning algorithms. Overall accuracy of 85% and a Kappa value of 0.85 were achieved for classifying 55 species. The results demonstrate that fine spatial resolution hyperspectral data is promising for wetland species mapping. However, due

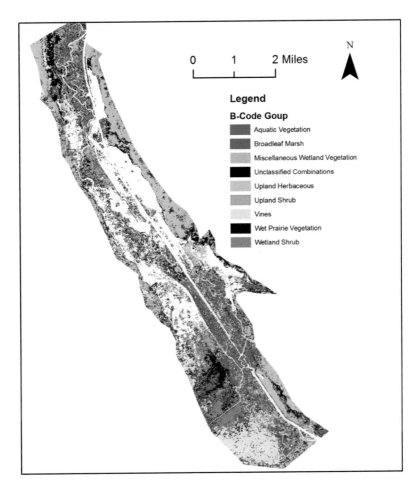

FIGURE 13.8
B-Code group-level mapping aggregated from species-level classification.

TABLE 13.4

Classification accuracies for each individual species

| B-Codes | NFS | SVM | | RF | | B-Code | NFS | SVM | | RF | |
		PA	UA	PA	UA			PA	UA	PA	UA
1. H.AF	6	100	90	90	81	29. H.PR	6	100	91	90	71
2. H.AG	5	87	100	87	87	30. H.PS	10	96	83	100	64
3. H.CD	5	71	41	42	100	31. H.PS-CO	6	90	71	63	63
4. H.CJ	5	62	83	25	100	32. H.PS-HG	6	100	90	70	63
5. H.CS	5	100	100	85	85	33. H.PS-PH	6	100	91	81	90
6. H.EC	6	81	90	100	84	34. H.PS-PH-CO	8	92	100	85	80

(*Continued*)

TABLE 13.4 (Cont.)

B-Codes	NFS	SVM PA	SVM UA	RF PA	RF UA	B-Code	NFS	SVM PA	SVM UA	RF PA	RF UA
7. H.EC-PST	4	80	57	100	71	35. H.PST	6	60	42	60	66
8. H.ES	6	90	100	80	100	36. H.RN	6	90	83	72	80
9. H.HA	6	90	100	90	90	37. H.SB	6	100	90	80	72
10. H.HG	6	100	100	100	83	38. H.SCF	6	90	76	72	80
11. H.HU	6	36	80	63	77	39. H.SS	6	100	100	100	83
12. H.JEd	5	100	100	85	100	40. H.TY	7	92	92	85	85
13. H.JEp	5	100	100	100	100	41. S.CO	7	92	92	85	75
14. H.LH	6	80	72	70	77	42. S.CO-PS	6	90	76	81	64
15. H.MFM	6	90	83	63	87	43. S.CO-PS-PH	6	80	66	60	100
16. H.MxE	6	70	100	80	100	44. S.HF	6	100	83	90	69
17. H.MxFA	4	80	66	80	100	45. S.LS	7	57	80	85	85
18. H.MxFN	6	90	100	81	90	46. S.LSF	6	63	77	36	80
19. H.MxM	5	42	100	00	00	47. S.MC	8	77	93	88	72
20. H.MxN	7	92	81	85	80	48. S.MCF	6	70	77	60	66
21. H.MxW	12	92	90	94	82	49. S.MxFS	7	92	86	92	68
22. H.MxWP	7	85	92	78	84	50. S.SC	6	100	78	90	90
23. H.MxWT	6	54	66	27	50	51. S.SR	4	100	100	80	100
24. H.NL	5	42	60	57	100	52. S.ST	4	40	50	20	100
25. H.PD	6	100	73	63	70	53. V.LM	7	85	92	85	66
26. H.PH	6	80	100	80	88	54. V.MxV	6	70	87	90	81
27. H.PN	6	63	100	81	75	55. XUNCL	6	90	100	81	100
28. H.PP	6	100	83	100	100						
Overall accuracy and Kappa value								Accuracy: 85% Kappa: 0.85		Accuracy: 79% Kappa: 0.79	

SVM: Support Vector Machine; RF: Random Forest; PA: Producer's Accuracy (%); UA: User's Accuracy (%); NFS: Number of Field Samples

to the high heterogeneity of spectral signatures within a class and a high degree of spectral similarity between classes, to improve the capability of hyperspectral data for species-level classification, more field samples should be collected. Here the capability of hyperspectral data might be biased due to limited training samples. Another issue is the high data cost in hyperspectral imagery collection, which may reduce its application for monitoring species change during different phases of the restoration. It is still challenging to use airborne hyperspectral sensors for monitoring purpose in a practical and operational manner. It is anticipated that this study can stimulate more research in wetland species mapping using remote sensing techniques, especially in the Greater Everglades ecosystem.

References

Adam, E., Mutanga, O., and Rugege, D., 2010. Multispectral and hyperspectral remote sensing for identification and mapping of wetland vegetation: A review. *Wetlands Ecology and Management*, 18, 281–296.

Artigas, F.J., and Yang, J.S., 2005. Hyperspectral remote sensing of marsh species and plant vigour gradient in the New Jersey Meadowlands. *International Journal of Remote Sensing*, 26, 5209–5220.

Belluco, E., Camuffo, M., Ferrari, S., et al., 2006. Mapping salt-marsh vegetation by multispectral and hyperspectral remote sensing. *Remote Sensing of Environment*, 105, 54–67.

Bousquin, S.G., Anderson, D.H., Williams, G.E., and Colangelo, D.J., 2005. *Establishing a Baseline: Pre-restoration Studies of the Channelized Kissimmee River*, South Florida Water Management District, West Palm Beach, FL, USA., Technical Publication ERA #432.

Demir, B., and Ertürk, S., 2009. Increasing hyperspectral image classification accuracy for data sets with limited training samples by sample interpolation. In: *4th International Conference Recent Advances in Space Technologies*, 367–369.

Judd, C., Steinberg, S., Shaughnessy, F., and Crawford, G., 2007. Mapping salt marsh vegetation using aerial hyperspectral imagery and linear unmixing in Humboldt Bay, California. *Wetlands*, 27, 1144–1152.

Li, L., Ustin, S.L., and Lay, M., 2005. Application of multiple endmember spectral mixture analysis (MESMA) to AVIRIS imagery for coastal salt marsh mapping, a case study in China Camp, CA, USA. *International Journal of Remote Sensing*, 26, 5193–5207.

Lowe Engineers and SAIC, 2003. *Kissimmee River Restoration Remote Sensing Pilot Study Project*. South Florida Water Management, West Palm Beach, Florida, USA.

Trimble, 2014. *eCognition Developer 9.0.1 Reference Book*. Trimble Germany GmbH, Arnulfstrasse 126, D-80636, Munich.

Wang, C., Menenti, M., Stoll, M., Belluco, E., and Marani, M., 2007. Mapping mixed vegetation communities in salt marshes using airborne spectral data. *Remote Sensing of Environment*, 107, 559–570.

Zhang, C., Denka, S., Cooper, H., and Mishra, D.R., 2018. Quantification of sawgrass marsh aboveground biomass in the coastal Everglades using object-based ensemble analysis and Landsat Data. *Remote Sensing of Environment*, 204, 366–379.

Zhang, C., and Xie, Z., 2013. Object-based vegetation mapping in the Kissimmee River watershed using HyMap data and machine learning techniques. *Wetlands*, 33, 233–244.

14

Benthic Habitat Mapping in the Florida Keys using EO-1/Hyperion

Caiyun Zhang

14.1 Introduction

Mapping benthic habitats in a coral reef ecosystem has long been the research focus in coastal remote sensing (Mumby et al., 2004; Hedley et al., 2016). As the "rain forests of the sea," coral reefs are the most diverse ecosystem with a variety of marine species and millions of undiscovered organisms. Their rich biodiversity is considered a living museum and key to finding new medicines in the 21st century. Reefs also offer valuable socio-economic resources worth billions of dollars each year to human societies. Benthic habitats are places where aquatic organisms live on or near the sea floor. These beds of seagrass, coral reef, mud, and sand provide shelter to a rich array of animals. There is a growing need to map benthic habitats of the reef environment in order to provide rapid assessment of health and stress response of these vulnerable ecosystems.

The Florida Keys, also known as the Florida Reefs and Florida Reef Tract, is the only living coral barrier reef in the continental US. It is the largest coral barrier reef system in the world after the Great Barrier Reef and Belize Barrier Reef (Florida Reef, 2019). It is about 6 to 7 km wide and extends 270 km from Fowey Rocks to south of the Marquesas Keys (Figure 14.1). Like other coral reef ecosystems, the Florida Reefs are not only threatened by global climate change and human activities, but also frequently damaged by hurricanes and tropical storms (Rohmann and Monaco, 2005). Since 2014, the Florida Reefs have experienced a multi-year outbreak of coral diseases, which have resulted in the mortality of millions of corals (Florida Reef Tract Coral Disease Outbreak, 2019). Since 2015, the Florida Department of Environmental Protection (DEP) and numerous partners from federal, state and local governments, universities, nongovernmental organizations, and South Florida have been making multifaceted efforts to understand the cause and contributing factors of this disease outbreak. Meanwhile, addressing other coral stressors (e.g., water quality) and understanding the change of corals will increase the ability of the corals to recover.

NOAA, in partnership with many other entities, has made decades of effort to map habitats in the Florida Reefs. Its mapping efforts have focused on the application of multiple data sources including aerial photography, satellite imagery from Ikonos, bathymetric data, and others using a manual interpretation procedure. Many of these maps are old, and updating them is difficult because the manual interpretation procedure is labor-intensive and time-consuming. Remote sensing has long been used to map reef benthic habitats in shallow waters (i.e., < 20 m). Based on the remote sensing data sources, research and application in reef benthic habitat mapping can be grouped into three categories (Zhang et al., 2013). The first is the application of multispectral sensors with a coarse spatial resolution (i.e., 20–30 m or larger), such as Landsat data (e.g.,

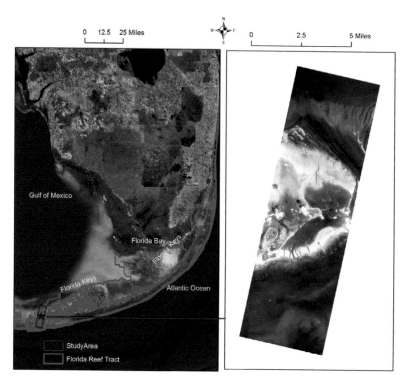

FIGURE 14.1
The Florida Reef Tract and study site shown as a color composite generated from the EO-1/Hyperion imagery.

Purkis and Pasterkamp, 2004). This type of data may have limited effectiveness in mapping heterogeneous benthos due to its relatively coarse spatial resolution. In addition, multispectral data are unable to discriminate more than seven different habitat classes because of their poor spectral resolution (Benfield et al., 2007). The second is the employment of multispectral data with a fine spatial resolution (i.e., 4 m or smaller), such as imagery collected from Ikonos and QuickBird (e.g., Mumby and Edwards, 2002; Andréfouët et al., 2003; Mishra et al., 2006; Phinn et al., 2012; Zapata-Ramírez et al., 2013). This type of data is able to produce higher accuracies than Landsat data for regions with low to intermediate numbers of habitat classes but again cannot be used reliably for mapping fine descriptive detail (e.g., more than 10 classes) (Mumby and Edwards, 2002). The third is the utilization of hyperspectral data, such as imagery collected from Earth Observation-1 (EO-1)/Hyperion, CASI, and HyMap (e.g., Lesser and Mobley, 2007; Mishra et al., 2007; Bertels et al., 2008; Phinn et al., 2008; Fearns et al., 2011; Pu et al., 2012; Botha et al., 2013). Studies have illustrated that this type of data yields higher accuracies than multispectral data in mapping areas with low to intermediate numbers of habitats, but evaluation of their potential in mapping fine descriptive habitats is limited (Lesser and Mobley, 2007). We have assessed the effectiveness of 17-m AVIRIS data for reef habitat mapping in the Florida Keys. We found that the 17-m AVIRIS imagery could produce an overall accuracy of 86.7% for mapping 12-class benthic habitats and application of hyperspectral sensors is promising for detailed reef mapping using automated and semi-automated digital imagery analysis (Zhang et al.,

2013). However, to date, acquisition of airborne hyperspectral data is costly, and application of this type of data is impractical for large area mapping. In this chapter, the spaceborne EO-1/Hyperion sensor was assessed for reef habitat mapping with an aim to looking into the potential of spaceborne hyperspectral sensors for reef habitat mapping in the Florida Keys.

14.2 Study Area and Data

The study area, with an approximate size of 148 km^2, is located in the lower Florida Keys (Figure 14.1). This area has a tropical climate, and its environment is similar to the Caribbean. It is characterized by spectacular coral reefs, extensive seagrass beds, and mangrove-fringed islands. The Florida Keys is one of the world's most productive ecosystems with more than 6000 species of marine life and 250 species of birds. The study area is mainly in a lagoon with water depth less than 10 m based upon the bathymetry data. It was selected because its reef habitats have a high degree of spatial heterogeneity and diversity compared with other regions, which makes it more challenging in mapping using manual or in-situ methods.

Data used in this chapter include an EO-1/Hyperion image and a reference reef habitat map. The EO-1/Hyperion imagery was collected on October 31, 2002, (path/row: 16/43). The EO-1 satellite was launched on November 21, 2000, as a one-year technology demonstration/validation mission. After the initial mission was completed, NASA and USGS agreed to continue the EO-1 program as an extended mission charted to collect and distribute Hyperion hyperspectral and Advanced Land Imager (ALI) multispectral products to users at no cost. EO-1/Hyperion collects imagery with 220 spectral channels from 0.357 to 2.576 μm with a 10 nm bandwidth. It operates in a pushbroom fashion with a spatial resolution of 30 m. A standard scene has a width of 7.7 km and length of 185 km. EO-1 satellite was decommissioned in 2017, but archived products including L1Gst are still available at USGS's EarthExplorer. The Hyperion L1Gst is terrain corrected and provided in 16-bit radiance values in GeoTIFF format.

The reef reference map was sourced from The Unified Florida Coral Reef Tract Map v2.0. In September 2014, the Florida Fish and Wildlife Conservation Commission and the Fish and Wildlife Research Institute published The Unified Florida Coral Reef Tract Map v1.0; this map integrated existing benthic habitat maps provided by multiple entities. This reef map has been updated as new data have become available. The current version (2.0) was released in 2017. Reef habitats were grouped into 5 levels using a unified classification system, with level 0 being the most general level and level 4 the most detailed. The unified classification system mainly follows the classification scheme developed by Zitello et al. (2009), in which the reef geographic zone, geomorphological structure type, and dominant biological cover are considered. Figure 14.2 shows the classification scheme used to define reef benthic habitats with 4 primary attributes and several hierarchical levels. For the selected study area, the reef map was manually interpreted from the Ikonos satellite imagery collected in 2006. The map was calibrated and validated through intensive in-situ methods and was reported to have an accuracy of 90 to 95%. In this chapter, the most general level (level 0) and the most detailed level (level 4) were mapped using Hyperion imagery. Three general reef classes are found based on the reference map: Coral Reef and Hardbottom, Seagrass, and Unconsolidated

Geographic Zone	Geomorphological Structure	Biological Cover
Land	Coral Reef and Hard Bottom	**Major Cover**
Salt Pond	Rock Outcrop	Algae
Shoreline Intertidal	Boulder	Live Coral
Lagoon	Aggregate Reef	Coralline Algae
Reef Flat	Individual Patch Reef	Mangrove
Back Reef	Aggregated Patch Reefs	Seagrass
Reef Crest	Spur and Groove	No Cover
Fore Reef	Pavement	Unknown
Bank/Shelf	Pavement with Sand Channels	**Percent Major Cover**
Bank/Shelf	Reef Rubble	10% - <50%
Escarpment	Rhodoliths	50% - <90%
Channel	Unknown	90% - 100%
Dredged	Unconsolidated Sediment	Unknown
Unknown	Sand	
	Mud	**Coral Cover**
	Sand with Scattered	**Percent Coral Cover**
	Coral & Rock	0% - <10%
	Unknown	10% - <50%
	Other Delineations	50% - <90%
	Land	90% - 100%
	Artificial	Unknown
	Unknown	

FIGURE 14.2

The classification scheme defines the benthic habitats with primary attributes described by separate boxes and several hierarchical levels of the classification scheme (from Zitello et al., 2009).

Sediment, as described in Table 14.1. Land, artificial features, and mangroves are also present in the study area but are masked out as others and not considered in the mapping procedure. For level 4, 17 classes are found, as shown in Table 14.1. The high degree of habitat diversity and spatial heterogeneity of the selected site provided an opportunity to assess the effectiveness of Hyperion for reef habitat mapping.

14.3 Methodology

Object-based image analysis (OBIA) and machine learning classifiers were combined to map reef habitats here, as shown in Figure 14.3. For benthic habitat mapping, 3 image corrections are often conducted first, including atmospheric correction, sun-glint correction, and water column correction. Zhang et al. (2013) systematically assessed the effects of these 3 corrections on reef habitat mapping in the Florida Keys and found that their effects were small if OBIA was applied. Thus, 3 corrections were not carried out for the Hyperion imagery. The EO-1/Hyperion products were georeferenced, but for the

TABLE 14.1

Description of level 0 and level 4 classes to be mapped in this chapter (Zitello et al., 2009)

Level 0	Description
1. Coral Reef and Hard Bottom	Areas of both shallow and deep-water seafloor with solid substrates including bedrock, boulders, and deposition of calcium carbonate by reef-building organisms. Substrates typically have no sediment cover, but a thin veneer of sediment may be present at times especially on low-relief hard-bottoms. Detailed structure classes include Rock Outcrop, Boulder, Spur and Groove, Individual Patch Reef, Aggregated Patch Reefs, Aggregate Reef, Reef Rubble, Pavement, Pavement with Sand Channels, and Rhodoliths.
2. Seagrass	Habitats dominated by any single species of seagrass or a combination of several species.
3. Unconsolidated Sediment	Areas of the seafloor consisting of small particles (<.25 m) with less than 10% cover of large stable substrate. Detailed structure classes of softbottom include Sand, Mud, and Sand with Scattered Coral and Rock.

Level 4	Description
1. Aggregate Reef, Algae	Continuous: 90-100% major biological cover type with nearly continuous coverage of the substrate.
2. Aggregate Reef, Patchy Live Coral	Patchy: 50-90% cover of the major biological type with breaks.
3. Aggregate Reef, Patchy Macroalgae	Sparse: 10-50% cover of the major biological type with breaks.
4. Aggregated Patch Reef, and Discontinuous Algae	Live coral: Substrates colonized with 10% or greater live reef building corals and other organisms.
5. Pavement	Seagrass: Habitats dominated by any single species of seagrass or a combination of several species.
6. Pavement, Algae	Algae: Substrates with 10% or greater distribution of any combination of numerous species of red, green, or brown algae.
7. Pavement, Discontinuous Algae	
8. Pavement, Sparse Live Coral	
9. Pavement, Patchy Macroalgae	
10. Pavement, Discontinuous Seagrass	
11. Sand, Patchy Macroalgae	
12. Sand, Continuous Seagrass	
13. Sand, Patchy Seagrass	
14. Sand, Sparse Seagrass	
15. Sand, Uncolonized	
16. Continuous Seagrass	
17. Unconsolidated Sediment	

selected study area, a big misalignment was observed between the Hyperion and Ikonos images. Since the reference reef map was generated from a manual procedure using the Ikonos imagery collected in 2006, thus the Hyperion imagery was re-georeferenced using the Ikonos imagery to correct the misalignment. Spectral channels with low signal-to-noise were dropped, leading to 30 visible bands and 9 near-infrared bands for further analysis. The reduced imagery was then clipped to the study area, and a color composite generated from this image was shown in Figure 14.1. Hyperspectral data contains a tremendous amount of redundant spectral information. The Minimum Noise Fraction (MNF) method (Green et al., 1988) is commonly used to reduce the high

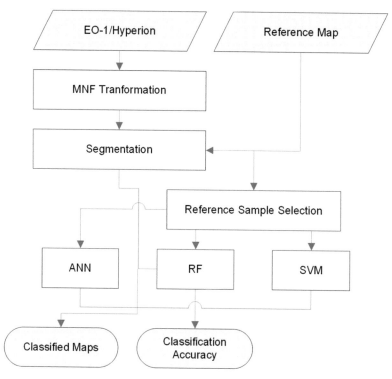

FIGURE 14.3
Methodology flowchart for reef benthic habitat mapping using EO-1/Hyperion data.

dimensionality and inherent noise of hyperspectral data. The MNF transformation was conducted to the 39-band Hyperion imagery and the most useful and spatially coherent eigenimages remained to generate a 10-band MNF imagery for classification. Zhang et al. (2013) have shown that MNF transformation is valuable for reef mapping when hyperspectral data are used. The 10-band MNF imagery was segmented using a relatively larger scale parameter in the multi-resolution segmentation algorithm (Trimble, 2014) for the 3-class level 0 classification, and a smaller scale parameter was used in the segmentation for the 17-class level 4 classification. Other parameters remained unchanged so that the detailed level (level 4) nested within the general level (level 0). The weights of the MNF layers were set based on their eigenvalues. Color and shape weights were set to 0.9 and 0.1 so that spectral information would be considered more heavily for segmentation. Smoothness and compactness weights were set to 0.5 and 0.5 so as to not favor either compact or non-compact segments. Part of the segmentation results is displayed in Figure 14.4.

Following segmentation, object boundaries and spatial and spectral attributes of each object (mean and standard deviation) were exported from the eCognition software package into ArcGIS to be used for object-based classification and reference sample selection. In total, 424 reference objects were labeled based on the reference map to be used as the training and testing data for the 3-class classification, and 1354 objects were labeled as training and testing samples for the 13-class detailed reef habitat classification. The selection of these samples was based upon a random sampling strategy. Three machine learning classifiers (Artificial Neural

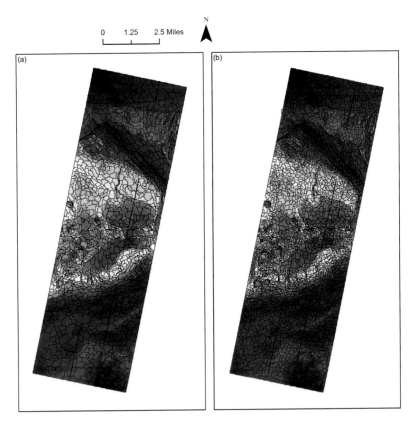

FIGURE 14.4
Segmentation results for (a) 3-class general classification (level 0), and (b) 17-class detailed reef benthic habitat classification (level 4).

Network (ANN), Random Forest (RF), and Support Vector Machine (SVM)) were examined for the classification and the best classifier was selected for final mapping. These machine learning algorithms have been extensively applied in remote sensing classification in the past decades, as reviewed by Mas and Flores (2008) for ANN, Belgiu and Drăguţ (2016) for RF, and Mountrakis et al. (2011) for SVM. More recently, Maxwell et al. (2018) gave an applied review of machine learning methods in remote sensing. Compared with the traditional classifiers such as Maximum Likelihood method, these non-parametric algorithms are more efficient because they do not require data normalization. Classification accuracy was assessed using traditional error matrix techniques based on the k-fold cross-validation approach.

14.4 Results and Discussion

Three classifiers were tuned in the open source software package WEKA, and the best results from each classifier are displayed in Table 14.2. For the level 0 classification with 3 classes, 3

TABLE 14.2

Classification accuracies for 2 reef habitat detail levels using different classifiers

	3-class classification		
	RF	**SVM**	**ANN**
Overall Accuracy (%)	80.4	80.0	77.8
Kappa Value	0.70	0.69	0.66
	17-class classification		
Overall Accuracy (%)	78.0	72.2	71.7
Kappa Value	0.73	0.65	0.66

classifiers produced comparable accuracy in terms of the overall accuracy (OA) and Kappa values. RF achieved the highest accuracy with an OA of 80.4% and Kappa value of 0.7. For the detailed level, 4 classification with 17 classes, again, RF generated the best results among 3 classifications with an OA of 78.0% and Kappa value of 0.73. The RF algorithm has been recognized for its high speed in data processing and less parameters to be specified compared with ANN and SVM. Here, RF was selected to produce the reef benthic habitat classification maps at 2 detail levels, as shown in Figures 14.5 and 14.6. In general, the classified maps are consistent with the spatial pattern of reef benthos presented in the reference map. Based on the level 0 mapping result, the study domain is dominated by seagrass and unconsolidated sediment. Coral reefs are observed as stripes or small patches. The level 4 map further details the presence of algae, patchy algae, patchy seagrass, etc. The accuracy assessment results of the classified maps are displayed in Tables 14.3 and 14.4. As can be seen, seagrass is easier to identify with the highest classification accuracy among 3 classes. For the 17-class classification, the user's accuracies are in the range of 0.0% and 93.9%, while producer's accuracies vary from 0.0% to 91.4%. Note that class 3 of level 4 (Aggregate Reef, Patchy Macroalgae) has only 5 samples, and the 0.0% accuracy might not be reliable due to limited reference samples. An exclusion of the minor classes may improve accuracy.

Application of EO-1/Hyperion successfully revealed the reef patterns for the selected study site, and high accuracy was produced. EO-1/Hyperion has the same spatial resolution of Landsat but a much finer spectral resolution. Previous studies have shown that Landsat could not achieve adequate accuracy for detailed reef mapping. Andréfouët et al. (2003) reported a relationship between the OA and the number of reef benthic habitats to be classified as $Y=-3.90X+86.38$ ($R^2=0.63$), where Y is the OA, and X is the number of classes using Landsat data. Thus, if Landsat multispectral imagery was used for the study domain, an approximate 75% and 20% of OA were expected for the 3-class and the 17-class classification, respectively. Here, for the 3-class classification, the improvement was smaller using EO-1/Hyperion data (80.4% vs. 75%) compared with the 17-class classification (78.0% vs. 20%), suggesting hyperspectral data are more valuable for detailed reef benthic habitat mapping. A previous study using 17-m AVIRIS data for mapping benthic habitat in the Florida Keys showed that an OA of 84.3% and 86.7% were produced to classify 3-class and 12-class benthic habitats, respectively (Zhang et al., 2013). This demonstrates that a higher spatial resolution can also improve classification. Like Landsat, the 30-m

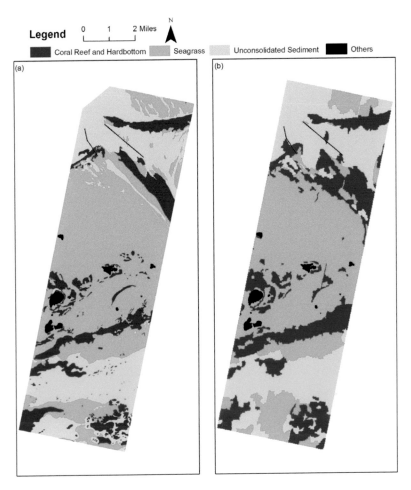

FIGURE 14.5
(a) Reference reef map of level 0 with 3 classes and (b) the classified map using the RF classifier and EO-1/ Hyperion imagery.

EO-1/Hyperion data cannot delineate small patches and narrow linear reef features due to coarse spatial resolution. A fusion of EO-1/Hyperion and fine spatial resolution data collected from Ikonos-type satellites or aerial photography may mitigate this issue. Zhang (2015) has combined 17-m AVIRIS hyperspectral imagery and 1-m aerial photography to improve reef mapping in the Florida Keys and achieved good results. The designed fusion framework can be also used to integrate EO-1/ Hyperion imagery and fine spatial resolution imagery for a better mapping result.

Though EO-1/Hyperion has shown a potential for detailed reef mapping, this sensor was decommissioned in February 2017. Presently, spaceborne multispectral sensors play the major roles in reef mapping. However, several hyperspectral missions were planned to launch soon, such as PRecursore IperSpettrale della Missione Applicativa (PRISMA) Italian mission (30-m; 400–2505 nm) (http://prisma-i.it/index.php/it/), the Environmental Mapping and Analysis Program (EnMAP) German mission (30-m; 400–2500 nm)

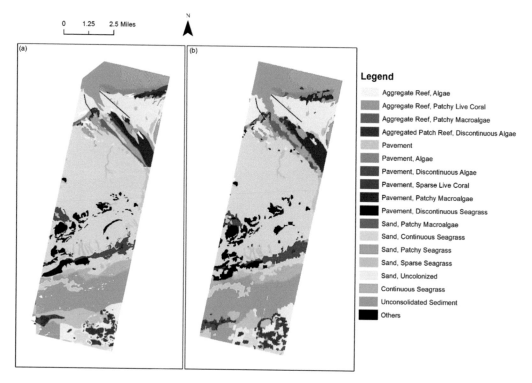

FIGURE 14.6
(a) Reference reef map of level 4 with 17 classes and (b) the classified map using the RF classifier and EO-1/ Hyperion imagery.

TABLE 14.3

Accuracy assessment of the classified map at level 0 from the RF classifier

Class	1	2	3	RT	UA
1	90	20	16	126	71.7
2	13	168	12	193	87.0
3	16	6	82	104	78.8
CT	119	194	110	OA (%): 80.4	
				Kappa value: 0.70	
PA	75.8	86.6	74.5		

RT: Row Total; CT: Column Total; UA (%): User's Accuracy; PA (%): Producer's Accuracy; OA (%): Overall Accuracy. The names of classes 1 to 3 are listed in Table 14.1.

(www.enmap.org/), and the Hyperspectral Imager SUIte (HISUI) Japanese mission (30-m; 400–2500 nm) (ISS Utilization: HISUI, 2019). The launch of these hyperspectral missions and others will increase the application of spaceborne hyperspectral sensors for reef mapping and monitoring.

TABLE 14.4

Accuracy assessment of the classified map at level 4 from the RF classifier

Class	1	2	3	4	5	6	7	8	9	10	11	12	13	14	15	16	17	RT	UA
1	4			3											3		2	12	33.3
2		2	1															5	40.0
3							1	2					3					8	0.0
4	1			21			7			4		1	4		1			36	58.3
5					8												11	19	40.0
6						2		1	6		2			3				14	14.3
7		3		5			41			6		2	7	1	1		3	66	62.1
8			1					2		1	2		1	1			1	12	16.7
9						3			29					2	1	1	1	37	78.4
10							2	1		65			44			1		114	57.0
11						1					6		1	8	1	1	1	17	35.3
12				4								23	1		2		6	36	63.9
13				5	1		1			5	3	1	440	5	2	4	38	499	88.2
14							1		4				8	39	2	6	3	66	59.1
15	1			1			2		1	3		2	31	2	38	6	4	89	42.7
16									3	1			4	2		169	1	180	93.9
17					3							3		2		1	132	141	93.6
CT	6	5	3	39	12	6	55	6	43	85	13	30	543	65	51	185	204	OA (%): 78.0	
PA	66.7	40.0	0.0	53.8	66.7	33.3	74.5	33.3	67.4	76.5	46.2	76.7	81.0	60.0	74.5	91.4	78.6	Kappa: 0.73	

RT, CT, UA, PA, and OA are the same abbreviations as in Table 14.3.

14.5 Summary and Conclusions

In this chapter, a spaceborne hyperspectral sensor, EO-1/Hyperion, was assessed to map benthic habitats in the Florida Keys. OBIA and machine learning classifiers were combined to generate accurate and informative reef benthic habitat maps. The results show that the EO-1/Hyperion sensor is valuable for reef mapping. Application of spaceborne hyperspectral sensors can significantly improve detailed reef habitat classification compared with the application of spaceborne multispectral sensors. Currently, reef habitat mapping in the Florida Reef Tract is mainly from manual interpretation of fine spatial resolution aerial photography or Ikonos-type satellite data. Updating the existing maps using the traditional field and manual methods is difficult. Hyperion-type spaceborne sensors can be used as an alternative to the field and manual methods by using modern digital image processing techniques. For the broad area mapping using optical images, image preprocessing (atmospheric correction, sun-glint correction, and water column correction) should be conducted. With the planned spaceborne hyperspectral missions, the study presented in this chapter is expected to stimulate the application of these sensors in reef environments for global reef conservation and preservation and regional Florida reef mapping.

References

Andréfouët, S., Kramer, P., Torres-Pulliza, D., et al., 2003. Multi-sites evaluation of Ikonos data for classification of tropical coral reef environments. *Remote Sensing of Environment*, 88, 128–143.

Belgiu, M., and Drăguţ, L., 2016. Random forest in remote sensing: A review of applications and future directions. *ISPRS Journal of Photogrammetry and Remote Sensing*, 114, 24–31.

Benfield, S.L., Guzman, H.M., Mair, J.M., and Young, J.A.T., 2007. Mapping the distribution of coral reefs and associated sublittoral habitats in Pacific Panama: A comparison of optical satellite sensors and classification methodologies. *International Journal of Remote Sensing*, 28, 5047–5070.

Bertels, L., Vanderstraete, T., Coillie, S.V., et al., 2008. Mapping of coral reefs using hyperspectral CASI data; A case study: Fordata, Tanimbar, Indonesia. *International Journal of Remote Sensing*, 29, 2359–2391.

Botha, E.J., Brando, V.E., Anstee, J.M., et al., 2013. Increased spectral resolution enhances coral detection under varying water conditions. *Remote Sensing of Environment*, 131, 247–261.

Fearns, P.R.C., Klonowski, W., Babcock, R.C., et al., 2011. Shallow water substrate mapping using hyperspectral remote sensing. *Continental Shelf Research*, 31, 1249–1259.

Florida Reef, 2019. https://en.wikipedia.org/wiki/Florida_Reef. Accessed on 8 May 2019.

Florida Reef Tract Coral Disease Outbreak, 2019. https://floridadep.gov/rcp/coral/content/florida-reef-tract-coral-disease-outbreak. Accessed on 8 May 2019.

Green, A.A., Berman, M., Switzer, P., and Craig, M.D., 1988. A transformation for ordering multispectral data in terms of image quality with implications for noise removal. *IEEE Transactions on Geoscience and Remote Sensing*, 26, 65–74.

Hedley, J.D., Roelfsema, C.M., Chollett, I., et al., 2016. Remote sensing of coral reefs for monitoring and management: A review. *Remote Sensing*, 8, 118.

ISS Utilization: HISUI, 2019. https://eoportal.org/web/eoportal/satellite-missions/content/-/article/iss-utilization-hisui-hyperspectral-imager-suite-#launch. Accessed on 8 May 2019.

Lesser, M.P., and Mobley, C.D., 2007. Bathymetry, water optical properties, and benthic classification of coral reefs using hyperspectral remote sensing imagery. *Coral Reef*, 26, 819–829.

Mas, J.F., and Flores, J.J., 2008. The application of artificial neural networks to the analysis of remotely sensed data. *International Journal of Remote Sensing*, 29, 617–663.

Maxwell, A.E., Warner, T.A., and Fang, F., 2018. Implementation of machine-learning classification in remote sensing: An applied review. *International Journal of Remote Sensing*, 39, 2784–2817.

Mishra, D., Narumalani, S., Rundquist, D., et al., 2007. Enhancing the detection and classification of coral reef and associated benthic habitats: A hyperspectral remote sensing approach. *Journal of Geophysical Research*, 112, C08014.

Mishra, D., Narumalani, S., Rundquist, D., and Lawson, M., 2006. Benthic habitat mapping in tropical marine environments using QuickBird multispectral data. *Photogrammetric Engineering & Remote Sensing*, 72, 1037–1048.

Mountrakis, G., Im, J., and Ogole, C., 2011. Support vector machines in remote sensing: A review. *ISPRS Journal of Photogrammetry and Remote Sensing*, 66, 247–259.

Mumby, P.J., and Edwards, A.J., 2002. Mapping marine environments with Ikonos imagery: Enhanced spatial resolution does deliver greater thematic accuracy. *Remote Sensing of Environment*, 82, 248–257.

Mumby, P.J., Skirving, W., Strong, A.E., et al., 2004. Remote sensing of coral reefs and their physical environment. *Marine Pollution Bulletin*, 48, 219–228.

Phinn, S.R., Roelfsema, C.M., and Mumby, P.J., 2012. Multi-scale, object-based image analysis for mapping geomorphic and ecological zones on coral reefs. *International Journal of Remote Sensing*, 33, 3768–3797.

Phinn, S.R, Roelfsema, C.M., Dekker, A., et al., 2008. Mapping seagrass species, cover and biomass in shallow waters: An assessment of satellite multispectral and airborne hyper-spectral imaging systems in Moreton Bay (Australia). *Remote Sensing of Environment*, 112, 3413–3425.

Pu, R., Bell, S., Meyer, C., et al., 2012. Mapping and assessing seagrass along the western coast of Florida using Landsat TM and EO-1 ALI/Hyperion imagery. *Estuarine, Coastal, and Shelf Science*, 115, 234–245.

Purkis, S.J., and Pasterkamp, R., 2004. Integrating in situ reef-top reflectance spectra with Landsat TM imagery to aid shallow-tropical benthic habitat mapping. *Coral Reef*, 23, 5–20.

Rohmann, S.O., and Monaco, M.E., 2005. *Mapping Southern Florida's Shallow-water Coral Ecosystems: An Implementation Plan*, NOAA Technical Memorandum NOS NCCOS 19. NOAA/NOS/NCCOS/CCMA, Silver Spring, MD.

Trimble, 2014. *eCognition Developer 9.0.1 Reference Book*. Trimble Germany GmbH, Arnulfstrasse 126, D-80636, Munich.

Zapata-Ramírez, P.A., Blanchon, P., Olioso, A., et al., 2013. Accuracy of Ikonos for mapping benthic coral-reef habitats: A case study from the Puerto Morelos Reef National Park, Mexico. *International Journal of Remote Sensing*, 34, 3671–3687.

Zhang, C., 2015. Applying data fusion techniques for benthic habitat mapping and monitoring in a coral reef ecosystem. *ISPRS Journal of Photogrammetry and Remote Sensing*, 104, 213–223.

Zhang, C., Selch, D., Xie, Z., et al., 2013. Object-based benthic habitat mapping in the Florida Keys from hyperspectral imagery. *Estuarine, Coastal and Shelf Science*, 134, 88–97.

Zitello, A.G., Bauer, L.J., Battista, T.A., et al., 2009. *Shallow-Water Benthic Habitats of St. John, U.S. Virgin Islands*, NOAA Technical Memorandum NOS NCCOS 96, Silver Spring, MD, USA.

Part IV

Lidar Remote Sensing Applications

15

Vulnerability Analysis of Coastal Everglades to Sea Level Rise using SLAMM

Hannah Cooper and Caiyun Zhang

15.1 Introduction

The coastal parks of the National Park Service (NPS) provide critical habitat areas for nesting shorebirds and many threatened or endangered species and are home to historic forts and lighthouses. Sea level rise (SLR) presents one of the greatest challenges of the 21st century to national parks (Caffrey and Beavers, 2013). This complicates natural resource managers' ability to monitor wildlife habitats, manage park infrastructure, and inform restoration efforts within the parks. SLR is an important conservation challenge already threatening wildlife habitats in Everglades National Park (ENP) (Stabenau et al., 2011).

ENP is projected to experience the highest rate of SLR by the end of the century (Caffrey et al., 2018). ENP is especially vulnerable to the physical impacts of SLR because physiographic regions are characterized by flat land with elevations at or below sea level. The physical impacts of SLR in ENP are being observed as saltwater intrusion into previously freshwater areas (Stabenau et al., 2011; Park et al., 2017a, 2017b), leading to peat collapse in the coastal sawgrass and mangrove ecosystems. Peat loss can affect the integrity and elevation of the soil that supports these habitats. In addition, saltwater inundation will likely cause a decline in wading birds and other wildlife dependent on freshwater habitat in ENP (Pearlstine et al., 2010). More research is needed to assess how these changes may affect the parks (Caffrey et al., 2018).

In this chapter, we evaluated the potential of habitat change and loss in coastal ENP under a high SLR scenario by year 2050 using the Sea Level Affecting Marshes Model (SLAMM). SLAMM simulates the dominant processes involved in wetland conversions and shoreline modifications due to SLR (e.g., inundation, erosion, accretion). It uses a decision tree that incorporates geometric and qualitative relationships to simulate habitat change (Clough et al., 2012). SLAMM has been applied in Waccasassa Bay, FL, to assess coastal marsh and forest loss (Geselbracht et al., 2011), but to date, no studies have been conducted in ENP. Light detection and ranging (lidar) Digital Elevation Models (DEMs) are frequently applied in SLR modeling and mapping due to their better vertical accuracy (e.g., <0.15 m) and high horizontal resolution (e.g., <5 m) (Gesch, 2009). However, it is a challenge for lidar to obtain accurate measures in ENP due to the complexity of coastal marshes and swamps. Since DEMs are essential baseline datasets used in SLAMM, model predictions are prone to the vertical accuracy and errors of DEMs (Clough et al., 2012). Here, we first applied a correction procedure to improve lidar-DEM and then used the improved DEM in SLAMM with an aim to better analyze the vulnerability of coastal Everglades to SLR.

15.2 Study Area and Data

Our study area is located in the coastal ENP with an approximate size of 2220 km^2 (Figure 15.1). It is at the interface of fresh and brackish water with unique habitats including sawgrass freshwater prairies, mangroves, ridge-and-slough habitat, and estuarine habitats. These habitats will likely be lost or converted to other habitat types due to SLR inundation.

DEMs are frequently used for characterizing the current land elevation relative to the sea. The current best available lidar-DEM for the study area was created by the South Florida Water Management District (SFWMD) using the Florida Division of Emergency Management lidar classified bare-earth returns collected in August 2007, and January to February 2008. The SFWMD included breaklines to generate a Triangular Irregular Network and create a 3 m lidar-DEM using the natural neighbor interpolation method. This lidar-DEM is available to the public at www.sfwmd.gov/science-data/gis. Coastal Everglades' land elevations are often submerged and obscured by thick vegetation, which makes it difficult for lidar to penetrate to the ground. For this reason, the US Geological Survey (USGS) developed an approach using differential Global Positioning Systems (GPS) onboard either an airboat or helicopter, depending on accessibility, resulting in the High

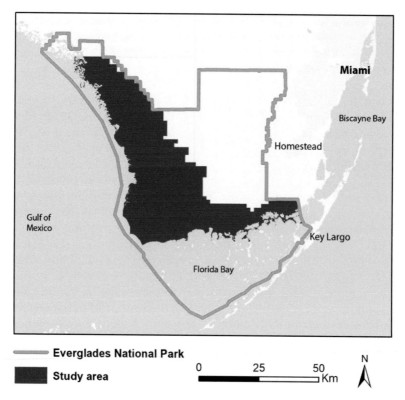

FIGURE 15.1
Study area of coastal Everglades National Park.

Accuracy Elevation Dataset (HAED). The Airborne Height Finder (AHF) is a combination of a GPS platform and surveyor's plumb bob mounted on a helicopter (hovering 3 to 6 m above ground) used for collecting HAED elevations in areas inaccessible by airboat. The ground surface was considered bare earth in non-swamp areas, whereas the layer of muck that supported a 1-pound weight on a bearing surface of a 13.5 cm circle was considered bare earth in swamp areas. The HAED data were collected from 1995 to 2007 on a 400 m grid. Using 17 NGS vertical order of 1 benchmarks at 2 independent helicopter hover heights, a Root Mean Square Error (RMSE) of 4.1 cm was reported for the HAED dataset (Desmond, 2003). The HAED data are available at: https://sofia.usgs.gov/projects/index.php?project_url=elev_data. In this chapter, the HAED dataset was used to calibrate and validate the lidar-DEM. The National Wetlands Inventory (NWI) is used to characterize different land cover types. The NWI land cover information was derived from imagery collected in 2010. This dataset is published by the US Fish and Wildlife Service and is available at: www.fws.gov/wetlands/Data/Data-Download.html.

15.3 Methodology

Multiple steps were required to set up the SLAMM model for the coastal ENP. We preprocessed the lidar-DEM first. We transformed the lidar-DEM to lidar-DEM Mean Tidal Level (MTL), a vertical datum used in SLAMM. Since lidar is known to overestimate marsh elevations, we thus applied a bias-correction procedure to improve the accuracy of the lidar-DEM for a better vulnerability analysis. After correction, a slope raster was generated from the corrected lidar-DEM, which was then used in SLAMM to calculate cell conversions. We also converted the NWI wetland categories from vector format to raster format wetland categories to be used by the SLAMM. SLAMM also requires the input of accretion and erosion rates of dominant land cover types, which were derived from past studies in the Everglades. In this chapter, we used the RCP 8.5 99th percentile SLR projection of 0.03 Mean Tide Level (MTL) (0.25 m NAVD 88) by 2050 for ENP (Park et al., 2017a). We calibrated the SLAMM using the current conditions of the landscape before modeling potential future habitat change. Major steps in the modeling procedure are detailed in the following subsections.

15.3.1 Lidar-DEM Transformation, Correction, and Derivation of Slope

To adapt the lidar-DEM for the SLAMM, we converted the lidar-DEM to Universal Transverse Mercator (UTM) zone 17 N in meters for horizontal units, and MTL in meters for vertical units using the NOAA's vertical and horizontal datum transformation tool VDATUM (https://vdatum.noaa.gov/). MTL is the arithmetic mean of mean high water and mean low water, which is the vertical datum used by SLAMM. This procedure resulted in a new lidar-DEM vertically referenced to MTL, limited to areas with a proximity to the shoreline, so further processing is needed to convert the original lidar-DEM for the entire study area. To achieve this, a transformation grid was created using a process similar to Gesch (2013). First, points were generated for each grid cell in the lidar-DEM referenced to MTL that is limited to areas with a proximity to the shoreline. A first order polynomial was used to fit a least-squares surface to the points matching the geographic extent of the study area, which resulted in a transformation grid. The transformation grid was then subtracted from the original lidar-DEM referenced to NAVD 88 to bring all grid cell elevations down to MTL (at Vaca Key tide station, MTL is below NAVD 88). As a result, the lidar-DEM is vertically referenced to MTL for the entire study area.

Lidar-DEM often has large errors in the coastal wetlands due to the penetration issues. Correction is thus required before using the lidar-DEM products. Researchers frequently calibrate lidar-DEMs of coastal marshes with correction factors such as the Mean Bias Error (MBE) approach (e.g., Hladik and Alber, 2012; McClure et al., 2016). This is a simple and attractive approach when lower computational intensity is required for a large study area. A vertical accuracy assessment must be performed on the lidar-DEM before it can be adjusted by the Mean Bias Error (MBE) approach. Once corrected, an independent vertical accuracy assessment is conducted on the corrected lidar-DEM. To accomplish this, the HAED data is first split into a calibration dataset for adjusting the lidar-DEM by the MBE and a validation dataset for accessing the vertical accuracy of the corrected lidar-DEM. From the total 12,806 HAED points, a criterion was placed where those with the shortest distance allowed between any 2 randomly placed points was 500 m. In total, 10,245 HAED points or 80% were then randomly selected as the calibration dataset used to assess vertical accuracy and to correct the uncorrected lidar-DEM. The remaining 2581 HAED points or 20% were then used to validate the corrected lidar-DEM. Vertical accuracy measures include descriptive statistics (e.g., minimum, maximum, standard deviation, etc.), estimates of the average error is estimated as the Mean Absolute Error (MAE) and vertical RMSE, and bias is estimated as MBE. These indices were calculated by:

$$MAE = \frac{\sum_{i=1}^{n}|P_i - O_i|}{n}, \tag{15.1}$$

$$RMSE_z = \sqrt{\sum_{i=1}^{n}(P_i - O_i)^2/n}, \tag{15.2}$$

$$MBE = \frac{\sum_{i=1}^{n}(P_i - O_i)}{n}, \tag{15.3}$$

where P_i is the lidar-DEM grid cell value, $\overline{P_i}$ is the mean of the lidar-DEM grid cell values, O_i is the HAED value, $\overline{O_i}$ is the mean of the HAED values, n is the number of matched lidar-DEM grid cells and HAED data samples, and i is an integer from 1 to n. Once the vertical accuracy measures for the lidar-DEM were calculated using the calibration dataset (10,245 HAED points or 80%), the lidar-DEM was adjusted by subtracting the MBE from the original lidar-DEM, resulting in a corrected lidar-DEM. Validation was then conducted to the corrected lidar-DEM using the HAED validation dataset (2561 HAED points or 20%).

In addition to the corrected lidar-DEM, the slope in degrees is used when calculating cell conversions in SLAMM. We employed ArcGIS version 10.4 Spatial Analyst Slope tool to calculate the maximum rate of change in elevation from each cell to its neighbors. As a result, the slope percent raster contains the same vertical and horizontal units, datums, and cells sizes and alignments.

15.3.2 Land Cover Preparation

In order to run the SLAMM, the NWI wetland categories in vector format need to be converted to SLAMM wetland categories in raster format. Several steps were taken using ArcGIS version 10.4 to prepare the NWI data for SLAMM. First, the NWI vector data were projected to match the horizontal units and datum of the corrected lidar-DEM and slope raster. A numeric field was then added to the table of the NWI feature class before using a lookup table found in Clough et al. (2012) to assign SLAMM wetland categories to each NWI polygon. The NWI vector data were then converted to a raster dataset using the SLAMM categories value field to assign values to the output raster where the cell sizes and alignments matched those of the corrected lidar-DEM and slope raster.

15.3.3 Accretion and Elevation Change Rates

The rate of soil elevation change is necessary for projecting the current biological components of the landscape into future SLAMM modeling. In the Everglades, accretion is measured using the radioactive isotope dating method or Maker Horizon technique. The anthropogenic occurring radioactive isotope Cesium-137 (^{137}Cs) with a half-life of 30.1 years is a product of nuclear fission that began during the mid-20th century. ^{137}Cs is deposited into soils as wet or dry fallout from the atmosphere. The known time deposition of ^{137}Cs atmospheric fallout to the environment (e.g., years 1954 and 1964) was used to compare depths of different ^{137}Cs horizons in sediment cores to calculate accretion rates for sawgrass marsh in the Everglades (Craft and Richardson, 1993; Reddy et al., 1993). Here, inland fresh marsh accretion was set to a constant 0.5 mm/yr based on data by Reddy et al. (1993). The Marker Horizon technique was combined with the Soil Elevation Table approach to measure vertical accretion and elevation change allowing for the calculation of shallow subsidence for mangroves in the Everglades (Cahoon and Lynch, 1997; Whelan et al., 2009). We set mangrove swamp elevation change as 0.6 mm/yr based on data presented by Cahoon and Lynch (1997) because accretion alone has shown to overestimate soil elevation rates for mangroves in the Everglades.

15.3.4 Erosion Rates

The rate of erosion is used to help SLAMM identify areas vulnerable to the conversion of open water. For swamp, Clough et al. (2016) set the most likely long-term horizontal erosion rates to 1 m/yr, which is also adopted for mangroves in this study. It is noted that marshes are typically less sensitive to the erosion parameter in SLAMM because a fetch of >9 km at the marsh to open water interface is needed for erosion to occur (Clough et al., 2016). Nonetheless, marsh erosion rates were set to 0.23 m/yr (Hine and Belknap, 1986; Geselbracht et al., 2011).

15.3.5 Sea Level Rise Projections

As the leading international organization for climate change assessment, the Intergovernmental Panel on Climate Change (IPCC) conducts scientific reviews on published literature to assess our current understanding of climate change. IPCC SLR projections have changed throughout the years due to distinct emissions scenarios and contributions of the individual inputs to SLR along with improvements to their respective modeling. The most recent assessment report at the time of this study was the fifth assessment report (AR5) published in 2013. New emissions scenarios were developed for the AR5, which are referred to as Representative Concentration Pathways (RCPs). The term representative refers to an RCP being representative of several different scenarios with similar radiative forcing (a measure of the influence a factor has on altering the balance of incoming and outgoing energy in the atmosphere) and emissions, and the term pathway refers to the long-term trajectory of emissions and radiative forcing outcomes (Moss et al., 2007). The RCPs have 4 pathways: RCP8.5, RCP6, RCP4.5, and RCP2.6; the numbers represent forcing for each RCP. For example, the RCP scenario where radiative forcing reaches >8.5 W/m^2 in the year 2100 is the RCP8.5 scenario. The various IPCC emissions scenarios are used to force the model simulations of IPCC future SLR, resulting in various projections. Although rapid ice flow from Greenland and Antarctic ice sheets was recently included in AR5, the IPCC warns that their projections could still be low due to loss of ice sheets. For this reason, the RCP8.5 scenario is a reasonable conservative decision support metric used for SLR risk assessment in the Everglades. It is important to note that local processes of ocean current and land uplift or subsidence are not accounted for in the AR5.

An estimate of the local processes for the RCP8.5 scenarios has been provided by Kopp et al. (2014); it was reasonably adopted as the basis of SLR scenarios for ENP (Park et al., 2017a, 2017b). In order to account for local oceanographic processes that influence sea levels around ENP, it is necessary to examine local tidal datum defined by the nearest tide stations (e.g., Vaca Key, Naples, and Key West). The Vaca Key tide station was found to best represent local oceanographic influences, so it was used with the RCP 8.5 50th and 99th percentiles to develop SLR scenarios for ENP. We converted the projections by Park et al. (2017a) from the North American Vertical Datum of 1988 (NAVD 88) to MTL 1983–2001 epoch for SLAMM by taking the difference between NAVD 88 (0.41 m) and MTL (0.16 m) at the Vaca Key tide station (0.41–0.16 = 0.25 m) and subtracting the resulting value (0.25 m) from their NAVD 88 projections. The RCP8.5 50th and 99th percentile SLR projections referenced to MTL for ENP are shown in Figure 15.2. We used the RCP 8.5 99th percentile SLR projection referenced to MTL of 0.03 m (0.25 m NAVD 88) for year 2050 to simulate potential land cover change.

15.3.6 SLAMM Setup and Calibration

We calibrated the SLAMM using the current conditions of the landscape before modeling potential future habitat change. This was achieved by using a "time-zero" time step with the current Great Diurnal Tide Range (GT) of 0.297 m for Vaca Key, Florida Bay, Florida tide station. GT is the difference in height between mean higher high water and mean lower low water. In the calibration, no SLR, accretion, or erosion was considered because the initial conditions of elevation, land cover, tidal ranges, and hydrologic connectivity were examined. Cells can become converted to a different category due to the course horizontal resolution of the land cover data, DEM accuracy and error, and simplifications within the SLAMM model (Clough et al., 2016). A threshold tolerance of ≤5% change is generally acceptable for all SLAMM land cover categories combined (New York State Energy Research and Development, 2014). The results of the SLAMM calibration for ENP show a 4% change, thus the current conditions model is satisfactory for estimating potential habitat change or loss due to future SLR, accretion, and erosion (Table 15.1). Once calibrated, the time step for the SLR scenario (RCP 8.5 99th percentile) in addition to accretion and erosion are simulated to estimate land cover change from 2010 to 2050.

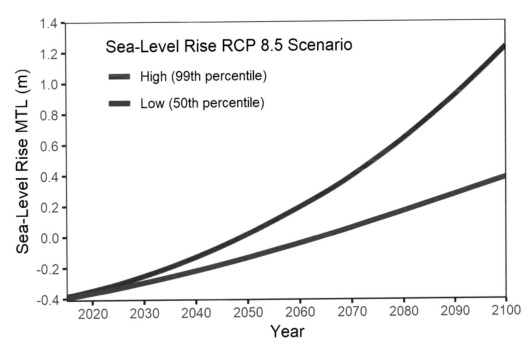

FIGURE 15.2
Everglades National Park sea-level rise projections using the RCP 8.5 scenario modified from Park et al. (2017a) to reference meters Mean Tide Level (MTL).

TABLE 15.1

Initial condition land cover categories and SLAMM predicted using "time-zero" time step

Land cover	Current (ha)	Time zero (ha)	Change (ha)	% change
Swamp	807.00	807.18	0.18	0%
Cypress Swamp	116.57	116.55	−0.02	0%
Inland-Fresh Marsh	10,522.60	10,521.85	−0.75	0%
Tidal-Fresh Marsh	3,031.48	3,031.97	0.49	0%
Trans. Salt Marsh	1,071.64	1,071.61	−0.04	0%
Regularly Flooded Marsh	5,211.92	5,211.60	−0.32	0%
Mangrove	142,303.65	142,320.77	17.12	0%
Estuarine Beach	5,632.47	5,636.37	3.90	0%
Tidal Flat	534.32	537.64	3.33	1%
Ocean Beach	62.72	63.14	0.42	1%
Ocean Flat	13.53	13.78	0.25	2%
Inland Open Water	19.35	19.38	0.03	0%
Estuarine Open Water	37,921.83	37,922.26	0.43	0%
Open Ocean	93.64	93.58	−0.05	0%
Irreg.-Flooded Marsh	14,215.81	14,214.23	−1.58	0%
Tidal Swamp	447.07	447.06	0.00	0%
			Total	**4%**

15.4 Results and Discussion

15.4.1 Lidar-DEM Correction and Vertical Accuracy in SLR Applications

The result of the corrected lidar DEM is shown in Figure 15.3, and the results of the vertical accuracy assessment performed on the uncorrected and corrected lidar-DEM s are shown in Table 15.2. The calibration dataset of 10,245 HAED points was used to assess the accuracy of the uncorrected lidar-DEM and provided a correction factor with MBE of 0.29. Once the uncorrected lidar-DEM was reduced by the MBE (on average this is the amount the lidar-DEM overestimates the ground surface), the validation dataset of 2561 HAED points was used to calculate new summary measures of the corrected lidar-DEM. In terms of the MAE and RMSE, a meaningful distinction can be made between the uncorrected and corrected lidar-DEMs. The MAE and RMSE are the lowest (better) for the uncorrected lidar-DEM. However, some caution should be used when considering the RMSE as a difference measure. The uncorrected lidar-DEM 's RMSE was reduced from 0.40 to 0.27 m, which seems to demonstrate that the correction reduced the error. It should be noted that the RMSE squares the errors, which inflates the influence of outliers. A better measure of error may be the standard deviation (σ), which is a measure of precision that represents a range of errors around the mean (accuracy) between the lidar-DEM and HAED. While the MBE was reduced from 0.29 to 0.01 m, the corrected lidar-DEM RMSE of 0.27 is not an improvement because the uncorrected and corrected lidar-DEM's

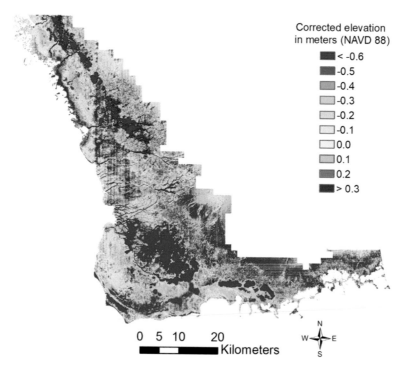

FIGURE 15.3
Map of the corrected lidar-DEM.

TABLE 15.2

Descriptive statistics for the uncorrected and corrected lidar-DEM s. Where MBE = mean bias error between uncorrected or corrected lidar -DEM and HAED, MAE = Mean Absolute Error, RMSE = Root Mean Square Error between uncorrected or corrected lidar-DEM and HAED, n = number of HAED, all in meters (except n)

Experiment	n	Min	Max	Median	Skew	Kurtosis	Standard deviation	MBE	MAE	RMSE
Uncorrected LiDAR DEM	10,245	−1.51	1.97	0.29	0.28	2.08	0.27	0.29	0.32	0.40
Corrected LiDAR DEM	2,561	−1.37	1.49	−0.02	0.45	1.48	0.27	0.01	0.21	0.27

standard deviations are 0.27 m (Table 15.2). Methods are available for reducing both bias and error in DEMs. Machine learning (Rogers et al., 2018) and object-based machine learning correction techniques (Cooper et al., 2019) reduce both bias and error in DEMs; however, these approaches are more complex and computationally intensive than the simple bias-correction approach applied in this study. For this reason, we did not apply object-based machine learning techniques for our study area of coastal ENP, but these methods have great potential to improve future SLR and storm surge modeling.

Application of the commonly used RMSE is limited, especially for SLR vulnerability assessment. The RMSE is not robust and thus is strongly influenced by outliers. As the MBE increases, the RMSE will overestimate the range of errors (Table 15.2), therefore overestimating potential inundation areas and resulting impacts (see review article by Cooper et al., 2013). This may result in incorrect conclusions about the possible impacts of future SLR, especially in wetlands where vegetation patterns are found at limited elevation gradients (e.g., a few centimeters). Note that the American Society for Photogrammetry and Remote Sensing (ASPRS, 2014) states that the MBE is often not equal to zero (as in the case of wetland habitats), thus it must be accounted for. This is important for existing methods that rely on the RMSE to propagate lidar error in SLR vulnerability mapping (Gesch, 2009, 2013; Clough et al., 2016) to ensure that the range of errors is not over-estimated. A more robust estimate is based on the mean error and σ, thus the ASPRS recommends that vendors calculate the mean error, σ, skew, kurtosis, and RMSE in lidar error analyses. This is especially important for wetland vegetation land covers where lidar elevations have a positive bias. For example, although Sadro et al. (2007) found the relationship between marsh canopy height and RMSE significant, the RMSE was not a robust measure to use in correcting lidar elevations. Instead, the MBE was used as an elevation adjustment measure. In this study, the MBE was accounted for so that the range of errors is not overestimated. This allows for a normal distribution (the MBE = 0.01 and is not significantly different from 0) specified by the RMSE because it is the same as the standard deviation (σ) (RMSE = 0.27 and σ = 0.27) to be utilized for when accounting for elevation uncertainty in SLAMM. Although not addressed in this study, future work in the Everglades may want to examine the confidence of the SLAMM results considering the uncertainty in the underlying data (Clough et al., 2016).

15.4.2 SLAMM Results

The quantitative results of potential habitat change under the RCP8.5 99th percentile scenario of 0.03 m MTL (0.25 NAVD 88) by 2050 are shown in Table 15.3. All land cover types are predicted to experience a loss except open ocean and estuarine open water. Ocean beach has the greatest potential of habitat loss (99%), while cypress swamp has the least potential of habitat loss (5%) for the study area. The current estimated inland-fresh marsh of ~10,5000 ha is predicted to lose 59% of habitat by 2050. In the maps (Figure 15.4), it is illustrated that the transition zone between the mangrove ridge and inland-fresh marsh has moved further inland by 2050. Mangroves are shown to replace inland-fresh marsh, which may lead to a decline in wading birds and other wildlife dependent on the freshwater habitat in ENP. Mangrove swamps are also predicted to lose 25% or ~32,500 ha of their current habitat by 2050 (Table 15.3). Figure 15.4 shows that mangroves will be replaced with estuarine open water, and so inland-fresh marsh areas may be more vulnerable to erosion and saltwater inundation from storm surges in 2050. Estuarine open water is estimated to expand an additional ~51,000 ha from its current ~40,000 ha, a 135% gain (Table 15.3) that is also illustrated in the maps (Figure 15.4). Less noticeable in the maps is the conversion of inland-fresh marshes to estuarine open water. This is often accompanied by peat collapse releasing large amounts of sequestered carbon into the atmosphere (Park et al., 2017a).

While the SLAMM results provide some useful information on the potential of habitat change and loss due to SLR in the coastal Everglades, some shortcomings should be noted. First, the current conditions of elevation and historical rates of accretion and erosion are expected to remain static into the future. Therefore, habitat change and loss

TABLE 15.3

Land cover change by 2050 compared to initial conditions in 2010 using the RCP 8.5 99th percentile sea-level rise projections

Land cover	Hectares in 2010	2010 to 2050 land cover change (ha)	2010 to 2075 land cover change (ha)
Swamp	807.18	−110.75	−694.67
Cypress Swamp	116.55	−5.83	−104.17
Inland-Fresh Marsh	10,521.85	−6,172.43	−10,425.93
Tidal-Fresh Marsh	3,031.97	−583.38	−2,720.79
Trans. Salt Marsh	1,071.61	−70.52	−692.14
Regularly Flooded Marsh	5,211.60	−320.08	−3,686.55
Mangrove	142,320.77	−35,521.87	−101,529.26
Estuarine Beach	5,636.37	−3,672.51	−5,447.78
Tidal Flat	537.64	−135.88	−503.95
Ocean Beach	63.14	−62.79	−63.14
Ocean Flat	13.78	−1.90	−9.42
Inland Open Water	19.38	−16.50	−18.59
Estuarine Open Water	37,922.26	51,104.14	139,173.03
Open Ocean	93.58	64.08	71.94
Irreg.-Flooded Marsh	14,214.23	−4,446.32	−13,026.93
Tidal Swamp	447.06	−70.82	−345.03

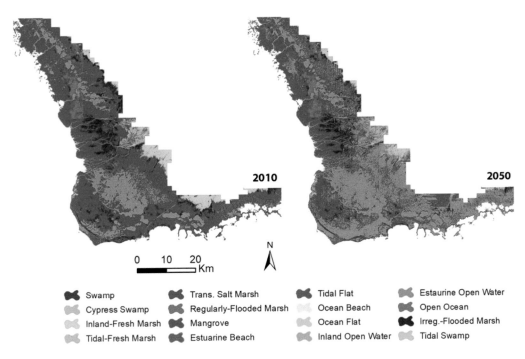

FIGURE 15.4
Initial conditions in 2010 and results of SLAMM forecasts 2010 to 2050 using the RCP8.5 99th percentile scenario of 0.03 m MTL (0.25 NAVD 88).

may be overestimated because ecogeomorphic feedbacks (e.g., where moderate increases in inundation boosts marsh productivity contributing to vertical accretion) are disregarded (Kirwan et al., 2016). These feedbacks are also poorly understood for mangroves where the effects of flooding on root production to maintain soil elevations may not be easily predicted (Krauss et al., 2013).

Incorporating predictions of ecogeomorphic processes under accelerated SLR conditions will improve future forecasts of habitat change and loss due to SLR (Kirwan et al., 2016). Second, changes in topography due to hurricanes such as elevation loss (peat collapse) or gain (deposition from storm sediments) along with influences on vegetation are not considered in the modeling. While hurricanes can cause extensive breakage in the mangrove community, rootlets have been shown to stabilize storm deposits that become a part of the soil profile (Whelan et al., 2009). Third, in low-lying coastal areas with gentle slopes such as the Everglades, there will likely be a rise in the groundwater table with sea level. The floor of the southern tip of the Florida peninsula and Florida Bay is formed by the Miami Limestone, which is highly permeable transmitting rainfall and water from the Everglades. Roughly 12 m thick, this surficial aquifer known as the Biscayne aquifer, allows easy water exchange between surface and groundwater (Lodge, 2010). It is also important to note that although the mangrove ridge will become permanently inundated with seawater, if fresh water is flowing downstream, the impacts of saltwater inundation will be mitigated (Park et al., 2017a). ENP is a special case when conducting SLAMM simulations, and future work should examine the inclusion of potential inland-fresh marsh surface water changes due to water management practices and groundwater changes due to SLR.

15.5 Summary and Conclusions

We evaluated the potential of habitat change and loss in coastal ENP under a high SLR scenario of 0.25 m NAVD 88 (0.03 m MTL) by year 2050 using SLAMM. An important baseline dataset used in SLAMM is a light detection and ranging (lidar) DEM because of improved vertical accuracy and resolution. Despite this, the best available lidar-DEM overestimated elevations in ENP by 0.29 m (MBE), so a simple bias-correction procedure was applied reducing the MBE to 0.01 m before performing SLAMM simulations. Since lidar-DEM correction provides a better representation of the current landscape, it may increase the integrity in SLR predictions. For coastal ENP, the SLAMM modeling predicted a 59% loss of inland-fresh marsh and 25% loss of mangroves by 2050. It is likely that these predictions provide an overestimate of marsh and swamp vulnerability to future SLR because dynamic feedbacks in vertical accretion and fresh surface water flowing downstream and groundwater change due to SLR are not captured. Future work should focus on incorporating these feedbacks to increase the integrity in SLR vulnerability assessments.

References

American Society for Photogrammetry and Remote Sensing (ASPRS), 2014. Positional accuracy standards for digital geospatial data. Edition 1, Version 1.0. *Photogrammetric Engineering & Remote Sensing*, 81(3): A1–A26.

Caffrey, M., and Beavers, R., 2013. Planning for the impact of sea-level rise on. *U.S. National Parks. Park Science*, 30, 6–13.

Caffrey, M.A., Beavers, R.L., and Hoffman, C.H., 2018. Sea level rise and storm surge projections for the National Park Service. Natural Resource Report NPS/NRSS/NRR—2018/1648. National Park Service, Fort Collins, CO.

Cahoon, D., and Lynch, J., 1997. Vertical accretion and shallow subsidence in a mangrove forest of southwestern Florida. *U.S.A. Journal of Mangroves and Salt Marshes*, 1, 173–186.

Clough, J., Park, R., and Fuller, R., 2012. SLAMM 6 Beta Technical Documentation. http://warren pinnacle.com/prof/SLAMM6/SLAMM6_Technical_Documentation.pdf

Clough, J., Polaczyk, A., and Propato, M., 2016. Modeling the potential effects of sea-level rise on the coast of New York: Integrating mechanistic accretion and stochastic uncertainty. *Environmental Modeling & Software*, 84, 349–362.

Cooper, H.M., Fletcher, C.H., Chen, Q., and Barbee, M.M., 2013. Sea-level rise vulnerability mapping for adaptation decisions using lidar-DEMs. *Progress in Physical Geography*, 37, 745–766.

Cooper, H.M., Zhang, C., Davis, S., and Troxler, T., 2019. Object-based correction of lidar-DEMs using RTK-GPS data and machine learning modeling in the coastal Everglades. *Environmental Modelling & Software*, 112, 179–191.

Craft, C.B., and Richardson, C.J., 1993. Peat accretion and N, P, and organic C accumulation in nutrient-enriched and unriched everglades peatlands. *Ecological Applications*, 3, 446–458.

Desmond, D.G., 2003. *Measuring and Mapping the Topography of the Florida Everglades for Ecosystem Restoration*, US Geological Survey Fact Sheet 021-03, USGS, Reston, VA.

Gesch, D., 2009. Analysis of Lidar elevation data for improved identification and delineation of lands vulnerable to sea-level rise. *Journal of Coastal Research*, 53, 49–58.

Gesch, D., 2013. Consideration of vertical uncertainty in elevation-based sea-level rise assessments: Mobile Bay, Alabama case study. *Journal of Coastal Research*, 6, 197–210.

Geselbracht, L., Freeman, K., Kelly, E., et al., 2011. Retrospective and prospective model simulations of sea level rise impact on Gulf of Mexico coastal marshes and forests in Waccasassa Bay, Florida. *Climatic Change*, 107, 35–37.

Hine, A.C., and Belknap, D.F., 1986. Recent geological history and modern sedimentary processes of the Pasco, Hernando, and Citrus county coastline: West-Central Florida. Florida Sea Grant Report 79. Florida Sea Grant, Gainesville, Florida.

Hladik, C., and Alber, M., 2012. Accuracy assessment and correction of a lidar-derived salt marsh digital elevation model. *Remote Sensing of Environment*, 121, 224–235.

Kirwan, M.L., Temmerman, S., Skeehan, E.E., et al., 2016. Overestimation of marsh vulnerability to sea level rise. *Nature Climate Change*, 6, 253–260.

Kopp, R.W., Horton, R.M., Little, C.M., et al., 2014. Probabilistic 21st and 22nd century sea-level projections at a global network of tide gauge sites. *Earth's Future*, 2, 383–406.

Krauss, K.W., McKee, K.L., Lovelock, C.E., et al., 2013. How mangrove forests adjust to rising sea level. *New Phytologist*, 202, 19–34.

Lodge, T.E., 2010. *The Everglades Handbook: Understanding the Ecosystem*, 3rd Edition, Taylor & Francis Group, Boca Raton, FL.

McClure, A., Liu, X.H., Hines, E., and Ferner, M.C., 2016. Evaluation of error reduction techniques on a lidar-derived salt marsh digital elevation model. *Journal of Coastal Research*, 32, 424–433.

Moss, R., Babiker, M., Brinkman, S., et al., 2007. Towards new scenarios for analysis of emissions, climate change, impacts, and response strategies. Technical Summary. Intergovernmental Panel on Climate Change, Geneva, 25. www.ipcc.ch/pdf/supporting-material/expert-meeting-ts-scenarios.pdf

New York State Energy Research and Development Authority, 2014. Application of Sea-Level Affecting Marshes Model (SLAMM) to Long Island, NY and New York City, Report Number 14-29. www.nyserda.ny.gov//media/Files/Publications/Research/Environmental/SLAMM%20report.pdf

Park, J., Stabenau, E., and Kotun, K., 2017a. Sea-level rise and inundation scenarios for national parks in South Florida. *Park Science*, 33, 63–73.

Park, J., Stabenau, E., Redwine, J., and Kotun, K., 2017b. South Florida' encroachment of the sean and environmental transformation over the 21st century. *Journal of Marine Science and Engineering*, 5, 31.

Pearlstine, L.G., Pearlstine, E.V., and Aumen, N.G., 2010. A review of the ecological consequences and management implications of climate change for the Everglades. *Journal of the North American Benthological Society*, 29, 1510–1526.

Reddy, K.R., Delaune, R.D., Debusk, W.F., and Koch, M.S., 1993. Long-term nutrient accumulation rates in the Everglades. *Soil Science Society of America*, 57, 1147–1155.

Rogers, J.N., Parrish, C.E., Ward, L.G., and Burdick, D.M., 2018. Improving salt marsh digital elevation model accuracy with full-waveform and nonparametric predictive modeling. *Estuarine, Coastal and Shelf Science*, 202, 193–211.

Sadro, S., Buhl-Gastil, M., and Melack, J., 2007. Characterizing patterns of plant distribution in a Southern California salt-marsh using remotely sensed topographic and hyperspectral data and local tidal fluctuations. *Remote Sensing of the Environment*, 110, 226–239.

Stabenau, E., Engel, V., Sadle, J., and Pearlstine, L., 2011. Sea-level rise: Observations, impacts, and proactive measures in Everglades National Park. *Park Science*, 28, 26–30.

Whelan, K.T., Smith, T.J., Anderson, G.H., and Ouellette, M.L., 2009. Hurricane Wilma's impact on overall soil elevation and zones within the soil profile in a mangrove forest. *Wetlands*, 29, 16–23.

16

Enhancing Lidar Data Integrity in the Coastal Everglades

Hannah Cooper and Caiyun Zhang

16.1 Introduction

The vertical accuracy of light detection and ranging (lidar) elevation measurements refers to how close those measurements are to the "true" elevations of an independent data source of better accuracy (e.g., Real Time Kinematic-Global Navigation Satellite Systems (RTK-GNSS) measurements). lidar elevation measurements are often mistakenly considered better when they have "high" vertical accuracy. However, high vertical accuracy (e.g., >0.15 m) means that the lidar obtains larger differences between the measured and "true" elevations. Lidar elevation measurements should be considered better with "low" vertical accuracy (e.g., <0.15 m) with smaller differences between the measured and "true" elevations. Accordingly, high accuracy lidar elevation measurements have larger errors compared to low accuracy lidar elevation measurements that have smaller errors. Since error is associated with all discreet measurements in science, it should be noted that uncertainty or error in lidar elevation measurement is not a careless mistake. Error or uncertainty in lidar elevation measurements may be expressed numerically by the standard deviation (σ), yet there are many standards that use a modified σ approach for assessing elevation measurement error, and the most commonly used is the National Standard for Spatial Data Accuracy (NSSDA) (FGDC, 1998). NSSDA uses statistical procedures to calculate the vertical Root Mean Square Error ($RMSE_z$), or the square root of the average squared differences between the measured and "true" elevations of the same locations. It is important to conduct an error analysis on lidar elevation measurements in order to keep uncertainty to a minimum so that a reliable estimate is made of how large or small those measurement errors may be. There are many standards for assessing elevation measurement error, and a review of the most commonly used NSSDA and its applicability to lidar elevation can be found in Cooper et al. (2013). It should also be noted that the American Society for Photogrammetry and Remote Sensing (ASPRS) accuracy standards (ASPRS, 2014) provide new recommendations that accommodate state-of-the-art technologies such as lidar. While marshes are excluded in the ASPRS (2014) guidelines, the acknowledgment of impenetrable vegetation (i.e., mangroves) is a good start for critical habitats such as those in the coastal Everglades.

The high point density and low (better) vertical accuracy of lidar data reported by commercial vendors makes it especially attractive for low-relief coastal wetland habitats. However, the acceptance of lidar presented as-is could potentially lead to false conclusions because commercial vendors may interpret land cover classes freely resulting in the collection of little to no ground control points in difficult regions such as coastal marshes (Schmid et al., 2011). As a result, researchers should be cautioned that a vertical accuracy assessment conducted on the marsh and mangrove ecosystems is likely not included in the commercial vendor's accuracy assessment. It should also be noted that the vertical accuracy of the commonly included land cover classes of tall weeds, grass, scrub-shrub and crops cannot be assumed equal to that of marsh habitats (Schmid et al., 2011). The exclusion of wetland environments in the data provider's accuracy assessment is further complicated by the laser's inability to penetrate dense marsh plants that are inundated with rich organic soils (Figure 16.1) and the commercial classification and filtering of lidar into ground returns. This has led to researchers to develop lidar-DEM correction techniques for wetlands.

Lidar-DEM errors tend to increase with marsh vegetation density and height (e.g., Rosso et al., 2006). It is common for lidar-DEMs to overestimate marsh ground elevations, for example, as much as 0.65 m (Medeiros et al., 2015). Therefore, several efforts have been made to address this dilemma. One approach is to apply a minimum binning procedure where the minimum lidar ground elevation point within a grid cell is assigned to this grid cell when more than one lidar point falls within this grid cell (Schmid et al., 2011; Medeiros et al., 2015; Buffington et al., 2016). The minimum binning procedure is a simple and reasonable approach to reduce the uncertainty of lidar-DEMs. A second approach is to use RTK-GNSS to calculate a species-specific bias used to correct the lidar-DEM (e.g., Hladik and Alber, 2012; McClure et al., 2016). This type of correction assumes that lidar-DEM errors are consistent within a plant species, but plant characteristics influence lidar uncertainty, presenting a varying elevation within a species rather than

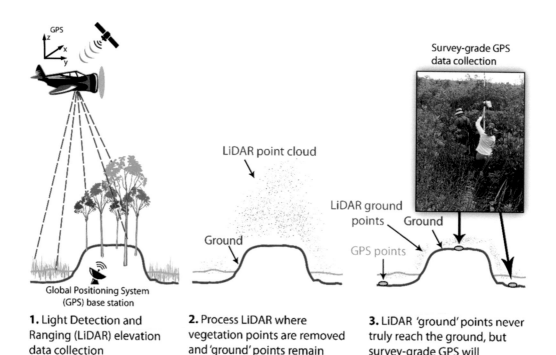

FIGURE 16.1
Illustration of the difficulties lidar has in penetrating dense marsh plants and inundated soils typical of the Everglades, while RTK-GPS can measure the ground truth.

a constant (Rogers et al., 2016, 2018). While the minimum binning and bias correction techniques typically work well if plant structure is homogeneous and the error is even, they are not a good option if the ecosystem has a complex landscape (Cooper et al., 2019). Recent studies have shown that applying modern machine learning techniques is promising for reducing errors in lidar-DEMs of coastal marshes (Rogers et al., 2018; Cooper et al., 2019). Compared to the straightforward minimum binning and bias-correction techniques, application of machine learning techniques is more complex. However, this new technique is attractive and more effective for improving the accuracy of lidar-DEMs when the datasets do not follow a normal distribution, contain a bias, and are highly spatially autocorrelated. Our recent study (Cooper et al., 2019) demonstrates that by combining machine learning with Object-Based Image Analysis (OBIA) techniques, the spatial bias and uncertainty of lidar-DEM in the coastal Everglades can be largely reduced. The object-based machine learning lidar-DEM correction approach can successfully address the correction issue when the lidar spatial bias has a high degree of heterogeneity within a plant community and across communities. In this chapter, we extended and applied this newly developed technique to correct lidar-DEM in more land cover types including sawgrass, tall mangroves, short mangroves, and coastal prairies in the coastal Everglades.

16.2 Study Area and Data

Our study area is located in the southern coastal region of the Flamingo district of ENP (Figure 16.2). On the southeastern edge of the park, Main Park Road (also known as Flamingo Road) runs from the Ernest Coe Visitor Center to the Flamingo Visitor Center for 38 miles. Main Park Road contains National Geodetic Survey (NGS) Benchmarks located roadside that are useful for our RTK-GNSS and total station surveys. Here, we selected 4 sites based on the domiant vegetation communities along Main Park road. The first site is located just below the Nine Mile Pond turnoff from Main Park Road. This site is historically a freshwater marsh with dominative vegetation of sawgrass (*Cladium jamaicense*) and thus is referred to as the sawgrass site in this chapter. The average hydroperiod for sawgrass at this site ranges anywhere from 6 months to continuous flooding (Lodge, 2010). When conducting fieldwork for this study, flooding was present. The ecological impact of hurricane winds can be seen within the plant communities just southwest of the sawgrass study site. Our second site is dominated by short mangroves with a low density of trees and heights less than 5 m. Red mangrove (*Rhizophora mangle*) and buttonwood (*Conocarpus erectus*) are present at this site. Hurricanes Donna (1960) and Betsy (1965) resulted in the establishment of the red mangrove seedlings in this freshwater marsh, thus the red mangrove does not require its tidal habitat to survive (Lodge, 2010). The vegetation substrate for the sawgrass and short mangroves study sites are defined as Everglades peat soils that overlay a porous limestone rock layer.

Just southwest of the short mangroves study site is the third study site referred to here as tall mangroves because it is dominated with high-density tall stands of trees with heights more than 5 m including red mangrove, black mangrove (*Avicennia germinans*), white mangrove (*Laguncularia racemosa*), and buttonwood. The ecological impact of strong hurricane winds and storm surges also changed the vegetation and

FIGURE 16.2
The study area in the coastal Everglades and field pictures collecting RTK-GPS elevation data.

terrain. Southwest of the tall mangroves site hurricanes resulted in a patchwork of coastal prairies dominated by three species: saltwort (*Batis maritima*), glasswort (*Salicornia spp.*), and saltgrass (*Distichlis spicata*). They all have a height less than 1 m. This is our fourth site, referred to as coastal prairies hereinafter, where field RTK-GPS elevation data were collected. Pictures of these 4 sites for field data collection are in Figure 16.2. The vegetation substrate for the tall mangroves and coastal prairies study sites are defined by mangrove peat soils and hurricane deposited marl that overlays a porous limestone rock layer. Underlying all 4 study areas is the Biscayne aquifer, an unconfined aquifer dominantly limestone and highly permeable. Water flows toward the southwest, which is limited to current water management practices.

Data sources include RTK-GNSS data for uncertainty assessment of lidar elevation and correcting lidar-DEM, lidar for DEM generation, and aerial imagery for generating image objects. RTK-GNSS data were collected in February and March 2016 to be seasonally consistent with the acquisition of lidar and aerial imagery. We used the Leica 1200 system with reported real time errors of 0.01 m in the horizontal and 0.02 m in the vertical (RMSE at 68% confidence interval) to collect the RTK-GPS data (Leica Geosystems, 2008). Elevations were surveyed near low tide with nominal precisions ≤0.01 m in horizontal and ≤0.03 m in vertical positions (RMSE at 68% confidence interval). The total number of RTK-GNSS measures used in this study is 62 for sawgrass, 138 for short mangroves, 131 for tall mangroves, and 151 for coastal prairies. The current best available lidar data for the study area were collected by the Florida Division of Emergency Management in February 2008 using the Leica ALS50 Airborne Laser Scanner system, which was reported to have a horizontal error of 0.07–0.64 m (1 standard deviation or 1 σ) and vertical error of 0.08 to 0.24 m (1 σ) after post-processing (Leica Geosystems, 2007). The average lidar point density reported is 2 points/m^2 for unobscured areas (no trees or buildings). The vendor filtered the lidar point cloud into classified ground (bare-earth) and non-ground returns (vegetation, structures, and buildings). The lidar bare-earth were used in this study and transformed from NAVD 88 using Geoid 03 to NAVD 88 using Geoid 12B to be consistent with our RTK-GNSS data using a vertical datum transformation tool from NOAA (https://vdatum. noaa.gov/). Fine spatial resolution aerial photography with 3 spectral channels (red, green, and blue) were collected in January 2016 with a spatial resolution of 0.25 m for the sawgrass and short mangroves site, while aerial imagery collected in January 2012 with a spatial resolution of 0.30 m was used for the coastal prairie and tall mangrove sites. The lidar and aerial imagery data are available at National Oceanic and Atmospheric Administration (NOAA)'s data access viewer (https://coast.noaa.gov/dataviewer).

16.3 Methodology

In this chapter, we applied the object-based machine learning correction approach developed in Cooper et al. (2019) to improve the lidar-DEM in the coastal Everglades. The flowchart of this methodology is displayed in Figure 16.3. Objects were first generated from the fine spatial resolution aerial photography using an image segmentation method; then, lidar descriptors were determined for each object using the bare-earth returns. The field surveyed RTK-GPS data were spatially matched with the objects in which lidar descriptors were available. A machine learning algorithm, Random Forest, was applied to develop a correction model. If an acceptable accuracy was produced, the model would be then used to generate a corrected lidar-DEM. The object-based machine learning correction approach was also compared with the bias-based method. Major steps of the methodology in this chapter are detailed in the following subsections.

16.3.1 Image Segmentation

To conduct the object-based lidar-DEM correction, image objects need to be created. Here, image objects were generated from the aerial imagery using the multi-resolution segmentation algorithm in eCognition Developer 9.3 (Benz et al., 2004; Trimble, 2014). The algorithm generates image objects by segmenting individual pixels before merging neighboring segments together until a heterogeneity threshold is reached (Benz et al., 2004). We followed Cooper et al. (2019) to determine the heterogeneity threshold including scale (parameter set to 50), color/shape (parameter set to 0.9/0.1 for all three bands to weight spectral information most heavily), and smoothness/compactness (parameter set to 0.5/0.5 for all three bands so compact and non-compact segments were favored equally). After segmentation, lidar statistical metrics (minimum, mean, maximum, standard deviation, range, and count or number of lidar points) were derived for each object using the original lidar bare-earth returns as independent variables to predict the true elevation (i.e., the RTK-GPS elevation).

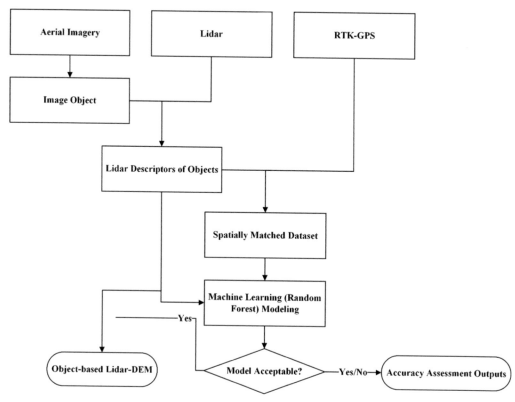

FIGURE 16.3
Methodology flowchart of object-based machine learning lidar-DEM correction.

16.3.2 Data Matching

The limited survey measures for each of the 4 study sites were spatially matched with lidar measures at the object and grid levels. This allowed for predictive modeling development and a comparison between the object-based and grid-based modeling for correcting lidar-DEMs. It is expected that matching a survey measure to relatively homogenous image objects with varying shape and size can produce a better result in both lidar elevation correction and DEM generation because an object is more representative than a grid cell in space for characterizing plant structures. The uniform size of a grid cell may include other species in this cell, leading to a higher heterogeneity and more lidar uncertainties.

16.3.3 Random Forest-based Lidar Data Correction

Predicting the elevation of survey data measurements as a function of nearby lidar classified ground return descriptive statistics may be thought of as a regression problem in which RTK-GPS surveyed elevation is the dependent variable while lidar elevation descriptors are the independent variables. Here, we applied a machine learning algorithm, Random Forest, to conduct the regression. This algorithm makes specifying the input parameters easier, it has a fast speed, and it is effective for lidar correction compared with other machine learning algorithms such as support vector machine and artificial neural network. Random Forest regression performs predictions using the following steps (Breiman, 2001). First, m variables are selected at random from the calibration data also known as the root node (the entire sample is represented in the root node before getting split into two homogeneous datasets or terminal nodes). In this bootstrapped or subsampled dataset, roughly one-third of the instances are left out to get out-of-bag estimates of the error rate. Although the out-of-bag technique is like cross-validation for statisticians, Breiman (2001) notes an important difference: out-of-bag estimates are unbiased when enough trees are grown. Second, the bootstrapped dataset is used to grow an independent CART regression tree by repeating a random selection of variables, picking the best variable, and splitting the node for each terminal node until the minimum node size is reached. The purpose of building independent regression trees is to split any complex nonlinear regression problem dataset into a set of smaller datasets. This allows for the smaller datasets consisting of continuous random and independent variables to be better managed by simple linear models when making predictions (Criminisi et al., 2011). Third, several more independent regression trees (*ntrees*) are grown to form a random forest where the predictions of each tree are combined by bagging or averaging over the predictions to get an unbiased result (Breiman, 2001). The idea is that combining the predictions of several weaker models or independent regression trees into one stronger model or random forest prediction helps reduce variance and overfitting while providing more likely generalizations. An illustration of the Random Forest regression algorithm is shown in Figure 16.4. This algorithm can be used for both classification and regression, as demonstrated in the previous chapters for sawgrass biomass modeling and mapping and vegetation classification and mapping.

Before implementing the Random Forest technique, it is necessary to split the survey datasets into two: (1) a calibration dataset for optimizing and training the model and (2) a validation dataset to judge model performance. The validation dataset was determined by randomly selecting 20% of the features from each respective dominant vegetation community. The remaining 80% of the features was used as the calibration dataset. We employed R with the "caret" package (Kuhn, 2016) train function to fit the Random Forest-based model over different tuning parameters. For our calibration datasets, the train control (p) was set to 10-fold cross-validation (Kohavi, 1995). In randomly splitting the calibration datasets into 10 separate training sets to build the prediction models and 10 separate testing sets to calculate the average RMSE using 10-fold cross-validation, our goals were to avoid (1) tuning the predictive function too closely to the calibration dataset and (2) a RMSE that is biased by a single split. We set *ntrees* in the package function based on a trial and error procedure to determine the lowest average RMSE. The number of randomly selected variables (*mtry*) was then set at a maximum of our total number of predictors where "caret" performs the tuning parameter automatically by fixing *ntrees* and testing various *mtrys* to identify the best model for our datasets. For repeatability, the seed is set to 123 for all model runs. Variable importance was determined using the "caret" package importance function. Once the Random Forest-based model is optimized and trained on the calibration datasets, the validation datasets were used to judge the performance of the predictions on unseen data.

16.3.4 Object-based Lidar-DEM Generation

We developed an object binning technique to calculate lidar descriptors for each object created in Section 16.3.1. This technique can assign an object the respective statistic elevation feature value when more than 1 lidar ground point falls within this object (null

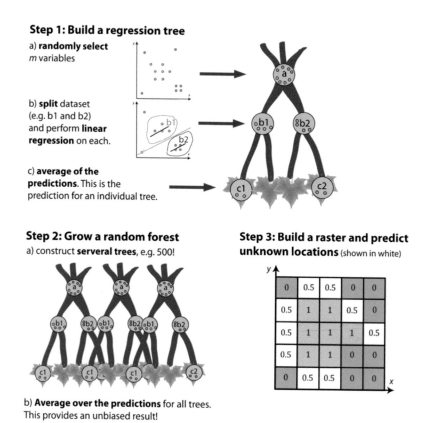

FIGURE 16.4
Illustration of the Random Forest regression algorithm. The decision tree is better shown upside down with the roots at the top and the leaves at the bottom because the terminal nodes or leaves lie at the bottom.

assigned to no data). It requires executing many GIS system tools in a sequence, thus we developed a Python script tool in ArcGIS to automate this technique. The first step of this tool is to attribute the object features by performing an identity analysis with the lidar bare-earth points. For each attribute value of an object, the lidar descriptive statistic values are joined first and then assigned to the object vector dataset to be used for the object-based lidar-DEM generation. To generate the final DEM product, we employed R "stats" (Hijmans, 2016) package where the result of the best-fitted model from Section 16.3.3 was used with all object vector datasets to predict the survey data values of each object.

16.3.5 Interpolated Lidar-DEMs

For comparison with the object-based lidar-DEM generated from the machine learning prediction, interpolated DEMs were generated for this study. Estimating the elevation of unknown lidar locations as a function of nearby lidar ground return measurements may be thought of as an interpolation problem where the output is a continuous variable (Figure 16.5). All spatial interpolation approaches are relevant to the first law of geography where "everything is related to everything else, but near things are more

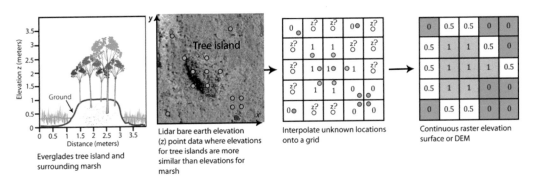

FIGURE 16.5
Illustration of the interpolation technique for generating raster DEMs in the Everglades.

related than distant things" (Tobler, 1970). In spatial analysis, this is referred to as spatial autocorrelation or evidence that data located near one another are likely to be more similar than data located further away from one another (O'Sullivan and Unwin, 2010). For example, elevation for an Everglades tree island will be more similar than elevation in the surrounding marsh (Figure 16.5). Therefore, many approaches to spatial interpolation apply weights to existing observations based on their proximities to the unsampled locations being estimated. Near points generally receive higher weights than points that are far away. Empirical Bayesian Kriging (EBK), an automatic geostatistical interpolator, produces near-accurate estimations along with their uncertainties based on statistical models that include autocorrelation. Here we applied the EBK interpolator to generate a lidar-DEM L for comparison purposes.

EBK performs estimations using the following algorithm (Krivoruchko, 2012). First, the data are tested for normality and transformed if necessary to more closely follow a Gaussian distribution. Second, a semi-variogram model, which is the function of the distance and direction separating two known locations, is estimated to quantify spatial dependency (autocorrelation). A new value is simulated at each known location. If the data was transformed, it is then transformed back at each of the known locations. Third, the new data is transformed (if necessary) and a new semi-variogram is estimated from the simulated data. The Bayes' rule, which is the likelihood that the observed data can be generated from the semi-variogram, is then used to calculate a weight for the semi-variogram. The second and third steps are repeated several times before the predictions and prediction standard errors are generated at unknown locations using the weights. If transformed, the data are back transformed with bias correction. The idea is that by creating a series of semi-variograms that are each an estimate of the true semi-variogram, predictions and prediction standard errors are more accurate.

We employed ArcGIS version 10.4 (www.esri.com/) Geostatistical Analyst to implement the EBK model over different tuning parameters. The following tuning parameters were defined: (1) the subset size for specifying the number of points in each subset, (2) the overlap factor for specifying the degree of overlap between subsets, (3) the number of simulations for specifying the number of semi-variograms to be simulated for each subset, (4) the transformation type for data that do not follow a Gaussian distribution, (5) the semi-variogram type for defining how the similarity of the observations diminishes over distances, (6) the neighborhood type for defining the searching ellipse,

(7) the maximum and minimum number of neighbors within the search ellipse to use when making predictions, and (8) the sector type for altering the number and orientation of sectors in the search neighborhood. All parameters were tested by trial and error. Once the EBK models were optimized, the validation datasets were used to judge the performance of the estimations on unseen data.

16.3.6 Lidar-DEM Accuracy Assessment

NSSDA uses statistical procedures to calculate the $RMSE_z$, or the square root of the average squared differences between the measured and "true" elevations of the same locations. The NSSDA $RMSE_z$ is computed as:

$$RMSE_z = \sqrt{\sum_{i=1}^{n} \left(z_{data,i} - z_{check,i}\right)^2 / n,}$$

(16.1)

where $z_{data,i}$ is the elevation measurement value of the ith checkpoint in the spatial dataset, $z_{check,i}$ is the elevation measurement value of the ith checkpoint in the independent source of better accuracy, n is the number of checkpoints, and i is an integer from 1 to n. Following Cooper et al. (2019), we also calculated each DEM's model-predicted (P) values and observed (O) RTK-GNSS measures using Pearson coefficient of correlation (r), Mean Absolute Error (MAE), and standard deviation (σ) using the following equations:

$$r = \frac{\sum_{i=1}^{n}\left(P_i - \overline{P_i}\right)\left(O_i - \overline{O_i}\right)}{\sqrt{\sum_{i=1}^{n}\left(P_i - \overline{P_i}\right)^2}\sqrt{\sum_{i=1}^{n}\left(O_i - \overline{O_i}\right)^2}},$$

(16.2)

$$MAE = \frac{\sum_{i=1}^{n}|P_i - O_i|}{n},$$

(16.3)

$$MBE = \frac{\sum_{i=1}^{n}(P_i - O_i)}{n},$$

(16.4)

where P_i is the model-predicted value, $\overline{P_i}$ is the mean of the model-predicted values, O_i is the observed value, $\overline{O_i}$ is the mean of the observed values, n is the number of matched test data samples, and i is an integer from 1 to n.

16.3.7 Lidar Bare-Earth Points Accuracy Assessment

It is necessary to determine the relationship between the RTK-GNSS surveyed data and lidar bare-earth points in order to assess the vertical accuracy of the lidar and to test the Random Forest and EBK methods. Here, proximity analysis and statistical approaches were used to determine this relationship. Proximity analysis was first carried out in Python version 2.7.10 (www.python.org/) using ArcGIS version 10.4 Arcpy Python site package to create an easy workflow. A search radius decision rule was applied where all lidar bare-earth points within 1 m distance of a survey point are spatially joined to the respective survey point. The 1 m search radius decision rule was chosen so that each survey point contains at least one lidar point. Statistical analysis was then carried out in

R version 3.3.2 (https://cran.r-project.org/). Descriptive statistics for all lidar points within a 1 m radius of a survey point was calculated.

16.3.8 Minimum Object-Based Bin (MOBB) and Bias Correction Lidar-DEMs

For comparison purposes, lidar-DEMs were generated using the object binning and bias-correction techniques. The object binning technique developed in this study was used where the lidar minimum elevation value was assigned to the object value when more than 1 lidar feature falls within that object (null assigned to no data) (see Section 16.3.4). This is referred to as the MOBB lidar-DEM. Since it is common for researchers to calibrate interpolated DEMs with bias correction procedures, the EBK lidar-DEM was calibrated by the MBE calculated from the calibration datasets (80% of the features; see Section 16.3.3). This dataset is referred to as the EBK bias-correction lidar-DEM.

16.4 Results and Discussion

16.4.1 Lidar Bare-Earth Errors

After a thorough literature review, no studies to date examined the accuracy of lidar bare-earth returns in the Everglades. To compare the errors of the lidar bare-earth returns used in this study for each dominant vegetation community, descriptive statistics are displayed for the uncorrected lidar in Table 16.1. The average magnitude of lidar errors in each dominant vegetation community tends to grow with vegetation height where the MAE for coastal prairies = 0.16 m, sawgrass = 0.19 m, short mangroves = 0.26 m, and tall mangroves = 0.30 cm. These results are conducive to other studies that found lidar errors tend to grow with vegetation species (e.g., *spartina alterniflora*) and height (Rosso et al., 2006; Sadro et al., 2007; Schmid et al., 2011). Therefore, it is important to take field measures to correct lidar error in coastal environments, especially the Everglades. The results also suggest that land cover type-based lidar correction is more reasonable because the errors show uneven distribution across the land cover types.

16.4.2 EBK vs. EBK Bias-Correction of Lidar-DEMs

For comparing the interpolated EBK lidar-DEM with the EBK bias-correction of lidar-DEM, descriptive statistics are displayed in Table 16.1. It is interesting to note that the unaltered MAE are no different from when the signs of the errors are not removed MBE for the EBK lidar-DEMs (MAE and MBE are the same for coastal prairies = 0.14, sawgrass = 0.26 m, short mangroves = 0.26 m, and tall mangroves = 0.30 m). This illustrates that the lidar measures overestimated the ground elevation, leading to a positive and consistent value in the difference between the prediction and observation. With respect to the average model bias, or in our case, average over-prediction, the EBK bias-correction produced a considerable improvement in lidar-DEM when compared to the EBK lidar-DEMs (MBE reduced from 0.14 to 0.00 m for coastal prairies, 0.26 to 0.06 m for sawgrass, 0.26 to 0.04 m for short mangroves, and 0.30 to 0.08 m for tall mangroves). The bias-correction technique is a simple approach to reducing errors in lidar-DEM and is preferred over simply interpolated lidar-DEMs. However, a RMSE less

TABLE 16.1

Descriptive statistics for each experiment. Where EBK = Empirical Bayesian Kriging, EBK bias-correction = Kriging DEM calibrated by the respective mean bias, MOBB = minimum object-based bin, RF = Random Forest, $\bar{P}$ = the mean of the model-predicted values, $\bar{O}$ = the mean of the observed RTK-GNSS, $\sigma_{\bar{P}-\bar{O}}$ = the standard deviation of the differences between $\bar{P}$ and $\bar{O}$, MBE = mean bias error between model-predicted and RTK-GNSS, MAE = Mean Absolute Error, RMSE = Root Mean Square Error between model-predicted and RTK-GNSS, n = number of RTK-GNSS, all in meters (except n)

Coastal Prairies

Experiment	n	$\bar{O}$	$\bar{P}$	$\sigma_{\bar{P}-\bar{O}}$	MBE	MAE	RMSE
Uncorrected lidar	31	0.06	0.22	0.07	0.16	0.16	0.17
EBK	31	0.06	0.02	0.07	0.14	0.14	0.16
EBK bias-correction	31	0.06	0.06	0.07	0.00	0.06	0.07
MOBB	31	0.06	0.01	0.08	−0.04	0.07	0.09
RF object-based	**31**	**0.06**	**0.05**	**0.02**	**0.01**	**0.02**	**0.02**

Mangrove > 5 meters in height

Uncorrected lidar	22	−0.08	0.21	0.28	0.28	0.30	0.40
EBK	22	−0.08	0.22	0.29	0.30	0.30	0.41
EBK bias-correction	22	−0.08	−0.16	0.29	0.08	0.24	0.29
MOBB	22	−0.08	−0.10	0.15	−0.02	0.09	0.14
RF object-based	**22**	**−0.08**	**−0.05**	**0.08**	**0.03**	**0.05**	**0.08**

Mangrove < 5 meters in height

Uncorrected lidar	28	−0.12	0.13	0.20	0.25	0.26	0.31
EBK	28	−0.12	0.14	0.18	0.26	0.26	0.31
EBK bias-correction	28	−0.12	−0.08	0.18	0.04	0.16	0.18
MOBB	28	−0.12	−0.15	0.19	−0.06	0.14	0.19
RF object-based	**28**	**−0.12**	**−0.17**	**0.12**	**−0.05**	**0.08**	**0.13**

Sawgrass

Uncorrected lidar	13	−0.14	0.05	0.14	0.19	0.19	0.24
EBK	13	−0.14	0.12	0.25	0.26	0.26	0.35
EBK bias-correction	13	−0.14	−0.08	0.25	0.06	0.16	0.24
MOBB	13	−0.14	−0.17	0.11	−0.03	0.08	0.11
RF object-based	**13**	**−0.14**	**−0.16**	**0.06**	**−0.02**	**0.05**	**0.06**

than the 0.15 m that is required for the Everglades projects (Jones et al., 2012) is met only for coastal prairies (EBK bias-correction RMSE = 0.07 m). Additional model comparisons are needed to determine the preferred approaches to meet this requirement.

16.4.3 EBK Bias-Correction vs. MOBB Lidar-DEMs

In terms of the MAE and RMSE, there is no significant difference between the EBK-bias correction and MOBB lidar-DEMs for coastal prairies and short mangroves (Table 16.1). However, the RMSE is considerably reduced from 0.29 to 0.14 m for the tall mangroves and 0.24 to 0.11 m for the sawgrass MOBB lidar-DEM. The tall mangroves and sawgrass

MOBB lidar-DEMs are therefore more appropriate for Everglades projects when compared to the EBK bias-correction of lidar-DEMs, at least for our study area. A major advantage to the MOBB technique in comparison to the bias-correction approach is that calibration data are not needed to calibrate the MOBB lidar-DEM. A disadvantage to the MOBB technique is that the segmentation algorithms such as the multi-resolution segmentation approach have not been available in traditional remote sensing software packages. So far, no approaches have been satisfactory at achieving a RMSE less than 0.15 m for sawgrass. An approach that can consistently achieve satisfactory results regardless of the dominant land cover type would increase efficiency in accurate and precise DEM generation in the coastal Everglades.

16.4.4 Object-based Lidar-DEMs from Machine Learning Models

In reviewing the difference measures of MAE and RMSE, a meaningful distinction can be made between the object-based lidar-DEM corrected from the RF algorithm and the EBK, EBK bias-correction, and MOBB lidar-DEMs (Table 16.1). The MAE and RMSE are lowest (best) for the RF object-based DEMs for each of the sawgrass, coastal prairies, short mangroves, and tall mangroves vegetation communities. The RF object-based is the only

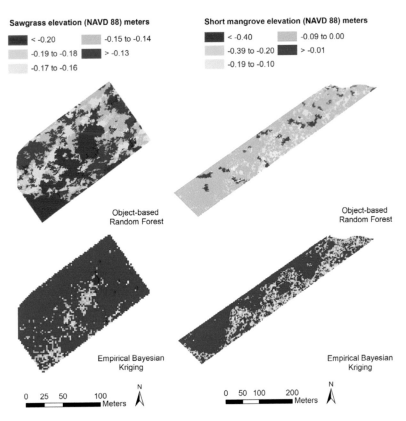

FIGURE 16.6
Object-based RF and Empirical Bayesian Kriging DEMs for sawgrass (left) and short mangroves (right).

approach that achieves a satisfactory RMSE less than 0.15 m (RMSE = 0.06 for sawgrass, 0.02 for coastal prairies, 0.13 for short mangroves, and 0.08 for tall mangroves). Overall, the object-based lidar-DEMs corrected from the RF performed best when compared to EBK, EBK bias-correction, and MOBB lidar-DEMs. This study demonstrates that the object-based machine learning technique is an optimal approach for generating DEMs in the Everglades, which is also consistent with the study by Cooper et al. (2019) who consolidated the vegetation communities.

For comparison purposes, the DEMs generated from RF and EBK are shown in Figures 16.6 and 16.7. In the maps, red represents lower elevations, while the dark blue represents higher elevations, all relative to NAVD 88 in meters. The EBK maps show that the EBK interpolation approach continuously overestimates the ground elevation (shown by the dark blue) regardless of land cover.

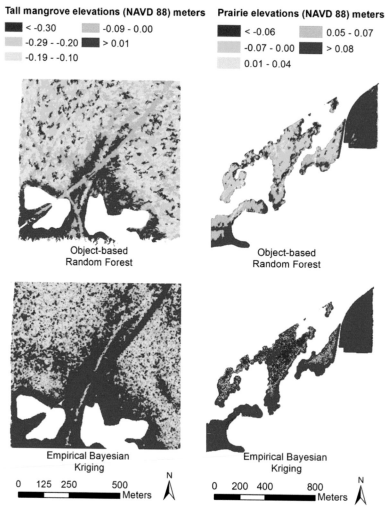

FIGURE 16.7
Object-based RF and Empirical Bayesian Kriging DEMs for tall mangroves (left) and prairies (right).

16.5 Summary and Conclusions

We evaluated 3 DEM correction techniques to determine which met the required vertical RMSE of <0.15 m for 4 different vegetation communities that are representative of the coastal Everglades (i.e., coastal prairies, sawgrass, short mangroves, and tall mangroves). We found that lidar bare-earth point errors tend to grow with both vegetation species and height in the coastal Everglades (i.e., the MAE for coastal prairies = 0.16 m, sawgrass = 0.19 m, short mangroves = 0.26 m, and tall mangroves = 0.30 cm), which confirmed the findings in the literature. It is typical for researchers to accept interpolated lidar-DEMs with an accuracy assessment provided by the vendor that does not capture the characteristics of wetland land covers. The results of this study demonstrated that the interpolated EBK lidar-DEM does not meet the required vertical RMSE of <0.15 m for any of the 4 vegetation communities, so researchers should be cautioned when using data presented as-is in wetland environments. On the other hand, the EBK bias-correction approach can produce DEM that meets this requirement for only the coastal prairies. We thus developed the MOBB technique that has a major advantage over the bias-correction approach because calibration data are not needed to calibrate the MOBB lidar-DEM, and the vertical RMSE of <0.15 m is met for all vegetation communities with one exception, the vertical RMSE is 0.19 m for short mangroves. The overall best approach that can consistently achieve satisfactory results regardless of the dominant land cover in the Everglades is the object-based machine learning approach using RF (vertical RMSE <0.15 m for all 4 vegetation communities). Overall, the MOBB and object-based machine learning approaches increase efficiency in accurate and precise DEM generation and should be considered in future applications that require reliable elevation data in the Everglades.

References

American Society for Photogrammetry and Remote Sensing (ASPRS), 2014. Positional accuracy standards for digital geospatial data. Edition 1, Version 1.0. *Photogrammetric Engineering & Remote Sensing*, 81 (3), A1–A26.

Benz, U.C., Hofmann, P., Willhauck, G., et al., 2004. Multiresolution, object-oriented fuzzy analysis of remote sensing data for GIS-ready information. *ISPRS Journal of Photogrammetry and Remote Sensing*, 58, 239–258.

Breiman, L., 2001. Random forests. *Machine Learning*, 45, 5–32.

Buffington, K.J., Dugger, B.D., Thorne, K.M., et al., 2016. Statistical correction of lidar-derived digital elevation models with multispectral airborne imagery in tidal marshes. *Remote Sensing of Environment*, 186, 616–625.

Cooper, H.M., Fletcher, C.H., Chen, Q., and Barbee, M.M., 2013. Sea-level rise vulnerability mapping for adaptation decisions using lidar-DEMs. *Progress in Physical Geography*, 37, 745–766.

Cooper, H.M., Zhang, C., Davis, S., et al., 2019. Object-based correction of lidar-DEMs using RTK-GPS data and machine learning modeling in the coastal everglades. *Environmental Modelling & Software*, 112, 179–191.

Criminisi, A., Shotton, J., and Konukoglu, E., 2011. Decision forests: A unified framework for classification, regression, density estimation, manifold learning and semi-supervised learning. *Foundations and Trends in Computer Graphics and Vision*, 7(2–3), 81–227.

FGDC, 1998. Geospatial positioning accuracy standards, Part 3: National standard for spatial data accuracy. US Geological Survey Report no. FGDC-STD 007.3-1998.

Hijmans, R.J., 2016. Package 'raster'. Available at: https://cran.r-project.org/web/packages/raster/raster.pdf

Hladik, C., and Alber, M., 2012. Accuracy assessment and correction of a lidar-derived salt marsh digital elevation model. *Remote Sensing of Environment*, 121, 224–235.

Jones, J.W., Desmond, G.B., Henkle, C., et al., 2012. An approach to regional wetland digital elevation model development using a differential global positioning system and a custom-built helicopter-based surveying system. *International Journal of Remote Sensing*, 33, 450–465.

Kohavi, R., 1995. A study of cross-validation and bootstrap for accuracy estimation and model selection. In: C.S. Mellish (ed.), *Proceedings of the 14th International Joint Conference on Artificial Intelligence*, 1137–1143.

Krivoruchko, K., 2012. "Empirical Bayesian kriging," *ArcUser Fall 2012*. www.esri.com/NEWS/ARCUSER/1012/files/ebk.pdf

Kuhn, M., 2016. Package 'caret': Classification and regression training. www.cran.r-project.org/pub/R/web/packages/caret/caret.pdf

Leica Geosystems, 2007. Leica ALS50 – II airborne laser scanner product specifications. www.nts-info.com/inventory/images/ALS50-II.Ref.703.pdf

Leica Geosystems, 2008. Leica GPS1200 + Series technical data. www.leica-geosystems.se/se/gps_1200_glonass_150dpi.pdf

Lodge, T.E., 2010. *The Everglades Handback: Understanding the Ecosystem*, 3rd Edition, Taylor and Francis Group, Boca Raton, FL.

McClure, A., Liu, X.H., Hines, E., et al., 2016. Evaluation of error reduction techniques on a lidar-derived salt marsh digital elevation model. *Journal of Coastal Research*, 32, 424–433.

Medeiros, S., Hagen, S., Weishampel, J., et al., 2015. Adjusting lidar-derived Digital Terrain Models in coastal marshes based on estimated aboveground biomass density. *Remote Sensing*, 7, 3507–3525.

O'Sullivan, D., and Unwin, D., 2010. *Geographic Information Analysis*, 2nd Edition, John Wiley & Sons Inc., Hoboken, NJ, 432.

Rogers, J.N., Parrish, C.E., Ward, L.G., et al., 2016. Assessment of elevation uncertainty in salt marsh environments using discrete-return and full waveform lidar. *Journal of Coastal Research Special Issue*, 76, 107e122.

Rogers, J.N., Parrish, C.E., Ward, L.G., et al., 2018. Improving salt marsh digital elevation model accuracy with full-waveform and nonparametric predictive modeling. *Estuarine Coastal and Shelf Science*, 202, 193–211.

Rosso, P.H., Ustin, S.L., and Hastings, A., 2006. Use of lidar to study changes associated with Spartina invasion in San Francisco Bay Marshes. *Remote Sensing of Environment*, 100, 295–306.

Sadro, S., Buhl-Gastil, M., and Melack, J., 2007. Characterizing patterns of plant distribution in a southern California salt marsh using remotely sensed topographic and hyperspectral data and local tidal fluctuations. *Remote Sensing of Environment*, 110, 226–239.

Schmid, K.A., Hadley, B.C., and Wijekoon, N., 2011. Vertical accuracy and use of topographic lidar data in coastal marshes. *Journal of Coastal Research*, 27, 116–132.

Tobler, W., 1970. A computer movie simulating urban growth in the detroit region. *Economic Geography*, 46(2), 234–240.

Trimble, 2014. *eCognition Developer 9.0.1 Reference Book*. Trimble Germany GmbH, Arnulfstrasse 126, D-80636, Munich.

17

Assessing the Effects of Hurricane Irma on Mangrove Structures in the Coastal Everglades using Airborne Lidar Data

Caiyun Zhang

17.1 Introduction

Mangrove forests are important ecosystems occupying intertidal settings along the land-sea interface in tropical and subtropical regions worldwide (Giri et al., 2011). They are ecologically significant to humans and the natural environment by providing a range of benefits such as stabilizing coastlines, protecting communities from storms, and storing vast amounts of carbon (Zhang et al., 2012; Beck et al., 2018). Despite their roles in providing goods and services, mangrove forests are threatened by human activities and natural disasters such as sea level rise and tropical cyclones (Gilman et al., 2008; Ward et al., 2016; Thomas et al., 2017; Sippo et al., 2018). The Florida Everglades has the largest contiguous tracts of mangrove forests in the United States, and most of them are located in Everglades National Park (ENP). Though they are in the protected park area with little effects from human activities, they are extremely vulnerable to hurricanes because Florida has been ranked the number one location hurricanes make landfall in the USA. Based on the data from National Oceanic and Atmospheric Administration (NOAA) since 1851, 40% of all US hurricanes hit Florida (NOAA/FAQ, 2019). Of 91 major hurricanes, 37 hit Florida directly.

Hurricane Irma, one of the strongest and costliest hurricanes on record in the Atlantic basin, made landfall on September 10, 2017 as a Category 3 hurricane in the southern coastal Everglades. Immediately after Irma, multiple agencies, such as National Aeronautics and Space Administration (NASA), have collected light detection and ranging (lidar) data and aerial photography over the margin of the coastal Everglades. NASA's preliminary analysis showed devastating mangrove damage. We have applied Landsat data to detect the large-scale damage of mangroves from Hurricane Irma and made the first effort to develop machine learning based mangrove hurricane risk models (Zhang et al., 2019). The mangrove damage detected from Landsat data is shown in Figure 17.1. We found that 332 km^2 of mangroves were severely damaged from Irma, and 635 km^2 would be devastated from a worst-case scenario Category 5 hurricane along the mangrove distribution in the southern coastal Everglades. Optical Landsat is effective for large-scale mangrove damage analysis but cannot provide information on the vertical structure dynamics of mangroves caused by hurricanes. Though NASA has conducted preliminary analyses of lidar data to reveal mangrove damage from Irma in the ENP, more studies are needed. Airborne lidar data has proven useful in the ENP for quantification of disturbances and lightning strikes from hurricanes Katrina and Wilma in 2005 (Zhang et al., 2008). In this chapter, pre-and post-Irma lidar data were applied to look at the effects of Hurricane Irma on mangrove structures.

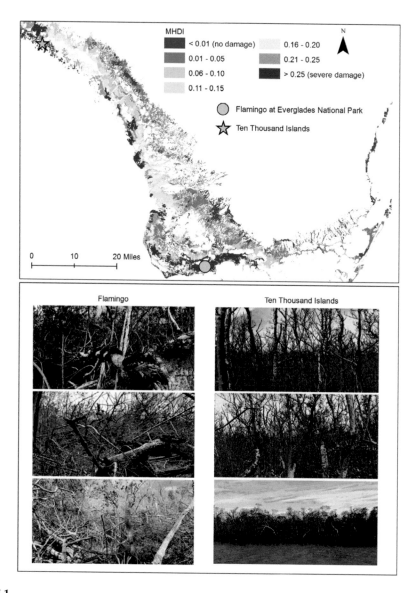

FIGURE 17.1
Large-scale mangrove damage detection from Landsat data collected before and after Hurricane Irma (Zhang et al., 2019). A high value in the Mangrove Damage Hurricane Index (MHDI) represents severe damage, while a low value refers to small damage. Field pictures shown in the bottom panel were collected immediately after Hurricane Irma.

17.2 Study Area and Data

The study area is located at the Shark River Slough in the ENP with an approximate size of 10 km^2 (Figure 17.2). Shark River Slough is a low-lying area of land that channels water through the Florida Everglades. It is the major path for water flow in the

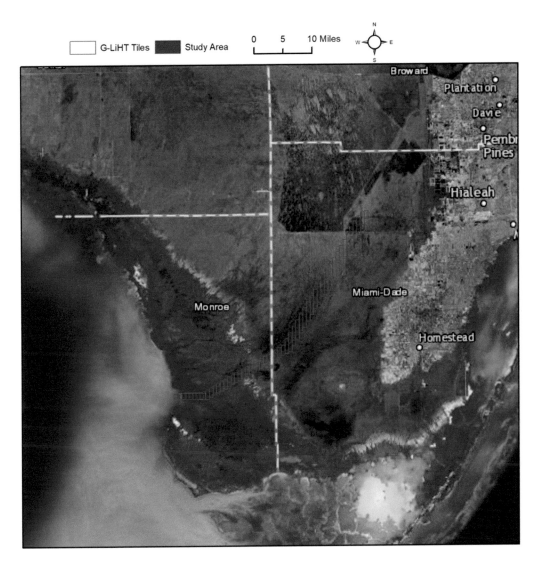

FIGURE 17.2
Study area located in the Shark River Slough and LiDAR data tiles for this area.

Everglades system. This area is a mixture of sawgrass marshes, tree islands, sloughs, and wet prairies. At the fringe of the slough where fresh water enters Florida Bay, the region is dominated by mangroves. NASA's Goddard's Lidar, Hyperspectral & Thermal Imager (G-LiHT) team collected pre- and post-lidar data over part of this region in March and December 2017. The G-LiHT team started a South Florida project to address how freshwater ecosystems such as sawgrass marshes of the Everglades are transitioning to saltwater ecosystems using remote sensing techniques, and collected lidar data in March 2017. After three months of Hurricane Irma, the team flied the same paths over the Everglades. Leafless trees, broken branches, and even uprooted mangrove trees were revealed by the team. G-LiHT is an airborne system that simultaneously collects lidar data, hyperspectral imagery, and thermal data in order to map the composition, structure, and function of terrestrial ecosystems (Cook et al., 2013). Given the complementary nature of lidar, optical, and thermal data, this airborne system aids data fusion studies by providing coincident data in time and space. G-LiHT has acquired a large volume of data across a broad range of ecoregions in the United States and Mexico since 2011; these data have been processed into standard data products, such as the 1-meter lidar derived Digital Terrain Model (DTM) and Canopy Height Model (CHM). The products have been freely distributed to the public at the G-LiHT Data Center through an interactive web map, and an FTP data portal (http://gliht.gsfc.nasa.gov/). Analysis of these products has been focused on providing new insights on photosynthetic functionality and vegetation productivity, characterizing fine-scale spatial and temporal heterogeneity in ecosystem structures, and creating new methods in data fusion to monitor ecosystem health (Cook et al., 2013).

We have applied G-LiHT products to map urban land cover types (Zhang et al., 2018). In this chapter, G-LiHT lidar products, DTM, CHM, and canopy rugosity, were applied to look at the effects of Hurricane Irma on the structures of mangroves. Thermal and hyperspectral imagery was not available when this study was conducted. G-LiHT collects both profiling and scanning lidar data. The scanning lidar data are collected by the VQ-480 airborne laser-scanning instrument, and the profiling lidar data are collected by an LD321-A40 multi-purpose laser distance meter. Based on the metadata of the lidar data used in this chapter, scanning lidar data were collected by the VQ-480i model using a wavelength of 1550 nm. Detailed data processing and product generation information were provided in Cook et al. (2013). The 1-meter lidar-DTM, CHM, and rugosity products were generated from the scanning lidar data. G-LiHT produces CHM by selecting the greatest return height in every 1-grid cell and using these points to interpolate the CHM raster grid. Canopy rugosity is determined as the standard deviation of canopy height. To generate DTM, G-LiHT uses a progressive morphological filter to classify the ground returns first, then creates a Triangulated Irregular Network using ground hits, and finally linearly interpolates DTM elevations to 1-m grid cell. Pre-Irma lidar data were collected on March 29, 2017, while a post-Irma lidar survey was conducted on December 2, 2017. The overlapping tiles for both dates are displayed in Figure 17.2.

17.3 Methodology

The mangrove DTM, CHM, and rugosity products were downloaded, mosaicked, and clipped to the study area highlighted in red in Figure 17.2. Noisy pixels were observed such as pixels falling over rivers and streams. To filter out these pixels, a mangrove mask was determined first by selecting pixels with a CHM less than 1 m. A vector mangrove mask layer was then created by converting the raster mask in ArcGIS. Further analysis was limited to pixels with a CHM more than 1 m. Meanwhile, some pixels were found to have a canopy height more than 30 m, which were also considered noisy pixels and were dropped in further analysis. Statistical metrics including mean and standard deviation were determined to look at the effects of Hurricane Irma on the mangrove canopy height. Change of CHM, DTM, and rugosity were mapped and analyzed. It has been reported that tropical cyclones and lightning can produce various sizes of openings or gaps in mangrove forests by causing the death of individual or groups of trees. Thus, the gaps of pre-Irma and post-Irma using lidar-derived CHM were identified and delineated. To achieve this, 4 major steps were followed: (1) a binary file was created by selecting pixels with a canopy height less than 5 m using the Con or reclassification function in ArcGIS; (2) the binary file was refined by selecting connected pixels of groups with at least 5 pixels, resulting in a new binary file without small noisy holes or gaps; (3) the refined binary file was imported into ENVI to conduct a morphological opening function with a window size of 5 to further refine the gaps. The morphological opening algorithm can eliminate small islands and sharp peaks or capes in an image. The opening of an image is defined as the erosion of the image followed by subsequent dilation using the same structure elements; and (4) the final refined raster gap file was imported into ArcGIS and converted into a vector gap file to be used for mapping and analysis. Mapping and geospatial analysis functions in ArcGIS were used to visualize and present the results.

17.4 Results and Discussion

17.4.1 Impacts on Mangrove Canopy Height

The mangrove CHMs of pre- and post-Irma are displayed in Figure 17.3 (a) and (b). The change is distinct. Hurricanes can cause severe defoliation and damage to mangrove trees and lower the canopy top surface. This has been confirmed by the analysis of lidar CHMs of mangroves over the Shark River Slough. More areas with tall canopies (height >15 m) were observed in the pre-Irma CHM than in the post-Irma CHM. The map of CHM became more heterogeneous after Hurricane Irma. The difference of pre- and post-Irma further revealed the damage of Hurricane Irma on mangrove canopies, as demonstrated in Figure 17.3 (c). For the selected study site, most canopies were shortened by 1–2 m. In contrast, an increase of CHMs was observed over small streams or big gap areas. This could be caused by the fallen branches or stems of mangroves over adjacent streams or gaps where no mangroves were present. The histogram of pre- and post-mangrove CHMs is shown in Figure 17.4, which further confirmed the change of CHMs caused by Hurricane Irma. The histogram also revealed that a threshold of 30 m to the CHMs was suitable to exclude noisy pixels, some of which had a height of more than 60 m. The mean and standard deviation of CHMs before and after Hurricane Irma are displayed in Table 17.1. The average of CHM was decreased from 15.77 m to 13.79 m, and standard deviation was changed from 5.39 m to 5.77 m. On average, mangrove canopy height was shortened by 1.98 m from Irma. The map of CHM change showed that the mean change was 2.69 m, and the standard deviation of CHM change was 4.66 m, as shown in Table 17.2. Canopies with a height more than 15 m but less than 30 m were considered tall canopies and were further analyzed because tall trees were more easily damaged than short canopies (Zhang et al., 2012). The mean and standard deviation of tall canopies did not have a significant change before and after Hurricane Irma (Table 17.1) with a mean pre-Irma CHM of 18.73 m and a mean post-Irma CHM of 18.33. Standard deviation was slightly increased. However, the coverage of tall canopies was largely reduced from a pre-Irma 5.71 km^2 to a post-Irma 3.99 km^2, a decrease of

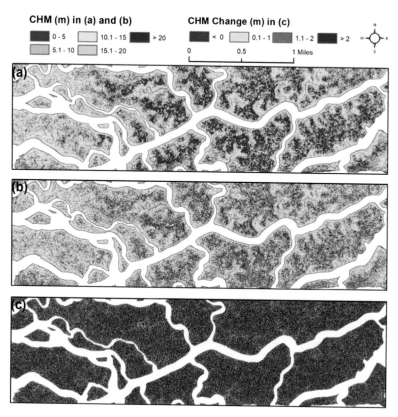

FIGURE 17.3
Mangrove canopy height models of (a) pre-Irma, (b) post-Irma and (c) change caused by Hurricane Irma.

1.72 km^2. On average, the tall canopies were shortened 2.8 m by Hurricane Irma (Table 17.2). The spatial height changes is displayed in Figure 17.5. In total, 1.3 km^2 was observed to have a change of more than 5 m in the heights of tall canopies. The damage of Hurricane Irma on mangrove canopies was also demonstrated by the change in canopy rugosity, as shown in Figure 17.6. The pre-Irma rugosity of mangrove canopies is smaller and canopies were smoother, while after Irma swept through, the rugosity became bigger and canopies became more complex and heterogeneous (Figure 17.6). Some regions had a change larger than 4 m (Figure 17.6 (c)).

17.4.2 Increased Canopy Gaps from Hurricane Irma

Lidar-derived fine resolution CHMs can effectively reveal canopy gaps within mangrove forests, as illustrated in Figure 17.7. A profile of CHM successfully identified a big gap within the pre-Irma CHM with a width of about 60 m. The spatial distribution of the detected mangrove canopy gaps with a size more than 50 m^2 from the pre-and post-Irma CHMs are displayed in Figure 17.8. Hurricane Irma significantly increased the number of canopy openings, as demonstrated. In total, 364 gaps were found in the pre-Irma CHM; the number of gaps increased to 3259 after the hurricane with 2895 new canopy openings, not counting those less than 50 m^2 (Table 17.3). Zhang et al. (2008)

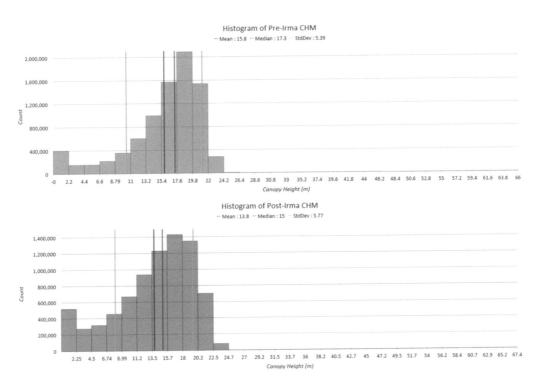

FIGURE 17.4
Histogram of pre- and post-Irma lidar-derived CHMs.

TABLE 17.1

Statistical metrics of mangrove canopy heights before and after Hurricane Irma

	All canopies		
	Pre-Irma	**Post-Irma**	**Difference**
Mean (m)	15.77	13.79	1.98
Standard deviation (m)	5.39	5.77	−0.38
	Tall canopies		
Mean (m)	18.73	18.33	0.40
Standard deviation (m)	2.0	2.02	−0.02
Area (km^2)	5.71	3.99	1.72

Canopies with a height more than 15 m were considered as tall canopies; canopies with a height more than 30 m were considered as noises and were excluded from analysis.

reported that the number of mangrove gaps per square kilometer in the ENP increased from about 400 or 500 to 4000 after hurricanes Katrina and Wilma. Note that in their study, gaps with a size of more than 10 m^2 were counted. The total area of gaps in the study area was about 0.057 km^2 (0.57% of the total study area). After Irma, the total area of gaps was increased to 0.89 km^2 (8.9% of the total study area). The average size of the gaps before Hurricane Irma was 156.4 m^2. After Irma, the average size increased to 272.6 m^2.

TABLE 17.2

Statistical metrics of canopy height change
caused by Hurricane Irma

All trees	
Mean change (m)	2.69
Standard deviation of changes (m)	4.66
Tall trees	
Mean change (m)	2.80
Standard deviation of changes (m)	4.35

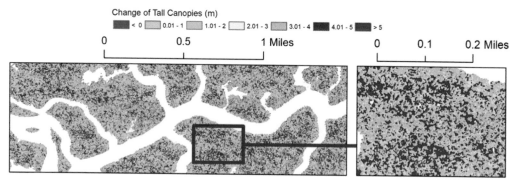

FIGURE 17.5
Damage of tall canopies detected from pre- and post-Irma lidar data. Canopies with a height more than 15 m were considered as tall canopies; canopies with a height more than 30 m were considered as noises and excluded from damage analysis.

The gap density (i.e., the number of gaps per square kilometer) was about 45 and 402, in pre-and post-Irma CHMs, respectively.

The mangrove ecosystem in the coastal Everglades is impacted by sea level rise, hurricanes, Everglades hydrological restoration, and canopy gaps caused by hurricanes and lightning strikes. The study shows that lidar is valuable for monitoring the dynamics of canopy gaps using fine-resolution CHMs. The geographic locations, scale of gaps, number of gaps, etc., can be determined from lidar products. Acquisition of this information is often difficult using boat, helicopter, or other in-situ methods. There were no in-situ gap data available before and after Hurricane Irma. Thus, the lidar-derived mangrove gap results were not assessed. However, the assessment from in-situ plot surveys for lidar-estimated gaps in Zhang et al. (2008) shows that lidar is an effective method for mangrove gap identification. Thus, the determined gap information is useful for mangrove preservation and conservation in the ENP.

It is known that hurricanes and other disturbances can cause severe damage to mangrove forests by defoliating the canopies, snapping branches, damaging barks, and uprooting trees, resulting in a large modification of the mangrove structure. This modification in mangrove structure can be revealed using airborne lidar data. A range of lidar parameters are useful for this purpose such as the number of returns, ratio

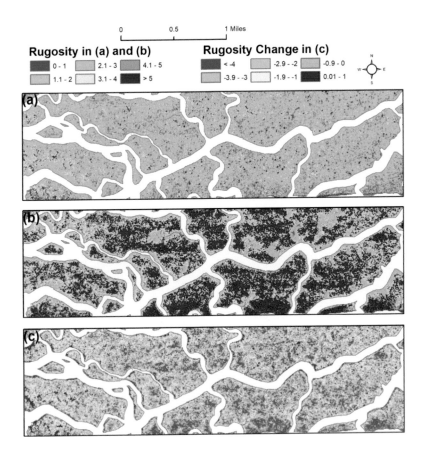

FIGURE 17.6
Mangrove canopy rugosity of (a) pre-Irma, (b) post-Irma, and (c) change caused by Hurricane Irma.

between the number of ground points and non-ground points, and canopy top surface height. The study shows that CHMs are valuable for assessing the damage from hurricanes by quantifying the decrease of canopy heights and tall mangrove canopies and detecting new openings within canopies. It is worth including other information in the analysis to gain insights into the mechanism of species-specific effects and varying impacts of storm surges.

17.4.3 Impacts on Terrains

The pre-and post-Irma DTMs and their differences are displayed in Figure 17.9. In general, the coastal Everglades mangrove ecosystem has a low-lying elevation with most regions less than 1 m. Small streams were clearly revealed in the (a) pre-Irma DTM. The (b) post-Irma DTM was generally consistent with the spatial pattern of the pre-Irma DTM, but most streams had disappeared. The post-Irma lidar DTM seems more reliable because more lidar pulses can pass through the mangrove canopy top and hit ground due to severe defoliation from the hurricane. In contrast, streams or gaps will be blocked from lidar due to the fallen branches, trunks, and other debris. The difference in the (c)

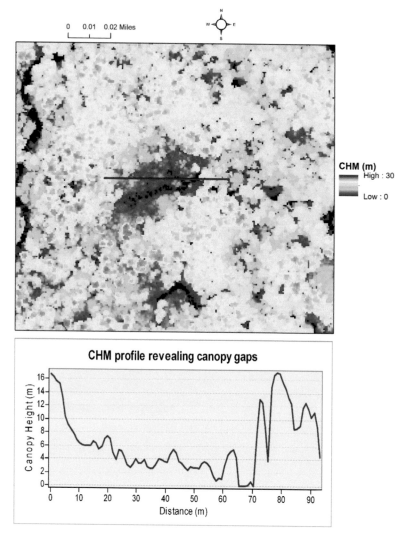

FIGURE 17.7
Illustration of canopy gaps revealed using 3-D Spatial Analyst function in ArcGIS 10.7 from the pre-Irma CHM.

pre-and post-Irma DTMs can delineate the deposition and erosion of soils caused by hurricanes. Blue represents deposition with negative values; orange and red, with positive values, refer to erosion. A majority of areas were observed to have an accretion in elevation after Hurricane Irma.

Hurricanes have been recognized to play an important ecological role in wetlands and coastal ecosystems by affecting soil elevation. Hurricanes can increase mangrove terrain elevation by depositing materials and folding peat layers. Hurricane storm surge can cause large-scale redistribution of sediments, resulting in sediment deposition, erosion, compaction, disruption of vegetated substrates, or some combination of these processes (Cahoon,

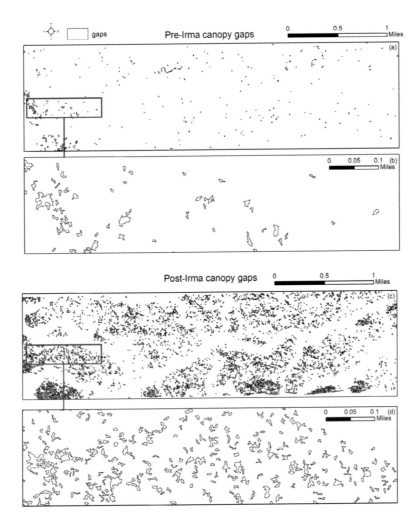

FIGURE 17.8
Canopy gaps detected from (a) and (b) pre-Irma and (c) and (d) post-Irma lidar-derived CHMs.

TABLE 17.3

Mangrove canopy gaps revealed from pre-and post-Irma CHMs

	Pre-Irma	Post-Irma	Difference
Number of gaps	364	3259	2895
Mean size (m²)	156.4	272.6	116.2
Total coverage (m²)	56,920	888,316	831,396
Density (number/km²)	45	402	357

Gaps are connected pixels with a canopy height less than 5 m and size more than 50 m².

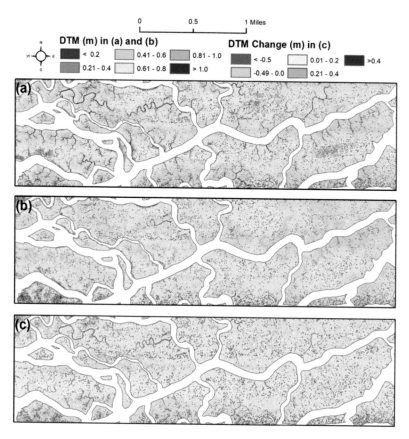

FIGURE 17.9
DTMs of (a) pre-Irma, (b) post-Irma, and (c) change caused by Hurricane Irma.

2006). For example, Hurricane Wilma deposited an average of 77 mm of sediment in the mangrove forests at Shark River in the ENP. Whelan et al. (2009) measured and analyzed the mangrove soil elevation change from Hurricane Wilma (6 months and 1 year post-Wilma) and also reported the direct deposit of sediments to mangrove terrains. There are pros and cons of mangrove terrain accretion from hurricanes. The accretion can mitigate the long-term sea level rise to reduce the risk for mangroves. However, the large deposition from the hurricane's storm surge can suffocate mangrove roots, leading to a delayed mortality of mangrove trees (Radabaugh et al., 2019). Most of these studies used in-situ soil elevation data, which had limited coverage. Redistribution of the sediment caused by hurricanes is difficult to characterize using the in-situ measurements. Lidar provides an attractive alternative to in-situ methods to understand the effects of hurricanes on soil elevations and is able to map the accretion and erosion of sediments at a landscape scale. However, for wetlands with thick vegetation covers, such as mangrove ecosystems in the coastal Everglades, the capability of lidar may be constrained because only limited pulses can penetrate the plants and hit the ground, leading to an overestimate of DTM in the coastal Everglades. To better apply lidar for studying mangrove terrain dynamics, lidar DTM should be corrected first, as discussed in Chapter 16.

17.5 Summary and Conclusions

In this chapter, pre-and post-Irma lidar datasets released by NASA's G-LiHT were applied to quantify the effects of Hurricane Irma on the structure of mangrove forests at the Shark River Slough in the ENP. Irma shortened the mangrove canopy top surface by ~2.0 m and reduced the coverage of tall mangrove canopies by 30%. Hurricane Irma also created more than 2000 new openings with a size more than 50 m^2 at the study area, which significantly increased the coverage of mangrove canopy gaps. In general, Hurricane Irma severely impacted the mangrove surface structure indicated from the variation of canopy heights and canopy rugosity, and mangrove vertical structure by opening a large number of new gaps. Mangrove terrain surface soil elevation was also changed. A large area of deposition from Irma was revealed from lidar-derived DTMs. The study indicates that lidar is a valuable tool for quantifying the dynamics of mangrove structures caused by hurricanes. However, a large-scale application of lidar is costly because, to date, lidar data is mainly collected from airborne platforms. Lidar also suffers from penetration issues in landscapes with a dense vegetation. The low-lying elevation and high inundation of mangroves make terrain characterization using lidar more challenging. A synergy of lidar, in-situ surveys, and optical sensors is beneficial.

References

Beck, M.W., Narayan, S., Trespalacios, D., et al., 2018. *The Global Value of Mangroves for Risk Reduction.* Summary Report, The Nature Conservancy, Berlin.

Cahoon, D.R., 2006. A review of major storm impacts on coastal wetland elevations. *Estuaries and Coasts*, 29, 889–898.

Cook, B.D., Corp, L.W., Nelson, R.F., et al., 2013. NASA Goddard's Lidar, Hyperspectral and Thermal (G-LiHT) airborne imager. *Remote Sensing*, 5, 4045–4406.

Gilman, E., Ellison, J., Duke, N.C., and Field, C., 2008. Threats to mangroves from climate change and adaptation options: A review. *Aquatic Botany*, 89, 237–250.

Giri, C., Ochieng, E., Tieszen, L.L., et al., 2011. Status and distribution of mangrove forests of the world using earth observation satellite data. *Global Ecology and Biogeography*, 20, 154–159.

NOAA/FAQ, 2019. www.aoml.noaa.gov/hrd/tcfaq/E19.html. Accessed on 20 May 2019.

Radabaugh, K.R., Moyer, R.P., Chappel, A.R., et al., 2019. Mangrove damage, delayed mortality, and early recovery following Hurricane Irma at two landfall sites in Southwest Florida, USA. *Estuaries and Coasts*. DOI:10.1007/s12237-019-00564-8.

Sippo, J.Z., Lovelock, C.E., Santos, I.R., et al., 2018. Mangrove mortality in a changing climate: A review. *Estuarine, Coastal and Shelf Science*, 215, 241–249.

Thomas, N., Lucas, R., Bunting, P., et al., 2017. Distribution and drivers of global mangrove forest change, 1996–2010. *PLoS One*, 12, e0179302.

Ward, R.D., Friess, D.A., Day, R.H., and MacKenzie, R.A., 2016. Impacts of climate change on mangrove ecosystems: A region by region review. *Ecosystem Health and Sustainability*, 2, e01211.

Whelan, K.R.T., Smith, T.J., Anderson, G.H., and Ouellette, M.L., 2009. Hurricane Wilma's impact on overall soil elevation and zones within the soil profile in a mangrove forest. *Wetlands*, 29, 16–23.

Zhang, C., Durgan, S., and Lagomasino, D., 2019. Modeling risk of mangroves to tropical cyclones: A case study of Hurricane Irma. *Estuarine, Coastal, and Shelf Science*, 224, 108–116.

Zhang, C., Smith, M., and Fang, C., 2018. Evaluation of Goddard's Lidar, hyperspectral, and thermal data products for mapping urban land-cover types. *GIScience & Remote Sensing*, 55, 90–109.

Zhang, K., Liu, H., Li, Y., et al., 2012. The role of mangroves in attenuating storm surges. *Estuarine, Coastal and Shelf Science*, 102–103, 11–23.

Zhang, K., Simard, M., Ross, M., et al., 2008. Airborne laser scanning quantification of disturbances from hurricanes and lightning strikes to mangrove forests in Everglades National Park, USA. *Sensors*, 8, 2262–2292.

Part V

Fusing Multiple Sensors for Everglades Applications

18

Integrating Aerial Photography, EO-1/Hyperion, and Lidar Data to Map Vegetation in the Coastal Everglades

Caiyun Zhang

18.1 Introduction

Mapping vegetation at different detail levels can document the changes of vegetation communities, types, and species to inform the progress and effects of restoration on the Everglades ecosystem. Several efforts have been made to automate vegetation mapping through digital image analysis. Such efforts have been grouped into 3 categories in Zhang et al. (2016). The first is the application of single-source remotely sensed data such as multispectral imagery (Rutchey and Vilchek, 1994, 1999; Jensen et al., 1995; Szantoi et al., 2015) and hyperspectral imagery (Hirano et al., 2003; Zhang and Xie, 2012, 2013). Chapters 4, 5, 6, 12, and 13 in this book also demonstrate the application of single-source remote sensing imagery for vegetation mapping in the Everglades. Studies have shown that most available remotely sensed imagery (e.g., Landsat) is not able to generate classifications with reasonable accuracies for detailed vegetation mapping in the Everglades due to limitations of spatial or spectral resolution. High spatial-resolution (i.e., 5 meters or smaller) airborne hyperspectral imagery is powerful in vegetation characterization in the Everglades, but data collection is costly. The second is the employment of a combination of single-source remotely sensed imagery with ancillary data such as environmental and hydrological variables and soil data (Griffin et al., 2011; Szantoi et al., 2013). The third is the application of multi-source data remotely sensed through data fusion techniques (Zhang et al., 2013, 2016, 2018; Zhang and Xie, 2014; Zhang, 2014) that integrate data and information from multiple sources to achieve refined/improved information for decision-making (Gómez–Chova et al., 2015). With the increasing availability of multi-sensor, multi-temporal, and multi-resolution images, data fusion has become a valuable tool for updating wetland inventory.

We have made efforts to combine multiple remote sensing data sources to map vegetation in the Everglades. We have examined a combination of 2 remote sensing data sources for mapping vegetation at different detail levels. For example, in Zhang et al. (2013), a fusion of lidar and aerial photography was assessed to map 7 forest communities in the Lake Okeechobee watershed, which produced an overall accuracy (OA) of 71% and Kappa value of 0.64. In Zhang (2014), a synergy of lidar and 12-m hyperspectral imagery collected from Airborne Visible/Infrared Imaging Spectrometer (AVIRIS) produced an OA of 86% and Kappa value of 0.82 for mapping 13 vegetation communities. A recent study explored a synergy of lidar-derived Digital Elevation Model (DEM) and 1-foot aerial photography for mapping freshwater marsh species in the wetland of Lake Okeechobee and achieved an OA of 85% and Kappa value of 0.83

(Zhang et al., 2018). The final classified vegetation maps (combining lidar with aerial photography or hyperspectral data) are displayed in Figure 18.1. The results illustrated that lidar data was valuable for vegetation mapping in the Florida Everglades. A synergy of 1-m aerial photography and 20-m AVIRIS hyperspectral imagery was also examined to map 9 vegetation types in the coastal Everglades and achieved an OA of 90% and Kappa value of 0.86 (Zhang and Xie, 2014).

A combination of 2 of the remote sensing data sources lidar, aerial photography, and hyperspectral imagery improves vegetation mapping. High spatial-resolution aerial photography can provide important spatial information for vegetation discrimination,

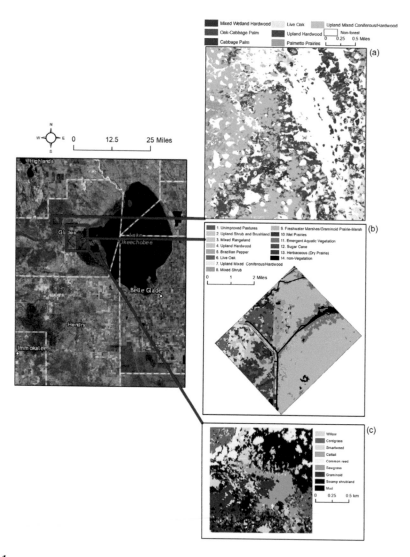

FIGURE 18.1
Vegetation mapping results in the Lake Okeechobee watershed; through a combination of (a) lidar and 1-m aerial photography (Zhang et al., 2013), (b) lidar and 12-m AVIRIS hyperspectral imagery (Zhang, 2014), and (c) lidar and 1-foot aerial photography (Zhang et al., 2018).

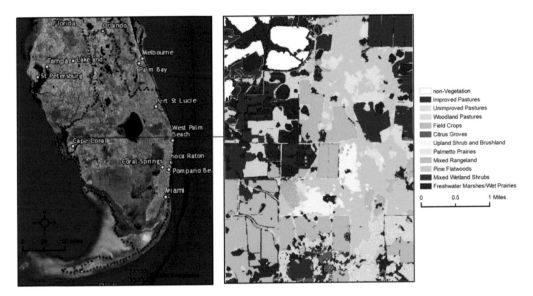

FIGURE 18.2
Classified land use/land cover-level vegetation map by fusing lidar, aerial photography, and EO-1/Hyperion for a portion of Caloosahatchee River watershed in the central Everglades.

but limited spectral resolution constrains its application for heterogeneous regions with many vegetation types or species. Fine spatial-resolution hyperspectral imagery is valuable for this application, but such data is commonly collected through airborne platforms, and acquisition of data is costly. Spaceborne hyperspectral sensors such as EO-1/Hyperion have high spectral resolution but cannot delineate small/linear features due to their coarse spatial resolution. Lidar contains useful elevation information complementary to optical sensors, which can improve vegetation characterization. Each remote sensing data source has its strengths and limitations; thus, a synergy of sources can improve vegetation classification through data fusion techniques. Zhang et al. (2016) have designed a framework to combine lidar, aerial photography, and EO-1/hyperspectral data for land use/land cover-level vegetation mapping in the Caloosahatchee River watershed and achieved an OA of 91% and Kappa value of 0.89 for mapping 11 land use/land cover vegetation types. The mapped location and final classified map using a fusion of three data sources is displayed in Figure 18.2. In this chapter, lidar, aerial photography, and EO-1/Hyperion data were combined to map detailed vegetation types/species in the coastal Everglades, where the ecosystem is different from in the central Everglades.

18.2 Study Area and Data

The study area is located in the coastal ENP covering approximate 55 km^2, a region known as Lostmans River (Figure 18.3). This area is dominated by typical plants found

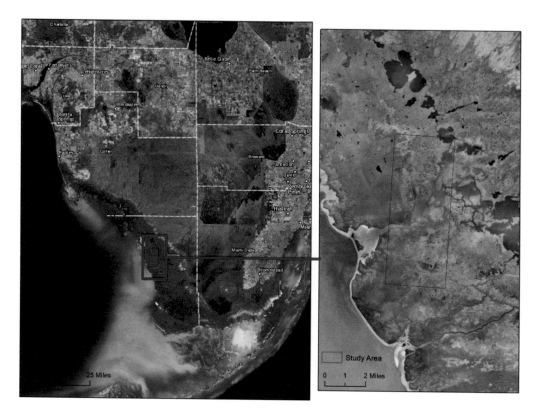

FIGURE 18.3
Study area located in the coastal ENP.

in the coastal Everglades including mangroves and cordgrass (*Spartina*). The area was selected due to the availability of data, especially an EO-1/Hyperion image with little cloud coverage. Data used in this chapter include EO-1/Hyperion imagery, digital aerial photography, lidar, and a vegetation species reference map. EO-1/Hyperion is the first spaceborne hyperspectral sensor acquiring imagery in 242 contiguous spectral bands (0.4 μm-2.5 μm) with a spatial resolution of 30 meters. The US Geological Survey (USGS) conducted the radiometric and systematic geometric corrections for the raw scenes. The preprocessed data are delivered to users as the Level 1 Gst and Level 1T products. In this chapter, the Level 1T product was used. Level 1T products have applied radiometric and systematic geometric corrections by incorporating ground control points. The Hyperion scene used in this chapter was acquired on February 18, 2005. Note that EO-1/Hyperion was decommissioned in February 2017, but historical data are still available through the USGS's Earth Explorer.

Lidar data were collected using the Leica ALS-50 system in January 2008 to support the Florida Division of Emergency Management. The Leica ALS-50 LiDAR system collects small footprint multiple returns and intensity at 1060 nm wavelength. The average point density for the study area is 4.2 pts/m². All the lidar point cloud data are available to the public at the International Hurricane Research Center and NOAA's Data Access Viewer. In total, 36 tiles (5000 × 5000 feet per tile) were downloaded and processed. Fine spatial resolution aerial photographs were collected on March 15, 1994, by the National Aerial Photography Program (NAPP). USGS processed these aerial photos into data products known as Digital Orthophoto Quarter Quads (DOQQs). The accuracy and quality of DOQQs meet National Map Accuracy Standards. Color infrared (CIR) DOQQs with a spatial resolution of 1 meter were used in this chapter. DOQQs are available to the public at USGS' Earth Explorer. Hyperspectral data and lidar and aerial photographs were all collected in the dry season of the Everglades (from November to April), which provides a good basis for evaluating vegetation mapping using data fusion techniques.

A reference vegetation dataset was used to identify which vegetation communities/ species would be mapped using remote sensing data. This reference dataset was produced through a project entitled Vegetation Map and Digital Database of South Florida's National Park Service (Welch et al., 1999). The University of Georgia and the South Florida Natural Resources Center at Everglades National Park created this database from NAPP CIR aerial photographs collected in 1994 and 1995 using photo-interpretation techniques. Note that the reference data were manually interpreted from the same DOQQs used in this chapter. As described in Chapter 3, a hierarchical Everglades Vegetation Classification System was used in this database. Vegetation classes were organized under 8 major headings: forest, scrub, savanna, prairies and marsh, shrub land, exotics, additional classes, and special modifiers. Each major group was further subdivided into classes corresponding to plant communities/species. The vegetation database was formatted to ArcGIS shape files with polygon features geographically aligned to the USGS DOQQs using the Universal Transverse Mercator projection. To accommodate the complex Everglades vegetation patterns, a 3-tiered classification scheme was developed by assigning a polygon with a dominant species that accounts for more than 50% of the vegetation in this polygon. Secondary and tertiary classes were added as required to account for mixed-plant communities. The vegetation database was validated by extensive GPS-assisted field observations including helicopter and automobile surveys. The database is currently serving as a baseline for monitoring vegetation changes in the ENP. The extracted vegetation reference data for the selected study area is shown in Figure 18.7 (b). In total, 13 dominant vegetation types/species were observed: bay-hardwood scrub, black mangrove (*Avicennia germinans*), red mangrove (*Rhizophora mangle*), white mangrove (*Laguncularia racemose*), mixed mangrove, black rush (*Juncus roemerianus*), Brazilian pepper (*Schinus terebinthifolius*), buttonwood (*Conocarpus erectus*), cabbage palm (*Sabal palmetto*), cordgrass (*Spartina* spp.), mixed graminoids, mixed scrub, and sawgrass (*Cladium jamaicense*). Descriptions of these types/species are listed in Table 18.1. Rivers, streams, and small ponds were also observed and mapped.

18.3 Methodology

EO-1/Hyperion data were provided as 224 individual TIF files by USGS. The individual TIF files were first clipped into the study area by using a Python script and then stacked as a single ENVI format file to be further processed in ENVI. Low-signal-noise ratio bands, uncalibrated bands, and severe stripping bands of the EO-1/Hyperion imagery were dropped, leaving 142 usable bands covering visible, near-infrared, and shortwave infrared for further analysis. Hyperspectral data has high dimensionality and contains

TABLE 18.1

Vegetation types/species to be mapped in this chapter, adapted from Welch and Madden (1999)

Vegetation types/species	Description
1. Bay hardwood scrub	Mixed association of bayhead swamp species, buttonwood scrub, and hardwood scrub species. Occurs in the transition zone between saline and fresh environments
2. Black mangrove (*Avicennia germinans*)	Low and high density black mangrove forest or scrub
3. Black rush (*Juncus roemerianus*)	Prairies and marshes dominated by black rush
4. Brazilian pepper (*Schinus terebinthifolius*)	An exotic plant species
5. Buttonwood (*Conocarpus erectus*)	Conocarpus erectus with variable mixtures of subtropical hardwoods
6. Cabbage palm (*Sabal palmetto*)	Cabbage Palm forest
7. Cordgrass (*Spartina* spp.)	Prairies and marshes dominated by cordgrass
8. Mixed graminoids	Prairies and marshes dominated by mixed graminoids
9. Mixed mangrove	Forest with mixed mangroves
10. Mixed scrub	Scrubs
11. Red mangrove (*Rhizophora mangle*)	Low and high density of red mangrove forest or scrub
12. Sawgrass (*Cladium jamaicense*)	Prairies and marshes dominated by sawgrass
13. White mangrove (*Laguncularia racemose*)	Low and high density of white mangrove forest or scrub
14. Water	Streams, rivers, ponds

a tremendous amount of redundant spectral information. The Minimum Noise Fraction (MNF) method (Green et al., 1988) is commonly used to reduce high dimensionality and inherent noise of hyperspectral data. Previous studies have shown that MNF transformed data can significantly improve the accuracy of vegetation mapping in the Everglades (Zhang and Xie, 2012; Zhang et al., 2016). Thus, here the MNF transformation was conducted for the hyperspectral imagery and selected the first 10 MNF eigen-images, which were the most useful and spatially coherent layers. For lidar data processing, a Digital Elevation Model (DEM) was created by using the last return of point with a minimum elevation within a 10-feet grid, and a Digital Surface Model (DSM) was produced by using the averaged elevation of first returns within the 10-feet grid in ArcGIS 10.7. A Canopy Height Model (CHM) was then produced by subtracting the DEM from the DSM to represent the height of plant canopies. Two DOQQ tiles were mosaicked and clipped for the study site. The final preprocessed results of DOQQ, Hyperion, and lidar data for the study area are displayed in Figure 18.4.

After data preprocessing, a framework was developed to combine 3 remote sensing data sources and the reference vegetation data for mapping vegetation types/species, as shown in Figure 18.5. This framework was adapted from the methodology developed in Zhang et al. (2016) but was more straightforward. Data fusion methods can be grouped into 3 categories: pixel-, feature-, and decision-level fusion (Gómez–Chova et al., 2015). Pixel-level fusion combines raw data from multiple sources into single resolution data to improve the performance of image processing tasks. Information may be lost during the data resampling procedure if the spatial resolution of input data sources is different (Solberg, 2006). For example, to fuse 1-m aerial photography with 30-m EO-1/Hyperion imagery, if a pixel-level fusion scheme was used, the 1-m aerial photography would be

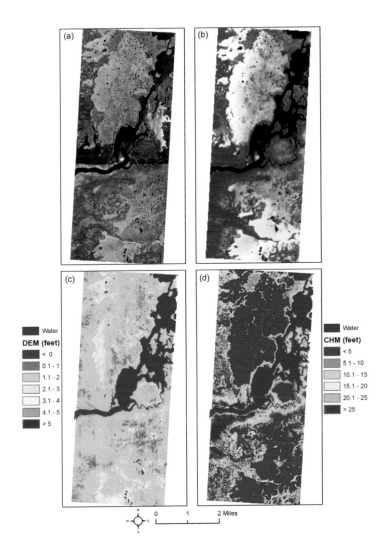

FIGURE 18.4
(a) Processed aerial photography, (b) EO-1/Hyperion imagery shown as a color infrared composite,
(c) lidar-derived DEM, and (d) CHM for the study area.

spatially resampled to 30 m to be consistent with the Hyperion data, or the 30-m Hyperion
imagery can be resampled into 1 m to be consistent with the aerial photography.
A resample of 1 m to 30 m will cause important spatial information loss, while 30 m to
1 m will largely increase data volume but will not add any spatial value in the classification. Feature-level fusion extracts features from each individual data source and merges
these features into one or more feature maps for further processing. Decision-level fusion
conducts a preliminary classification for each individual data source and then combines
the classification results into one outcome. Here, a pixel/feature-level fusion scheme was
applied. In the framework, the fine spatial resolution aerial photograph was segmented
first to generate image objects and extract features from aerial photography (mean and

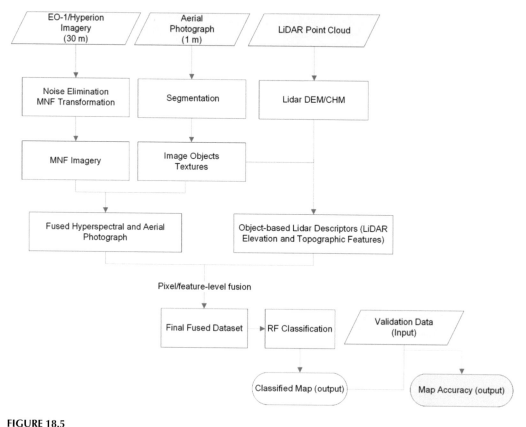

FIGURE 18.5
Methodology flowchart used to fuse 1-m aerial photography, 30-m Hyperion, and 10-feet lidar-DEM/CHM for detailed vegetation type mapping in the coastal Everglades.

standard deviation). The extracted features were then combined with pixel-level values of hyperspectral data. This was achieved by calculating a mean spectrum of hyperspectral pixels of each band within an image object and then integrating the mean spectrum with the extracted aerial features of this object. For the lidar data, the descriptive statistics (mean and standard deviation) of topography (from lidar DEM) and canopy height (from lidar CHM) within each image object were extracted and then merged with the pixel-level values of the hyperspectral imagery and feature-level of the aerial photograph, leading to a final fused dataset from 3 data sources. The fusion procedure combines pixel- and feature-level fusion methods at the object level, and thus this fusion scheme was referred to as the object-based pixel/feature-level fusion. A machine learning classifier, Random Forest (RF), was then applied to classify the fused dataset for vegetation mapping. RF has proven valuable for classifying a large volume of data due to its fast speed, fewer parameters to be specified, and high accuracy (Breiman, 2001). RF is a supervised classifier, and training data are thus required to calibrate the algorithm. Here, a spatially stratified data sampling strategy was followed to select reference object samples. The selected reference objects were manually labeled and refined by jointly checking the digital reference vegetation map, aerial photography, and hyperspectral imagery. In total, 4336 image objects were selected as reference samples to calibrate and validate the RF classifier, which accounted for approximately 10% of the total image objects produced from the aerial photography. The classification result was assessed using traditional error matrix techniques in remote sensing and the k-fold cross-validation approach. Image segmentation was conducted using the multi-resolution segmentation algorithm in eCognition Developer 9.0 (Trimble, 2014). An example of the segmentation result for part of the study area is displayed in Figure 18.6. As can be seen, small ponds were well delineated from the segmentation. Spectral mean of MNF transformed Hyperion imagery for each image object was determined through a Python script. Final data fusion, reference sample collection, and vegetation mapping were conducted in ArcGIS. WEKA (Frank et al., 2016) was used to implement and tune the RF classifier.

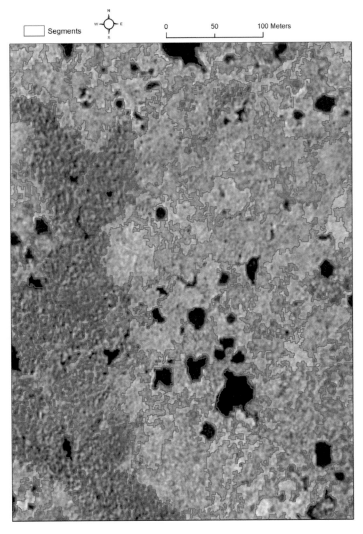

FIGURE 18.6
An illustration of the segmentation results for part of the selected study area.

18.4 Results and Discussion

To examine whether a data fusion is more effective for vegetation mapping, 4 experiments were conducted. Experiment 1 applied the aerial photography only, while experiments 2 and 3 applied a fusion of 2 data sources: aerial photography and lidar and aerial photography and Hyperion, respectively. Experiment 4 combined all 3 remote sensing data sources. The best result from each experiment is displayed in Table 18.2. Applying aerial photography alone produced an OA of 60.6% with a Kappa value of 0.51. A combination of aerial photography and Hyperion or lidar improved the OA to 64.9% and 70.3%, respectively. Kappa values also demonstrated the improvement of fusing 2 data sources. In Zhang and Xie (2014), a fusion of 1-m aerial photography with 20-m AVIRIS hyperspectral imagery, could increase 20% in the OA for a close region with 9 similar dominated vegetation types/species compared with the application of aerial photography alone. Here, less improvement was produced when 30-m EO-1/Hyperion imagery was combined. This suggests that the spatial resolution of the hyperspectral imagery is also important. In contrast, the inclusion of lidar improved accuracy more than the addition of Hyperion data. Lidar has proven valuable for vegetation classification in the Everglades, as demonstrated in Zhang et al. (2013), Zhang (2014), and Zhang et al. (2016, 2018). Lidar derived topography information is useful for vegetation classification in the Everglades. Topographical features are usually homogeneous within the same vegetation type, which can help reduce within-class variability among adjacent objects caused by shadows or gaps, thus increasing classification accuracy. Hydroperiod has been recognized as the main factor in determining the spatial distribution of plant species in a wetland. The topography impacts the hydroperiod. Higher elevation means a shorter hydroperiod, while lower elevations have longer hydroperiods. Inclusion of topographical data improved vegetation classification, as confirmed in this chapter. Lidar-derived CHM has proven valuable for vegetation characterization and commonly combined with optical imagery to improve vegetation classification. Note that here the lidar intensity has not been combined in the classification because intensity is similar to the reflectance of near infrared. Most airborne lidar collects data using near infrared wavelength. Here both aerial photography and hyperspectral imagery had near infrared reflectance.

Experiment 4 (i.e., applying a fusion of aerial photography, lidar, and hyperspectral imagery) achieved the highest accuracy with an OA of 73.4% and Kappa value of 0.67. This further confirms the benefit of fusing fine spatial resolution of aerial photography, hyperspectral data with a moderate spatial resolution (20–30 m), and lidar data (Zhang et al., 2016). Aerial photography has a fine spatial resolution but limited spectral resolution. The EO-1/Hyperion hyperspectral imagery has a fine spectral resolution

TABLE 18.2

Classification accuracies using different datasets

Dataset	Overall accuracy (%)	Kappa value
1. Aerial photography	60.6	0.51
2. Aerial photography and Hyperion	64.9	0.56
3. Aerial photography and lidar	70.3	0.64
4. Aerial photography, lidar, and Hyperion	73.4	0.67

but coarse spatial resolution. Hyperspectral imaging relies primarily on spectral features to discriminate vegetation, whereas fine spatial resolution aerial photography relies more on spatially invariant features to identify targets (Shaw and Burke, 2003). A combination of these two types of optical imagery improves the classification, as illustrated in Zhang and Xie (2014), Zhang et al. (2016), and this study. The gain from lidar has also been demonstrated in this study because of its complementary elevation and topographical features to the spectral and spatial features from the optical imagery. The developed object-based pixel/feature-level fusion scheme successfully combines the spatial features of aerial photography, rich spectral contents of hyperspectral imagery, and elevation and topographic features derived from lidar. It complements the shortages and takes advantage of the benefit of each individual data source. Application of data fusion is critical for the achieved result. Note that there is a large time gap in the acquisition of the 3 remote sensing data. A smaller time gap or simultaneous data collection can generate a better mapping result.

The final classified vegetation types/species from the fused dataset derived from 3 data sources is displayed in Figure 18.7 (a). The reference vegetation map produced from manual interpretation of aerial photography is displayed in Figure 18.7 (b). In general, the classified map well delineated the geospatial pattern of the 13 dominant

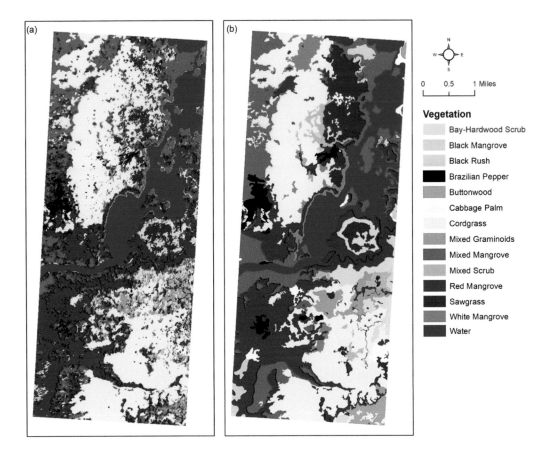

FIGURE 18.7
(a) Classified map fusing aerial photography, EO-1/Hyperion, and lidar data and (b) reference map manually interpreted from aerial photography.

vegetation types/species, and it was consistent with the reference map. A large coverage of red mangrove and cordgrass were revealed. Minor types/species such as bay-hardwood scrub and black mangrove were also delineated well. The reference map has a more homogeneous sawgrass region while the classified map shows that a large area occupied by sawgrass was encroached by cordgrass. Both cordgrass and sawgrass are marshes, and discrimination of these two marshes is challenging using either a manual procedure or automated digital image analysis procedure.

The error matrix of the classified map is displayed in Table 18.3. The user's accuracies (UAs) varied from 14.3% in classifying class 3 (black rush) to 93.8% in classifying class 7 (cordgrass). Discrimination of water from vegetation classes had the highest accuracy. For the producer's accuracy (PA), identifying cordgrass was the easiest with a high PA of 84.0%. Discriminating black mangrove and red mangrove was difficult. Note that classifying Brazilian pepper using the automated procedure was encouraging with a UA of 58% and PA of 75.8%. Brazilian pepper is one of the most aggressive and widespread invasive plant species in Florida. It invades aquatic and terrestrial habitats, greatly reducing the quality of native biotic communities in the state. Identifying and monitoring exotic plant species is an important component in the Everglades restoration.

18.5 Summary and Conclusions

A framework combining fine spatial resolution aerial photography, hyperspectral imagery with a moderate spatial resolution, and lidar data was developed to map more

TABLE 18.3

Error matrix of the classified map using a combination of aerial photography, EO-1/Hyperion, and lidar data

Class	1	2	3	4	5	6	7	8	9	10	11	12	13	14	CT	UA
1	72		1	3	6	2	6	39	9	7	11	8	2		166	43.4
2		14						4			39				57	24.6
3	2		8				24		7		5	8		2	56	14.3
4	2		1	69			1	27	2		16		1		119	58.0
5	13				14			22	2	1	19		6		77	18.2
6	8					7	1		5			1	3		25	28.0
7	2		1	0			1037	27	13	1	10	7	1	6	1105	93.8
8	8			12		1	1	643	2	3	194	1	3	2	870	73.9
9	5			1		1	46	8	74	1	5	12	4		157	47.1
10	14				2			31	1	22	12		1		83	26.5
11	6	8		4			8	134	8		755	2	7	2	934	80.8
12	3		1				101	1	6		1	137	1	2	253	54.2
13	4			2	1	1	0	42	8	2	28	4	29	1	122	23.8
14							9				2			301	312	96.5
RT	139	22	12	91	23	12	1234	978	137	37	1097	180	58	316	OA: 73.4%	
PA	51.8	63.6	66.7	75.8	60.9	58.3	84.0	65.7	54.0	59.5	68.8	76.1	50.0	95.3	Kappa: 0.67	

CT: Column Total; RT: Row Total; UA: User's Accuracy (%); PA: Producer's Accuracy (%).

detailed vegetation types/species in the coastal Everglades. The results demonstrate that applying data fusion is more effective for vegetation mapping and monitoring if the data complement each other. Using the pixel/feature level fusion strategy is more effective for integrating 3 types of remote sensing data sources than the pixel-level fusion scheme commonly used in the literature. A fusion of 3 data sources can produce an accuracy of over 70% for classifying detailed vegetation species. Since collecting the moderate spatial resolution hyperspectral imagery is more common for most projects, the designed framework assists with the manual methods for vegetation mapping in areas where lidar, aerial photography, and hyperspectral data are available.

References

Breiman, L., 2001. Random forests. *Machine Learning*, 45, 5–32.

Frank, E., Hall, M.A., and Witten, I.H., 2016. The WEKA Workbench. In: Witten, I. H., Frank, E., and Hall, M. A. (eds.), *Data Mining, Practical Machine Learning Tools and Techniques*, 4th Edition, Elsevier, Cambridge, UK.

Gómez-Chova, L., Tuia, D., Moser, G., and Camps-Valls, G., 2015. Multimodal classification of remote sensing images, a review and future directions. *Proc IEEE*, 103, 1560–1584.

Green, A.A., Berman, M., Switzer, P., and Craig, M.D., 1988. A transformation for ordering multi-spectral data in terms of image quality with implications for noise removal. *IEEE Transactions on Geoscience and Remote Sensing*, 26, 65–74.

Griffin, S., Rogan, J., and Runfola, D.M., 2011. Application of spectral and environmental variables to map the Kissimmee prairie ecosystem using classification trees. *GIScience & Remote Sensing*, 48, 299–323.

Hirano, A., Madden, M., and Welch, R., 2003. Hyperspectral image data for mapping wetland vegetation. *Wetlands*, 23, 436–448.

Jensen, J., Rutchey, K., Koch, M., and Narumalani, S., 1995. Inland wetland change detection in the Everglades Water Conservation Area 2A using a time series of normalized remotely sensed data. *Journal of Photogrammetric Engineering and Remote Sensing*, 61, 199–209.

Rutchey, K., and Vilchek, L., 1994. Development of an Everglades vegetation map using a SPOT image and the global positioning system. *Photogrammetric Engineering and Remote Sensing*, 60, 767–775.

Rutchey, K., and Vilchek, L., 1999. Air photointerpretation and satellite imagery analysis techniques for mapping cattail coverage in a northern Everglades impoundment. *Photogrammetric Engineering and Remote Sensing*, 65, 185–191.

Shaw, G.A., and Burke, H.K., 2003. Spectral imaging for remote sensing. *Lincoln Laboratory Journal*, 14, 3–28.

Solberg, A.H.S., 2006. Data fusion for remote sensing applications. In: Chen, C.H. (eds.) *Signal and Image Processing for Remote Sensing*, CRC Press, Boca Raton, FL, 515–537.

Szantoi, Z., Escobedo, F., Abd–Elrahman, A., Smith, S. and Pearlstine, L., 2013. Analyzing fine-scale wetland composition using high resolution imagery and texture features. *International Journal of Applied Earth Observation and Geoinformation*, 23, 204–212.

Szantoi, Z., Escobedo, F., Abd–Elrahman, A., Pearlstine, L., Dewitt, B. and Smith, S., 2015. Classifying spatially heterogeneous wetland communities using machine learning algorithms and spectral and textural features. *Environmental Monitoring and Assessment*, 187, 262.

Trimble, 2014. *eCognition Developer 9.0.1 Reference Book*. Trimble Germany GmbH, Arnulfstrasse 126, D-80636, Munich.

Welch, R., and Madden, M., 1999. Vegetation map and digital database of South Florida's National park service lands, final report to the U.S. department of the interior, National park service, cooperative agreement number 5280-4-9006. Center for Remote Sensing and Mapping Science, University of Georgia, Athens, 44.

Welch, R., Madden, M., and Doren, R., 1999. Mapping the Everglades. *Photogrammetric Engineering and Remote Sensing*, 65, 163–170.

Zhang, C., 2014. Combining hyperspectral and LiDAR data for vegetation mapping in the Florida Everglades. *Photogrammetric Engineering and Remote Sensing*, 80, 733–743.

Zhang, C., Denka, S., and Mishra, D.R., 2018. Mapping freshwater marsh species in the wetlands of Lake Okeechobee using very high-resolution aerial photography and lidar data. *International Journal of Remote Sensing*, 39, 5600–5618.

Zhang, C., Selch, D., and Cooper, H., 2016. A framework to combine three remotely sensed data sources for vegetation mapping in the central Florida Everglades. *Wetlands*, 36, 201–213.

Zhang, C., and Xie, Z., 2012. Combining object–based texture measures with a neural network for vegetation mapping in the Everglades from hyperspectral imagery. *Remote Sensing of Environment*, 124, 310–320.

Zhang, C., and Xie, Z., 2013. Object–based vegetation mapping in the Kissimmee River watershed using HyMap data and machine learning techniques. *Wetlands*, 33, 233–244.

Zhang, C., and Xie, Z., 2014. Data fusion and classifier ensemble techniques for vegetation mapping in the coastal Everglades. *Geocarto International*, 29, 228–243.

Zhang, C., Xie, Z., and Selch, D., 2013. Fusing LiDAR and digital aerial photography for object–based forest mapping in the Florida Everglades. *GIScience & Remote Sensing*, 50, 562–573.

19

Assessing a Multi-sensor Fusion Approach to Map Detailed Reef Benthic Habitats in the Florida Reef Tract

Caiyun Zhang

19.1 Introduction

Chapter 14 presents the capability of 30-m EO-1/Hyperion imagery for mapping benthic habitats. An encouraging result has been produced using this spaceborne hyperspectral sensor. However, a 30-m spatial resolution sensor cannot delineate small patches or linear reef features. Thus, NOAA, in partnership with other local and state agencies, mapped the reef benthic habitats in the Florida Reef Tract using the fine spatial resolution spaceborne sensor (Ikonos), aerial photography, and other surveyed data through a manual interpretation procedure. A reef-mapping product has been released, but updating this product is difficult because of the time-consuming and labor intensive manual work involved. Some of the maps are more than 20 years old. As demonstrated in Chapter 18, a combination of fine spatial resolution aerial photography, EO-1/Hyperion, and lidar can largely improve vegetation classification in species-level mapping in the coastal Everglades, as well as at land use land cover-level mapping in the central Everglades (Zhang et al., 2016). In this chapter, an integration of fine spatial resolution Ikonos imagery, EO-1/Hyperion imagery, and lidar bathymetry data was assessed to map reef benthic habitats at a detailed level.

Remote sensing has long been used for reef benthic habitat mapping, particularly the use of optical sensors such as Landsat and Ikonos (Mumby et al., 2004; Hedley et al., 2016). Airborne bathymetric lidar has also been applied to map benthic habitats (e.g., Collin et al., 2008, 2011; Tulldahl and Wikström, 2012; Zavalas et al., 2014) and proven valuable for classifying habitats into 3 to 4 categories. Airborne bathymetric lidar collects data using a single green wavelength with limited spectral information but can provide useful elevation data complementary to information provided by optical sensors. Thus, a combination of bathymetric lidar and optical sensor has the potential to improve benthic habitat mapping in a reef environment. Several efforts have been made to improve reef benthic habitat mapping by fusing bathymetric lidar and optical sensors. For example, Tulldahl et al. (2013) fused waveform lidar data and WorldView-2 imagery to map benthic habitats in a Swedish archipelago area. Torres-Madronero et al. (2014) combined lidar-derived bathymetric data into a bio-optical model to improve benthic habitat mapping using hyperspectral imagery. In the Florida Reef Tract, Walker et al. (2008) combined bathymetric lidar with acoustic ground discrimination, subbottom profiling, and aerial photography to map benthic habitats into 2- and 3- categories through a manual interpretation procedure in Broward County. Note that NOAA applied the same procedure to generate the reef benthic habitat map for most of the Florida Reef Tract.

A framework was developed in Zhang (2015) to combine 17-m AVIRIS hyperspectral imagery, 1-m aerial photography, and a 10-m bathymetric product created by the National Geophysical Data Center of NOAA for mapping 3-class and 9-class benthic habitats in the Florida Keys. This study suggests that a synergy of fine spectral resolution imagery with a moderate spatial resolution and fine spatial resolution imagery with limited spectral channels can significantly improve reef benthic habitat classification. However, whether a combination of the 2 spaceborne sensors Ikonos and EO-1/Hyperion can improve reef mapping remains unknown. This chapter presents a framework for combining airborne bathymetric lidar and 2 spaceborne sensors Ikonos and EO-1/Hyperion for mapping benthic habitats at a detailed level and evaluates whether a fusion of 3 systems or 2 systems is more effective for reef mapping compared with the application of 1 sensor system.

19.2 Study Area and Data

The same study area as Chapter 14 was selected due to the availability of EO-1/Hyperion, Ikonos, and lidar data for this site, as shown in Figure 19.1. The site is located in the lower Florida Keys with a size of 4 km^2 (~1.55 mile^2). The Florida Keys or the Florida Reef Tract is a coral cay archipelago beginning at the southeastern tip of the Florida peninsula and extending in a gentle arc south-southwest and then westward to the inhabited islands known as Key West. The Florida Reef Tract is the only living coral barrier reef in the continental US and the third largest coral barrier reef system in the world (after Great Barrier Reef and Belize Barrier Reef). The Florida Keys has a tropical climate, and its environment is similar to that of the Caribbean. It is characterized by spectacular coral reefs, extensive seagrass beds, and mangrove-fringed islands. This area is one of the world's most productive ecosystems with more than 6000 species of marine life and 250 species of nesting birds. The selected site is located in a lagoon with water depth varying from 0.6 m to 5.8 m based upon the lidar bathymetry data.

Data used in this chapter include 30-m EO-1/Hyperion imagery, 4-m Ikonos imagery, airborne lidar point cloud, and a reef benthic habitat reference map dataset. All these data are in digital format and available to the public at no cost. The EO-1/Hyperion imagery was collected on October 31, 2002 (path/row: 16/43). A detailed description of this dataset is provided in Chapter 14. Here, the Hyperion L1Gst product was used, which was terrain corrected. The 4-m Ikonos imagery was collected on October 7, 2006 by the Ikonos-2 sensor and georeferenced to Florida's 2004 Digital Orthophoto Quarter-Quadrangles (DOQQ) with the Universal Transverse Mercator (UTM) Zone 17N projection. NOAA purchased the Ikonos imagery collected between 2005 and 2006 over the Florida Keys under the company's Tier 3 license agreement and made these images available to public users. The coverage of these images in the Florida Keys is displayed in Figure 19.2. Bathymetric lidar data were collected by NOAA's National Geodetic Survey Remote Sensing Division using a Riegl VQ880G system during 08/29/2014-09/19/2014 to assist with NOAA's Coastal Mapping Program. The aim of this bathymetric lidar data acquisition was to support the "Map Once-Use Many Times" paradigm through a variety of applications such as inundation modeling, coastal engineering,

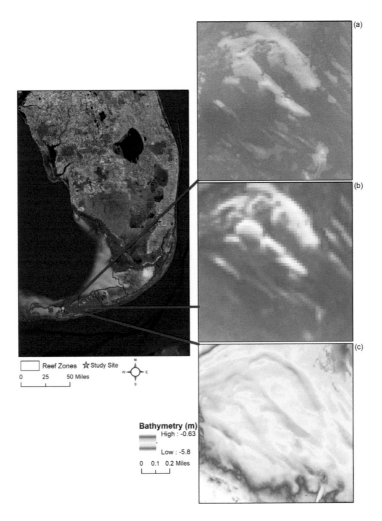

FIGURE 19.1
Study area located in the Florida Reef Tract and the site shown as a color composite of (a) 4-m Ikonos, (b) 30-m EO-1/Hyperion, and (c) 5-m lidar derived bathymetry.

and coral reef mapping. Currently, NOAA is providing this lidar dataset in point cloud LAS format to users with ground, submerged, and other features classified. It is available at NOAA's Data Access Viewer (https://coast.noaa.gov/dataviewer/#/). In total, 16 tiles covering the study site were downloaded and processed. An approximate 35 points/m^2 were estimated for the selected site. This high point density offers a good opportunity to evaluate bathymetric lidar for reef habitat characterization at a fine spatial scale but poses a challenge for using this dataset for broad reef mapping due to the large data volume.

The same reference map as Chapter 14 was used in this chapter. The reef reference map was sourced from the Unified Florida Coral Reef Tract Map v2.0. In September 2014, the Florida Fish and Wildlife Conservation Commission (FWC) and the Fish

FIGURE 19.2
The coverage of Ikonos image scenes purchased by NOAA in the Florida Keys. These images were collected during 2005 and 2006.

and Wildlife Research Institute (FWRI) published The Unified Florida Coral Reef Tract Map v1.0; this map integrated existing benthic habitat maps provided by multiple entities. Benthic habitats were mapped for 12 regions from Martin County to Dry Tortugas, as shown in Figure 19.3. The selected study site is within the Marquesas region. Based on the meta data, the reference map for the selected study site was interpreted from the fine resolution Ikonos pan-sharpened and color imagery with guidance from FWC-FWRI. All sea floor features were visible and mapped to the 0.4-ha (4000 m^2; ~1 acre) minimum mapping unit (MMU) specification, except for patchy reefs. Patchy reefs visible in the imagery were mapped using an approximately 0.0625-ha (625 m^2; ~0.154 acre) MMU. The map was validated and corrected through intensive field methods such as the application of video system, snorkel, and scuba. An over 90% thematic accuracy was reported for the reference map over the Marquesas region. Reef habitats in the Unified Florida Coral Reef Tract Map were grouped into 5 levels using a unified classification system with level 0 being the most general level and level 4 being the most detailed. The unified classification system mainly follows the classification scheme developed by Zitello et al. (2009) from NOAA, in which the reef geographic zone, geomorphological structure type, and dominant biological cover are considered. In this chapter, the most detailed level (level 4) was mapped. Based on the reference map, there are 9 benthic habitats appeared in the study site as level 4 and are listed in Table 19.1. The reference map was used to select training and testing data to calibrate and validate the remote sensing semi-automated mapping procedure.

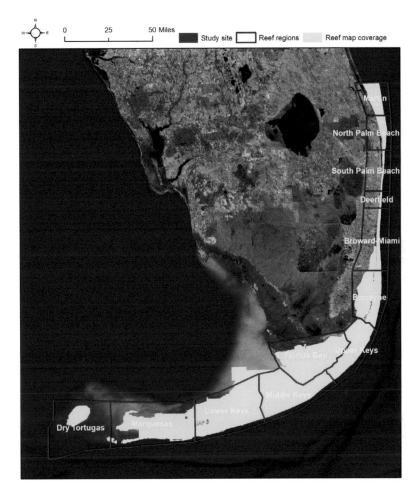

FIGURE 19.3
The coverage of mapped benthic habitats in the Unified Florida Coral Reef Tract Map v2.0.

19.3 Methodology

The original Ikonos, EO-1/Hyperion, and lidar point cloud data were pre-processed first. The Ikonos scene was clipped to the study site and only pixels within the study domain were considered. A natural color composite of Ikonos image for the study site is shown in Figure 19.1(a). For the EO-1/Hyperion imagery, a big misalignment was observed between the EO-1/Hyperion and Ikonos images, thus the EO-1/Hyperion was re-georeferenced using the Ikonos imagery to correct the misalignment. After the correction, spectral channels of EO-1/Hyperion with low signal-to-noise were dropped, leading to 30 visible bands and 9 near-infrared bands for further analysis. The reduced Hyperion imagery was then clipped to the study area, and a color composite generated from this image is shown in Figure 19.1(b). For the bathymetric lidar point cloud data,

TABLE 19.1

Description of benthic habitats classified in this chapter

Habitats	Description
1. Patchy live coral, aggregate reef	**Aggregate reef**: continuous, high-relief coral formation of variable shapes lacking sand channels, including linear reef formations that are oriented parallel to shore or the shelf edge.
2. Patchy macroalgae, aggregate reef	**Pavement**: flat, low continuous: Major biological cover type with nearly continuous (90-100%) coverage of the substrate.
3. Sparse live coral, pavement	**Continuous**: major biological cover type with nearly continuous (90-100%) coverage of the substrate.
4. Patchy macroalgae, pavement	**Patchy**: Discontinuous (50-90%) cover of the major biological type with breaks.
5. Sparse macroalgae, pavement	**Sparse**: Discontinuous (10-50%) cover of the major biological type with breaks. **Liver coral**: Substrates colonized with 10% or greater live reef building corals and other organisms.
6. Patchy macroalgae, sand	**Seagrass**: Habitats dominated by any single species of seagrass or a combination of several species.
7. Continuous seagrass, sand	**Algae**: Substrates with 10% or greater distribution of any combination of numerous species of red, green, or brown algae.
8. Patchy seagrass, sand	
9. Uncolonized, sand	

a 5-m Digital Elevation Model (DEM) (bathymetry) was generated to represent the elevation of bare sea floor from the lidar ground points using the Inverse Distance Weighted (IDW) interpolation technique, as shown in Figure 19.1(c). A normalized 5-m Digital Surface Model (DSM) was produced to represent the elevation of non-floor features using lidar submerged points. A 5-m intensity raster layer was also created from the lidar dataset.

After data pre-processing, the methodology for mapping the benthic habitats using a data fusion framework was applied, as shown in Figure 19.4. Similar to Chapter 18, an object-based data fusion scheme was applied here. Image objects were produced from 4-m Ikonos imagery using the multi-resolution segmentation algorithm, and then the Ikonos-based image features were extracted (mean and standard deviation). Similar to Chapter 14, the Minimum Noise Fraction (MNF) transformation (Green et al., 1988) was applied to the Hyperion imagery to reduce the redundant spectral information, and the first 10 coherent MNF image layers were selected for further analysis. After the MNF transformation, the spectral feature of Hyperion for each image object was calculated and merged with the Ikonos derived features. This was achieved by first calculating a mean spectrum of Hyperion pixels of each MNF band (within an Ikonos-derived image object) and then integrating the mean spectrum with the extracted Ikonos features of this object. Similar to Chapter 18, for the lidar data, the descriptive statistics (mean, and standard deviation) of DEM and DSM and the intensity of each object were extracted and then merged with Ikonos and Hyperion-derived features at the object level, leading to a final fused dataset from 3 data sources. A machine learning classifier, Random Forest (RF) was then applied to classify the fused dataset for benthic habitat classification and mapping. RF has proven valuable for classifying a large volume of data due to its fast speed, fewer parameters to be specified, and high accuracy (Breiman, 2001). RF is a supervised classifier, and training data are thus required to calibrate the algorithm. Here, a spatially stratified data sampling strategy was followed to select reference object samples for calibrating and

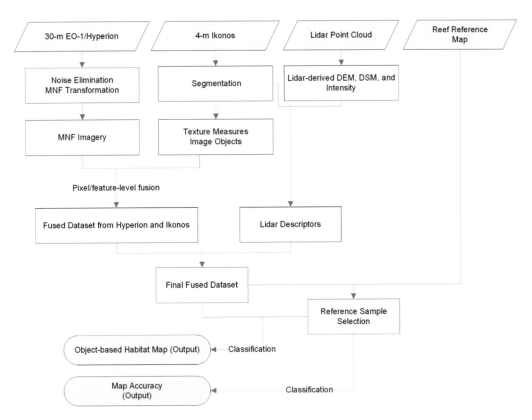

FIGURE 19.4
Methodology flowchart used to map detailed reef benthic habitats by fusing 30-m EO-1/Hyperion, 4-m Ikonos, and lidar data.

validating the classification. The selected reference objects were manually labeled and refined by jointly checking the digital reference reef map, Ikonos, and Hyperion imagery. In total, 659 image objects were selected as reference samples. The classification result was assessed using traditional error matrix techniques and the k-fold cross-validation approach. Image segmentation was conducted using the multi-resolution segmentation algorithm in eCognition Developer 9.0 (Trimble, 2014), and the segmentation result is shown in Figure 19.5. Spectral mean of MNF transformed Hyperion imagery for each image object was determined through a Python script. Final data fusion, reference sample collection, and vegetation mapping were conducted in ArcGIS. WEKA (Frank et al., 2016) was used to implement and tune the RF classifier.

19.4 Results and Discussion

To assess whether a combination of 2 or 3 remote sensing data sources was more effective than the application of Ikonos imagery alone, 4 experimental analyses were

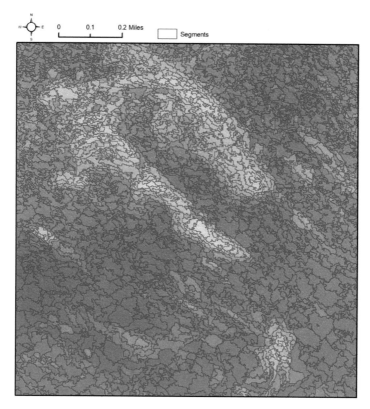

FIGURE 19.5
Image segments overlaid on the Ikonos imagery.

conducted, including the application of Ikonos image only, a combination of Ikonos and lidar data, a fusion of Ikonos and EO-1/Hyperion data, and a synergy of all 3 remote sensing data sources. The overall accuracy (OA) and the Kappa value based on a 5-fold cross-validation for each experiment is shown in Table 19.2. Application of Ikonos imagery only produced an OA of 66% and Kappa value of 0.52. A combination of Ikonos and lidar data improved the OA to 74% and Kappa value to 0.63, suggesting lidar is valuable for reef habitat mapping. A fusion of Ikonos and Hyperion was also promising, which improved the OA to 73% and Kappa value to 0.62. A synergy of all

TABLE 19.2

Classification accuracies using different datasets

Dataset	OA (%)	Kappa Value
Ikonos imagery	66	0.52
Ikonos and lidar data	74	0.63
Ikonos and EO-1/Hyperion	73	0.62
Ikonos, lidar, and EO-1/Hyperion	75	0.65

three remote sensing data sources only slightly increased the classification with an OA of 75% and Kappa value of 0.65 compared with the application of two data sources.

The previous study suggested that application of 4-m Ikonos imagery only could not produce adequate accuracy for detailed reef mapping using the automated or semi-automated classification procedure (Andréfouët et al., 2003). This study confirmed that low accuracy was produced using the Ikonos imagery alone. An inclusion of lidar data largely improved the detailed reef mapping when lidar was combined with Ikonos imagery. Lidar has proven valuable for reef mapping by providing complementary information (water depth, coral reef height, and intensity) to the optical sensors, which offer spectral information only. The study demonstrated that a fusion of Ikonos and lidar was beneficial for reef benthic habitat mapping. Another potential for improving reef mapping is to combine Ikonos with hyperspectral data. Ikonos sensors have high spatial resolution but limited spectral resolution, while spaceborne hyperspectral sensors such as EO-1/Hyperion have fine spectral resolution but relatively coarse spatial resolution. A fusion of Ikonos and EO-1/Hyperion sensors can take advantage of both sensors while reducing the constraints of each sensor. The study illustrated that a fusion of 4-m Ikonos and 30-m EO-1/Hyperion was useful for a detailed reef classification. In Chapter 14, application of EO-1/Hyperion alone produced an OA of 78% and Kappa value of 0.73. Note that though a high OA was produced based on the reference samples, small patches and linear features would not be delineated using the 30-m Hyperion data alone. In contrast, in this chapter, since the image objects were produced from the 4-m Ikonos imagery, this issue was mitigated due to the fine spatial resolution of Ikonos data. A fusion of Ikonos and EO-1/Hyperion type sensors are valuable for detailed reef mapping. A synergy of Ikonos, lidar, and Hyperion was expected to improve the classification compared with the application of 2 data sources. However, a slight improvement in the classification was produced. This needs further investigation.

The reef map was produced using the fused dataset of 3 remote sensing data sources, as shown in Figure 19.6(a). For comparison purpose, the reference map is displayed in Figure 19.6(b). In general, two maps are consistent in the spatial pattern of the distribution of 9 reef habitats. The selected study site was dominated by continuous seagrass, patchy seagrass, and uncolonized sand. Pavement with live coral, macroalgae, and aggregate reef with live coral and macroalgae were sparsely presented. The reference map is more smooth because the MMU is 4000 m^2 while the classified map had a MMU of 32 m^2 (2 Ikonos pixels). The classified map was thus able to provide more details in the distribution of reef benthic habitats. In contrast, small patches with a size less than 4000 m^2 were not delineated in the reference map.

The error matrix of the classified map based on the 5-fold cross-validation from a fusion of 3 remote sensing data sources is displayed in Table 19.3. Based on the reference samples, the digital procedure completely failed to classify sparse live coral on pavement with the User's Accuracy (UA) and Producer's Accuracy (PA) of 0% (class 3). However, on the classified map (Figure 19.6 (a)), sparse live coral on pavement (green) was partially delineated. Identification of sand covered with patchy seagrass (class 8) and uncolonized sand (class 9) had high accuracy. Minor habitats generally had lower accuracy in the classification due to their limited coverage and the small number of reference samples. This was expected in a digital classification procedure.

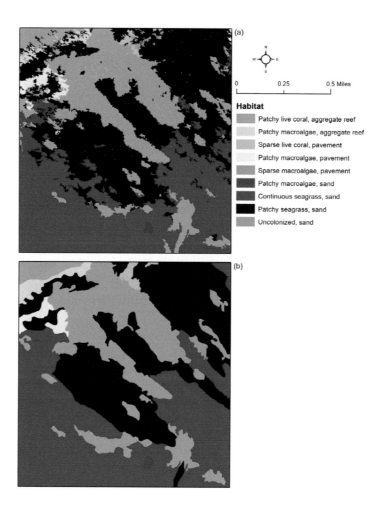

FIGURE 19.6
(a) Classified reef benthic habitat map combining three remote sensing data sources and (b) its comparison to the reference map manually interpreted from the 4-m Ikonos image.

A combination of Ikonos with lidar or Hyperion produced similar accuracy and Kappa value based on the reference samples. The classified maps from a fusion of Ikonos and lidar data and a fusion of Ikonos and Hyperion imagery are displayed in Figure 19.7(a) and 19.7(b), respectively. Three major habitats including continuous seagrass, patchy seagrass, and uncolonized sand were delineated well by both classifications. It was difficult to identify the difference between 2 classifications based on the classified maps. However, the difference between them could be quantified by defining a binary vector file, as shown in Figure 19.7(c). If the classifications from 2 datasets were the same for an input image object, an agreement was obtained. In contrast, if the classifications were different, a disagreement was assigned to the input image object. In this way, a binary file was produced showing the agreement and disagreement of 2 classifications. Though a combination of Ikonos with lidar or Hyperion produced similar accuracy based on the OAs and Kappa values from the reference samples, the classified maps could be different. In total, 481 image objects had different classifications using

TABLE 19.3

Error matrix of the classified map using the combined Ikonos, EO-1/Hyperion, and lidar data

Class	1	2	3	4	5	6	7	8	9	CT	UA (%)
1	5	1					1	10		17	29.4
2	2	2						13	1	18	11.1
3								10	3	13	0.0
4				5			1	7	4	17	29.4
5					6			3	6	18	33.3
6						2	1	9	2	14	14.3
7							104	34	3	141	73.8
8		1		3	1	1	11	208	11	236	88.1
9				1			5	15	164	185	88.6
RT	7	4	0	11	8	3	123	309	194	OA: 75% Kappa: 0.65	
PA (%)	71.4	50.0	0.0	45.5	75.0	66.7	84.6	67.3	84.5		

CT: Column Total; RT: Row Total; UA: User's Accuracy; PA: Producer's Accuracy; OA: Overall Accuracy. The name of each class is listed in Table 19.1.

2 datasets (shown in red on Figure 19.7(c)), accounting for 12% of the total size of the study area. This binary map can guide the field survey. Regions with a high uncertainty from classifications should be well sampled.

19.5 Summary and Conclusions

In this chapter, a data fusion framework was assessed to investigate whether a synergy of multiple sensors could improve a detailed reef mapping in the Florida Reef Tract to assist with Florida reef monitoring. A fusion of 4-m Ikonos, 30-m EO-1/Hyperion, and bathymetric lidar was expected to be more effective due to the complementary features of each sensor. Testing results in the lower Florida Keys were encouraging for applying data fusion techniques. A fusion of Ikonos imagery with either lidar or Hyperion data significantly improved reef classification, and a further minor improvement was obtained when all 3 data sources were combined. Data fusion techniques often require data acquisition to be simultaneous, especially for a dynamic environment. In this study, there is a large time gap among the data collection, resulting in uncertainties in the testing. In Florida, bathymetric lidar is frequently collected for multiple applications on the coasts. The fusion framework is promising for combining the valuable bathymetric lidar with any fine spatial resolution imagery (from spaceborne sensors or aerial photography) for updating reef maps in the Florida Keys. Although EO-1/Hyperion has been decommissioned, as discussed in Chapter 14, with the launch of other hyperspectral sensors, inclusion of these sensors in the reef mapping is valuable, using the data fusion techniques presented in the chapter.

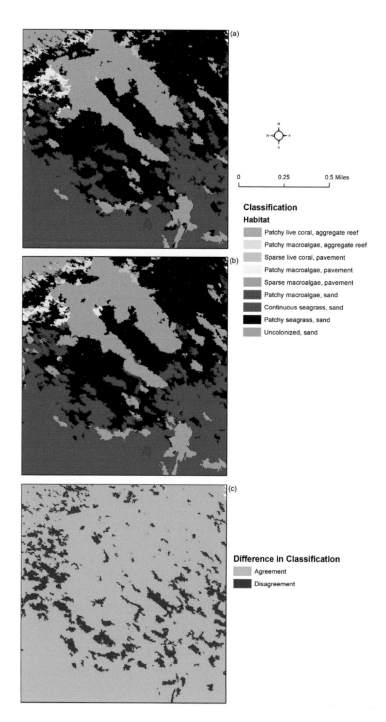

FIGURE 19.7
Classified reef benthic habitat map from (a) a combination of Ikonos and lidar data, (b) Ikonos and EO-1/ Hyperion data, and (c) the difference between the two classifications.

References

Andréfouët, S., Kramer, P., Torres-Pulliza, D., et al., 2003. Multi-sites evaluation of IKONOS data for classification of tropical coral reef environments. *Remote Sensing of Environment*, 88, 128–143.

Breiman, L., 2001. Random forests. *Machine Learning*, 45, 5–32.

Collin, A., Archambault, P., and Long, B., 2008. Mapping the shallow water seabed habitat with the shoals. *IEEE Transactions on Geoscience and Remote Sensing*, 46, 2947–2955.

Collin, A., Archambault, P., and Long, B., 2011. Predicting species diversity of benthic communities within turbid nearshore using full-waveform bathymetric Lidar and machine learners. *PLoS One*, 6. DOI:10.1371/journal.pone.0021265.

Frank, E., Hall, M.A., and Witten, I.H., 2016. The WEKA Workbench. In: Witten, I.H., Frank, E., and Hall, M.A. (eds.) *Data Mining, Practical Machine Learning Tools and Techniques*, 4th Edition, Elsevier, Cambridge, UK.

Green, A.A., Berman, M., Switzer, P., and Craig, M.D., 1988. A transformation for ordering multi-spectral data in terms of image quality with implications for noise removal. *IEEE Transactions on Geoscience and Remote Sensing*, 26, 65–74.

Hedley, J.D., Roelfsema, C.M., Chollett, I., et al., 2016. Remote sensing of coral reefs for monitoring and management: A review. *Remote Sensing*, 8, 118.

Mumby, P.J., Skirving, W., Strong, A.E., et al., 2004. Remote sensing of coral reefs and their physical environment. *Marine Pollution Bulletin*, 48, 219–228.

Torres-Madronero, M.C., Velez-Reyes, M., and Goodman, J.A., 2014. Subsurface unmixing for benthic habitat mapping using hyperspectral imagery and lidar-derived bathymetry. *Proc. SPIE*. DOI:10.1117/12.2053491.

Trimble, 2014. *eCognition Developer 9.0.1 Reference Book*. Trimble Germany GmbH, Arnulfstrasse 126, D-80636, Munich.

Tulldahl, H.M., Philipson, P., Kautsky, H., and Wikstrom, S.A., 2013. Sea floor classification with satellite data and airborne lidar bathymetry. *Proc. SPIE*. DOI:10.1117/12.2015727.

Tulldahl, H.M., and Wikström, S.A., 2012. Classification of aquatic macrovegetation and substrates with airborne lidar. *Remote Sensing of Environment*, 121, 347–357.

Walker, B.K., Riegl, B., and Dodge, R.E., 2008. Mapping coral reef habitats in southeast Florida using a combined technique approach. *Journal of Coastal Research*, 5, 1138–1150.

Zavalas, R., Ierodiaconou, D., Ryan, D., et al., 2014. Habitat classification of temperate marine macroalgal communities using bathymetric lidar. *Remote Sensing*, 6, 2154–2175.

Zhang, C., 2015. Applying data fusion techniques for benthic habitat mapping and monitoring in a coral reef ecosystem. *ISPRS Journal of Photogrammetry and Remote Sensing*, 104, 213–223.

Zhang, C., Selch, D., and Cooper, H., 2016. A framework to combine three remotely sensed data sources for vegetation mapping in the central Florida Everglades. *Wetlands*, 36, 201–213.

Zitello, A.G., Bauer, L.J., Battista, T.A., et al., 2009. Shallow-water benthic habitats of St. John, U.S. Virgin Islands. NOAA Technical Memorandum NOS NCCOS 96. Silver Spring, MD.

Index